Yu. V. Prokhorov · V. Sta

Limit Theorems of Pro

Springer-Verlag Berlin Heidelberg GmbH

Yu. V. Prokhorov V. Statulevičius
Editors

Limit Theorems of Probability Theory

Springer

Title of the Russian original edition:
Itogi Nauki i Tekhniki, Sovremennye Problemy Matematiki,
Fundamental'nye Napravleniya, Vol. 81, Teoriya Veroyatnostej 6.
Publisher VINITI, Moscow 1991

Library of Congress Cataloging-in-Publication Data
Limit theorems of probability theory / Yu. V. Prokhorov, V. Statulevicius, editors.
p. cm. Translation from the Russian series: Itogi nauki i tekhniki, sovremennye
problemy matematiki, fundamental'nye napravleniya, vol. 81,
teoriya veroyatnostej 6. Includes bibliographical references and index.
ISBN 978-3-642-08170-5 ISBN 978-3-662-04172-7 (eBook)
DOI 10.1007/978-3-662-04172-7
1. Limit theorems (Probability theory)
I. Prokhorov, Iu. V. (Iurii Vasil'evich) II. Statulevicius, V. A.
QA273.76.L56 2000 519.2-dc21 00-036582

Mathematics Subject Classification (2000):
60F05, 60F10, 60B12

ISBN 978-3-642-08170-5

Originally published by Springer-Verlag Berlin Heidelberg New York in 2000
Softcover reprint of the hardcover 1st edition 2000

Typesetting: Frank Herweg, Hirschberg-Leutershausen, using a Springer TEX macro-package.
Cover design: Erich Kirchner, Heidelberg

Printed on acid-free paper SPIN: 10064836 46/3142/Ko - 5 4 3 2 1 0

List of Editors, Authors and Translator

Editors

Yu.V. Prokhorov, Steklov Mathematical Institute, ul. Gubkina 8, 117966 Moscow, Russia, email: prohorov@class.mi.ras.ru

V. Statulevičius, Lithuanian Academy of Sciences, Institute of Mathematics and Informatics, Akademijos 4, 2600 Vilnius, Lithuania, email: stat@ktl.mii.lt

Authors

V. Bentkus, Lithuanian Academy of Sciences, Institute of Mathematics and Informatics, Akademijos 4, 2600 Vilnius, Lithuania, email: jurgita@ktl.mii.lt, bentkus@mathematik.uni-bielefeld.de

F. Götze, Universität Bielefeld, Fakultät für Mathematik, Universitätsstrasse 25, 33615 Bielefeld, Germany, email: goetze@mathematik.uni-bielefeld.de

P. Gudynas, Lithuanian Academy of Sciences, Institute of Mathematics and Computer Science, Akademijos 4, 2600 Vilnius, Lithuania, email: institutas@pi.elnet.lt

V. Paulauskas, Vilnius University, Department of Mathematics and Informatics, Naugarduko 24, 2600 Vilnius, Lithuania, email: vpaul@mif.vu.lt

V.V. Petrov, St. Petersburg University, Faculty of Mathematics and Mechanics, Stary Peterhof, 198904 St. Petersburg, email: petrov@stt.msu.edu

A. Račkauskas, Vilnius University, Department of Mathematics, Naugarduko 24, 2600 Vilnius, Lithuania, email: alfredas@ieva.maf.vu.lt

L. Saulis, Faculty of Fundamental Sciences, Department of Mathematics and Statistics, Vilnius Gediminas Technical University, Sauletekio a. 11, 2054 Vilnius, Lithuania, email: lsaulis@rasa.vtu.lt

V. Statulevičius, Lithuanian Academy of Sciences, Institute of Mathematics and Informatics, Akademijos 4, 2600 Vilnius, Lithuania, email: stat@ktl.mii.lt

J. Sunklodas, Lithuanian Academy of Sciences, Institute of Mathematics and Informatics, Akadmijos 4, 2600 Vilnius, Lithuania, email: jurgita@ktl.mii.lt

Translator

B. Seckler, 19 Ramsey Road, Great Neck, NY 11023-1611, USA, email: bersec@aol.com

Contents

Preface

This book consists of five parts written by different authors devoted to various problems dealing with probability limit theorems.

The first part, "Classical-Type Limit Theorems for Sums of Independent Random Variables" (V.V. Petrov), presents a number of classical limit theorems for sums of independent random variables as well as newer related results. The presentation dwells on three basic topics: the central limit theorem, laws of large numbers and the law of the iterated logarithm for sequences of real-valued random variables.

The second part, "The Accuracy of Gaussian Approximation in Banach Spaces" (V. Bentkus, F. Götze, V. Paulauskas and A. Račkauskas), reviews various results and methods used to estimate the convergence rate in the central limit theorem and to construct asymptotic expansions in infinite-dimensional spaces. The authors confine themselves to independent and identically distributed random variables. They do not strive to be exhaustive or to obtain the most general results; their aim is merely to point out the differences from the finite-dimensional case and to explain certain new phenomena related to the more complex structure of Banach spaces. Also reflected here is the growing tendency in recent years to apply results obtained for Banach spaces to asymptotic problems of statistics.

The third part, "Approximation of Distributions of Sums of Weakly Dependent Random Variables by the Normal Distribution" (J. Sunklodas), surveys known results on the normal approximation of sums of weakly dependent random variables and random fields defined on an integer lattice Z^d and gives some new results obtained by the author. Along with the general-purpose methods of Bernstein, cumulants and others, for studying limit laws and estimating their convergence rates for sums of weakly dependent random variables, the methods of Stein, Tikhomirov and Heinrich (for m-dependent random variables) prove to be very productive. Attention is therefore given mainly to the three methods of proof of Stein, Tikhomirov and Heinrich. It is shown how these methods work in estimating from above the convergence rate in the central limit theorem in various metrics (uniform, L_1, the Lipschitz bounded metric, and so on). A sequence of random variables is assumed to satisfy some weak dependency condition expressed in terms of a mixing coefficient (strong mixing, absolute regularity, uniformly mixing, ψ-mixing, m-dependent).

The fourth part, "Refinements of the Central Limit Theorem for Homogeneous Markov Chains" (P. Gudynas), discusses two of the most developed methods of proving limit theorems for sums of random variables connected in a homogeneous Markov chain. The first one is the spectral method. As the author showed in his papers, it enables one to apply the powerful analytic tools of perturbation theory of linear operators. The second one is the regenerative method. The crux of it is that one goes over to sums of independent random variables. On the one hand, the results obtained by the spectral method are more varied, while the regenerative method aids in weakening the conditions that have been imposed. A comparison of the two methods makes one think that there is a need for a synthetic approach that combines each of their advantages.

The fifth part, "Limit Theorems on Large Deviations" (L. Saulis and V. Statulevičius), is devoted to using the cumulant method in limit theorems with large deviations. The reader will be convinced that this method works well in studying the probabilities of large deviations for sums of both independent and dependent random variables, polynomial forms, multiple stochastic integrals of random processes and fields, polynomial statistics, and in functional limit theorems.

I. Classical-Type Limit Theorems for Sums of Independent Random Variables

V.V. Petrov

Contents

Introduction

This article presents a number of classical limit theorems for sums of independent random variables and more recent results which are closely related to the classical theorems. It concentrates on three basic subjects: the central limit theorem, the laws of large numbers and the law of the iterated logarithm for sequences of independent real-valued random variables. The author was restricted to an article of small size. Therefore many chapters of the classical theory of summation of independent random variables were omitted, particularly limit theorems with non-normal limit distributions, multidimensional limit theorems and local limit theorems. This article may be regarded as an introduction for the reader who wishes to become acquainted with the classical limit theorems for sums of independent random variables without spending much time. More detailed presentations may be found, for example, in the author's book (1987) and in its predecessors Csörgő and Révész (1981), Gnedenko and Kolmogorov (1949), Hall (1982), Ibragimov and Linnik (1965), Loève (1960), Petrov (1972, 1995), Révész (1967) and Stout (1974).

Chapter 1
The Central Limit Theorem and Its Generalizations

§1.1. Theorems of Lyapunov, Lindeberg and Feller

Let $X_1, X_2, \dots$ be a sequence of random variables. Write $S_n = \sum_{k=1}^n X_k$. Following the established terminology, we shall call a *central limit theorem* any assertion such that, under certain conditions, the distribution function of the properly centered and normalized sum S_n converges to the normal distribution function as $n \to \infty$. The first such results were obtained by DeMoivre and Laplace for the Bernoulli scheme and by Chebyshev and Lyapunov for much more general sequences of independent random variables.

Let $X_1, X_2, \dots$ be a sequence of independent random variables, $\mathbf{E}X_n = 0$ and $\mathbf{E}|X_n|^{2+\delta} < \infty$ for all n and some $\delta > 0$. Write

$$B_n = \sum_{k=1}^n \mathbf{E}X_k^2, \qquad F_n(x) = \mathbf{P}\left\{S_n < x\sqrt{B_n}\right\},$$

$$L_n = B_n^{-1-\delta/2} \sum_{k=1}^n \mathbf{E}|X_k|^{2+\delta}, \qquad \Phi(x) = \frac{1}{\sqrt{2\pi}} \int_{-\infty}^x \exp\{-t^2/2\}\, dt.$$

Lyapunov proved that if $L_n \to 0$ as $n \to \infty$, then

$$F_n(x) \to \Phi(x) \text{ for every } x. \tag{1}$$

The moment assumptions of this theorem were weakened by Lindeberg. He proved (1) for any sequence of independent random variables with zero means and finite variances satisfying the following condition (*Lindeberg's condition*):

$$\frac{1}{B_n} \sum_{k=1}^{n} \int_{|x| \geq \varepsilon \sqrt{B_n}} x^2 \, dV_k(x) \to 0$$

for every $\varepsilon > 0$, where $V_k(x)$ is the distribution function of X_k. It is easy to show that if *Lyapunov's condition* $L_n \to 0$ is satisfied, then Lindeberg's condition is satisfied. Lindeberg's theorem is therefore a generalization of Lyapunov's theorem. If $X_1, X_2, \ldots$ all have the same distribution function with a finite variance, then Lindeberg's condition obviously holds. We obtain the following *Lévy theorem* as a simple consequence of Lindeberg's theorem: If $X_1, X_2, \ldots$ is a sequence of independent and identically distributed random variables such that $\mathbf{E}X_1 = 0$, $\mathbf{E}X_1^2 = \sigma^2$ $(0 < \sigma^2 < \infty)$, then $\mathbf{P}\{S_n < x\sigma\sqrt{n}\} \to \Phi(x)$ for every x.

Feller showed that Lindeberg's condition is necessary for (1) and the relation $B_n^{-1} \max_{1 \leq k \leq n} \mathbf{E}X_k^2 \to 0$ as $n \to \infty$.

There are more general forms of the central limit theorem with no assumptions about the existence of moments. These more general forms of the central limit theorem involve a double sequence of random variables $\{X_{nk};\ k = 1, 2, \ldots, k_n;\ n = 1, 2, \ldots\}$ such that the variables of the nth row are mutually independent and either satisfy or do not satisfy the following infinitesimality condition:

$$\lim_{n \to \infty} \max_{1 \leq k \leq k_n} \mathbf{P}\{|X_{nk}| \geq \varepsilon\} = 0$$

for every $\varepsilon > 0$. Detailed presentations may be found, for example, in Gnedenko and Kolmogorov (1949), Loéve (1960), Petrov (1987) and Zolotarev (1986).

The central limit theorem may be deduced as a consequence of more general theorems which state necessary and sufficient conditions for the convergence of the distributions of sums of independent random variables to a given infinitely divisible distribution. The degenerate and normal distributions are infinitely divisible. Therefore it is possible to use the general theorems on convergence to an arbitrary infinitely divisible distribution to obtain general forms of the weak law of large numbers and the central limit theorem.

To prove the central limit theorem, Lyapunov and his successors applied the method of characteristic functions. Afterwards the convolution method and the operator method were also used.

§1.2. The Esseen and Berry-Esseen Inequalities

Any limit theorem becomes more valuable if it is accompanied by estimates for the rate of convergence. The estimates which are in some sense optimal are of special interest. The first estimates of the convergence rate in the central limit theorem were obtained by Lyapunov. We introduce the following additional notation:

$$\Delta_n = \sup_{x \in R} |F_n(x) - \Phi(x)|.$$

If the assumptions of Lyapunov's theorem are satisfied, then $\Delta_n = O(L_n)$ for $0 < \delta < 1$ and $\Delta_n = O(L_n|\log L_n|)$ for $\delta = 1$. These estimates were proved by Lyapunov. Attempts to refine them led to essential improvements in the method of characteristics functions and to significant progress in the area of limit theorems of probability theory.

Optimal estimates were obtained for Δ_n by Cramér for the special case of identical distributions satisfying the Cramér condition (C)

$$\limsup_{|t| \to \infty} |\mathbf{E}e^{itX_1}| < 1.$$

Later on, Berry (1941) and Esseen (1942) derived such estimates for independent and identically distributed random variables having finite absolute third moments.

Let $X_1, \ldots, X_n$ be mutually independent random variables with a common distribution function such that $\mathbf{E}X_1 = 0$, $\mathbf{E}X_1^2 = \sigma^2 > 0$ and $\mathbf{E}|X_1|^3 < \infty$. Then

$$\sup_x \left| \mathbf{P}\left\{ \frac{S_n}{\sigma\sqrt{n}} < x \right\} - \Phi(x) \right| \le \frac{A\mathbf{E}|X_1|^3}{\sigma^3\sqrt{n}},$$

where A is a positive absolute constant. This is the *Berry-Esseen inequality.*

In the general case of non-identical distributions, we have the following theorem of Esseen (1945). Let $X_1, \ldots, X_n$ be mutually independent random variables such that $\mathbf{E}X_k = 0$ and $\mathbf{E}|X_k|^3 < \infty$, $k = 1, \ldots, n$. Put $B_n = \sum_{k=1}^n \mathbf{E}X_k^2$,

$$L_n = B_n^{-3/2} \sum_{k=1}^n \mathbf{E}X_k^3,$$

$$\Delta_n = \sup_x \left| \mathbf{P}\{S_n < x\sqrt{B_n}\} - \Phi(x) \right|.$$

Then $\Delta_n \le AL_n$, where A is a positive absolute constant (the *Esseen inequality*).

In the special case of identical distributions, the Esseen inequality reduces to the Berry–Esseen inequality.

The theorems of this chapter assume that the means are zero. Of course, this assumption does not lessen the generality of the theorems. If it is not satisfied, then the variables $Y_k = X_k - \mathbf{E}X_k$, $k = 1, \ldots, n$, have zero means and we can apply the above results to these random variables.

The order of the Esseen and Berry–Esseen estimates is unimprovable with no additional assumptions about the distributions of our random variables. We shall give an elementary proof of this assertion. Consider a sequence of independent symmetric random variables taking the two values -1 and 1. Then $\mathbf{E}X_1 = 0$, $\mathbf{E}X_1^2 = 1$, $\mathbf{E}|X_1|^3 = 1$ and $L_n = n^{-1/2}$. (By means of Lyapunov's moment inequality, it is easy to show that $L_n \geq n^{-1/2}$ for any distribution.) The event $S_n = 0$ occurs if and only if half of the terms in the sum S_n take the value 1 and the other half the value -1. Therefore $\mathbf{P}\{S_n = 0\} = \binom{n}{n/2} 2^{-n}$ if n is an even number. Applying Stirling's formula, we easily find that $\mathbf{P}\{S_n = 0\} \approx \frac{2}{\sqrt{2\pi n}}$. Thus the function $F_n(x) = \mathbf{P}\{S_n < x\sqrt{n}\}$ has a jump of magnitude $\frac{2}{\sqrt{2\pi n}}\,(1 + o(1))$ at $x = 0$. It follows that in a neighborhood of $x = 0$, F_n cannot be approximated by a continuous function to an accuracy exceeding $\frac{1}{\sqrt{2\pi n}}\,(1 + o(1))$.

We now see that the absolute constant A in the Berry–Esseen inequality and in the Esseen inequality is not less than $1/\sqrt{2\pi}$. It is possible to show that both inequalities hold with $A = 0.8$. More exact information about the constants in the inequalities of this kind are found, for example, in Petrov (1987) and Zolotarev (1986).

§1.3. Generalizations of Esseen's Inequality

Esseen's inequality was stated under the assumption that the absolute third moments are finite. It is desirable to estimate the rate of convergence in the central limit theorem under weaker conditions.

Let G be the set of functions $g(x)$, defined for all x, that satisfy the following conditions: (a) $g(x)$ is non-negative, even, and non-decreasing for $x > 0$; (b) $x/g(x)$ is non-decreasing for $x > 0$.

Let $X_1, \ldots, X_n$ be mutually independent random variables such that $\mathbf{E}X_k = 0$ and $\mathbf{E}X_1^2 g(X_k) < \infty$ for $k = 1, \ldots, n$ and some $g \in G$. Then

$$\Delta_n \leq \frac{A}{B_n g(\sqrt{B_n})} \sum_{k=1}^{n} \mathbf{E}X_k^2 g(X_k),$$

where A is an absolute constant and Δ_n is the same as in Esseen's inequality.

If $0 < \delta \leq 1$, then the function $g(x) = |x|^\delta$ belongs to G. Therefore the last result implies Esseen's theorem and the following generalization of it: If $\mathbf{E}X_k = 0$ and $\mathbf{E}|X_k|^{2+\delta} < \infty$, $k = 1, \ldots, n$, for some positive $\delta \leq 1$, then

$$\Delta_n \le AB_n^{-1-\delta/2} \sum_{k=1}^{n} \mathbf{E}|X_k|^{2+\delta}.$$

This inequality strengthens Lyapunov's estimate.

Thus we have arrived at a bound for the convergence rate in the central limit theorem assuming the existence of moments of order $2+\delta$. It is possible to derive bounds for Δ_n without this last condition by assuming only the existence of second-order moments.

Let $X_1, \dots, X_n$ be independent random variables with zero means and finite variances. Put $B_n = \sum_{k=1}^n \mathbf{E}X_k^2$,

$$V_k(x) = \mathbf{P}\{X_k < x\}, \qquad \Lambda_n(\varepsilon) = \frac{1}{B_n} \sum_{k=1}^{n} \int_{|x|\ge\varepsilon\sqrt{B_n}} x^2\, dV_k(x),$$

and

$$l_n(\varepsilon) = \frac{1}{B_n^{3/2}} \sum_{k=1}^{n} \int_{|x|<\varepsilon\sqrt{B_n}} |x|^3\, dV_k(x).$$

Then

$$\Delta_n \le A(\Lambda_n(\varepsilon) + l_n(\varepsilon))$$

for every $\varepsilon > 0$, where A is a positive absolute constant.

This result follows for $\varepsilon = 1$ from the above generalization of Esseen's inequality if we take $g(x) = |x|$ when $|x| < \sqrt{B_n}$, and $g(x) = \sqrt{B_n}$ when $|x| \ge \sqrt{B_n}$. We have $\ell_n(\varepsilon) \le \varepsilon$ and therefore $\Delta_n \le A(\varepsilon + \Lambda_n(\varepsilon))$ for every $\varepsilon > 0$. Let us mention a simple consequence of the last inequality. If $\Lambda_n(\varepsilon) \to 0$ as $n \to \infty$ for every $\varepsilon > 0$, then $F_n(x) \to \Phi(x)$ (*Lindeberg's theorem*).

It is possible to go further and obtain estimates in the central limit theorem with no assumptions about the existence of moments.

Let $X_1, \dots, X_n$ be random variables (here we do not assume independence). Let $t_1, \dots, t_n$ be positive numbers. Introduce the truncated random variables

$$\overline{X}_k = \begin{cases} X_k, & \text{if } |X_k| < t_k, \\ 0, & \text{if } |X_k| \ge t_k, \end{cases}$$

where $k = 1, \dots, n$. Write

$$M_n = \sum_{k=1}^{n} \mathbf{E}\overline{X}_k = \sum_{k=1}^{n} \int_{|x|<t_k} x\, dV_k(x),$$

$$N_n = \mathbf{E}\left(\sum_{k=1}^{n} \overline{X}_k - M_n\right)^2,$$

$$\Delta_n = \sup_x \left| \mathbf{P}\left\{ N_n^{-1/2} \sum_{k=1}^{n} (\overline{X}_k - \mathbf{E}\overline{X}_k) < x \right\} - \Phi(x) \right|,$$

$$\Gamma_n = \sum_{k=1}^{n} \mathbf{P}\{|X_k| \geq t_k\}.$$

For all real numbers $a > 0$ and b, we have

$$\sup_x \left| \mathbf{P}\left\{ \frac{1}{a} \sum_{k=1}^{n} X_k - b < x \right\} - \Phi(x) \right|$$
$$\leq \Delta_n + \Gamma_n + \frac{|ab - M_n|}{\sqrt{2\pi N_n}} + \frac{1}{2\sqrt{2\pi e}} \left| 1 - \frac{N_n}{a^2} \right| \max\left(1, \frac{a^2}{N_n}\right).$$

This inequality was proved by Osipov and Petrov (1967). A close result was obtained later by Feller (1968). Heyde (1973) discovered the asymptotically optimum behavior of the Osipov–Petrov estimate.

Two preceding results of this section are due to Petrov (1965) and Osipov (1966), respectively. The first result is a generalization of a theorem of Katz (1963) on independent and identically distributed summands.

Osipov (1968) found upper and lower bounds of the same order for the remainder in the central limit theorem for independent and identically distributed random variables having finite variances. Later on more general results were obtained by Rozovskii (1975), (1978a), Hall (1980), (1982) and Heyde and Nakata (1984). In particular the following theorem was proved. If $X_1, X_2, \ldots$ is a sequence of independent and identically distributed random variables such that $\mathbf{E}X_1 = 0$ and $\mathbf{E}X_1^2 = 1$, then

$$\sup_x \left| \mathbf{P}\left\{ \sum_{k=1}^{n} X_k < x\sqrt{n} \right\} - \Phi(x) \right| + \frac{1}{\sqrt{n}} \asymp \psi_n + \frac{1}{\sqrt{n}},$$

where

$$\begin{aligned}\psi_n &= \mathbf{E}X_1^2 I(|X_1| \geq \sqrt{n}) \\ &\quad + n^{-1/2} |\mathbf{E}X_1^3 I(|X_1| < \sqrt{n})| + n^{-1} \mathbf{E}X_1^4 I(|X_1| < \sqrt{n}),\end{aligned}$$

and $I(A)$ is the indicator of event A.

§1.4. Asymptotic Expansions in the Central Limit Theorem

The concept of asymptotic expansions in the central limit theorem for sums of independent random variables is due to Chebyshev. It was developed by Cramér, Esseen and others. Let us state two results of Esseen for a sequence of independent and identically distributed random variables X_1, $X_2, \ldots$ with zero mean and finite positive variance σ^2. As usual, we put $F_n(x) = \mathbf{P}\left\{\sum_{k=1}^{n} X_k < x\sigma\sqrt{n}\right\}$. If $\mathbf{E}|X_1|^3 < \infty$ and X_1 has a non-lattice distribution (by definition, a random variable has a lattice distribution if, with probability 1, its values may be written in the form $a+kh$, where a and h are fixed real numbers and $k=0, \pm 1, \pm 2, \ldots$), then

$$F_n(x) = \Phi(x) + \frac{\alpha_3(1-x^2)\exp\{-x^2/2\}}{6\sigma^3\sqrt{2\pi n}} + o\left(\frac{1}{\sqrt{n}}\right) \quad (n \to \infty)$$

uniformly in x. Here $\alpha_3 = \mathbf{E}X_1^3$. If $E|X_1|^p < \infty$ for some integer $p \geq 3$ and Cramér's condition (C) is satisfied, then

$$F_n(x) = \Phi(x) + \sum_{\nu=1}^{p-2} \frac{Q_\nu(x)}{n^{\nu/2}} + o\left(\frac{1}{n^{(p-2)/2}}\right),$$

where $Q_\nu(x) = P_{3\nu-1}(x)\,\mathrm{e}^{-x^2/2}$, $P_{3\nu-1}(x)$ being a polynomial in x of degree $3\nu-1$ with coefficients depending only on the moments of X_1 up to order $\nu+2$. In particular, $Q_1(x) = \alpha_3(1-x^2)(6\sigma^3\sqrt{2\pi})^{-1}\,\mathrm{e}^{-x^2/2}$. Explicit formulas for $Q_\nu(x)$ were derived by Petrov (1962). The asymptotic expansion of the distribution function of a normalized sum of independent random variables having a common lattice distribution has some additional terms. Corresponding asymptotic expansions have been found for sequences of independent and non-identically distributed random variables. For the case of identical distributions, Osipov (1971) derived upper and lower bounds for the remainder in the asymptotic expansions of the same order. Osipov's methods and results were used and developed by Rozovskii, Hall and others (see Hall (1982) and Petrov (1987)).

§1.5. Nonuniform Estimates

Consider a sequence of independent and identically distributed random variables $X_1, X_2, \ldots$ such that $\mathbf{E}X_1 = 0$, $\mathbf{E}X_1^2 = \sigma^2 > 0$ and $\mathbf{E}|X_1|^3 = \beta_3 < \infty$. Put $F_n(x) = \mathbf{P}\{\sum_{k=1}^n X_k < x\sigma\sqrt{n}\}$. §§ 1.2 and 1.3 stated some estimates for the difference $F_n(x) - \Phi(x)$ which are uniform in x. More precise information is given by nonuniform estimates of this difference taking into account its dependence not only on n but also on x. A nonuniform strengthening of the Berry–Esseen inequality has the following form:

$$|F_n(x) - \Phi(x)| \leq \frac{A\beta_3}{\sigma^3\sqrt{n}(1+|x|^3)}$$

for every $x \in R$ (Nagaev (1965)). We now state a generalization of this result to the case of non-identically distributed summands.

Let $X_1, \ldots, X_n$ be mutually independent random variables, $\mathbf{E}X_k = 0$ and $\mathbf{E}|X_k|^{2+\delta} < \infty$ for some positive $\delta \leq 1$, $k = 1, \ldots, n$. Put

$$B_n = \sum_{k=1}^n \mathbf{E}X_k^2, \qquad F_n(x) = \mathbf{P}\left\{\sum_{k=1}^n X_k < x\sqrt{B_n}\right\},$$

$$L_n = B_n^{-1-\delta/2} \sum_{k=1}^{n} \mathbf{E}|X_k|^{2+\delta}.$$

Then

$$|F_n(x) - \Phi(x)| \leq \frac{AL_n}{1 + |x|^{2+\delta}}$$

for every $x \in R$, where A is a positive absolute constant (Bikelis (1966)).

There are nonuniform strengthenings of other inequalities of §§ 1.2 and 1.3 as well as nonuniform estimates of the remainder in the asymptotic expansions connected with the central limit theorem. One of the most important results of the last kind was proved by Osipov (1967). His general theorem about asymptotic expansions for distributions of sums of independent and identically distributed random variables implies the following result: If the assumptions of the Berry–Esseen theorem are satisfied and if $\mathbf{E}|X_1|^r < \infty$ for some $r \geq 3$, then

$$|F_n(x) - \Phi(x)| \leq \frac{C(r)}{1 + |x|^r} \left(\frac{\mathbf{E}|X_1|^3}{\sigma^3 \sqrt{n}} + \frac{\mathbf{E}|X_1|^r}{\sigma^r n^{(r-2)/2}} \right)$$

for every x, where $C(r)$ is a positive constant depending only on r.

The nonuniform bounds for the remainder in the central limit theorem and subsequent integration make it possible to obtain a so-called global form of the central limit theorem with assertions such as

$$\int_{-\infty}^{\infty} |F_n(x) - \Phi(x)|^p \, dx \to 0$$

for some $p > 0$ and $n \to \infty$. General global theorems were proved by Kruglov (1976).

§1.6. Convergence Rates: Necessary and Sufficient Conditions

We are interested in conditions which are necessary and sufficient for the distributions of sums of independent random variables to converge to the normal law at a given rate. Let $X_1, X_2, \ldots$ be a sequence of independent and identically distributed random variables. Let there exist numerical sequences $\{a_n\}$ and $\{b_n\}$ such that

$$F_n(x) = \mathbf{P}\left\{ \frac{1}{a_n} \sum_{k=1}^{n} X_k - b_n < x \right\} \to \Phi(x).$$

Put

$$r_n = \inf_{a_n, b_n} \sup_x |F_n(x) - \Phi(x)|,$$

$$V(x) = \mathbf{P}\{X_1 < x\}.$$

In order that $r_n = O(n^{-\delta/2})$, where $0 < \delta < 1$, it is necessary and sufficient that

$$\int_{-\infty}^{\infty} x^2\,dV(x) < \infty, \qquad \int_{|x|\geq z} x^2\,dV(x) = O(z^{-\delta})\ (z \to \infty).$$

If $\delta = 1$, then $r_n = O(n^{-1/2})$ if and only if these last conditions are satisfied for $\delta = 1$ and

$$\int_{-z}^{z} x^3\,dV(x) = O(1)\ (z \to \infty).$$

(Ibragimov (1966)).

It is interesting to obtain estimates of the remainder in the central limit theorem that are equivalent to the existence of the moments of a given order. Let $X_1, X_2, \ldots$ be a sequence of independent and identically distributed random variables, $\mathbf{E}X_1 = 0$ and $\mathbf{E}X_1^2 = 1$. Put $\Delta_n = \sup_x |\mathbf{P}\{\sum_{k=1}^{n} X_k < x\sqrt{n}\} - \Phi(x)|$. Heyde (1967) proved that the series $\sum_{n=1}^{\infty} n^{-1}\Delta_n$ converges if and only if $\mathbf{E}X_1^2 \log(1+|X_1|) < \infty$; when $0 < \delta < 1$, the convergence of the series $\sum_{n=1}^{\infty} n^{-1+\delta/2}\Delta_n$ is equivalent to the condition $\mathbf{E}|X_1|^{2+\delta} < \infty$.

Further results about the equivalence of certain moment conditions and the convergence of the series of weighted remainders in the central limit theorem were obtained by Heyde (1969, 1973) and Egorov (1973). They made no aprori assumptions about the existence of the moments of the random variables under consideration.

Chapter 2
Laws of Large Numbers

§2.1. The Weak Law of Large Numbers

Let $X, X_1, X_2, \ldots$ be a sequence of random variables defined on a probability space $(\Omega, \mathcal{A}, \mathbf{P})$. By definition, the sequence $\{X_n\}$ converges to X in probability if $\mathbf{P}\{|X_n - X| < \varepsilon\} \to 1$ for every fixed $\varepsilon > 0$ as $n \to \infty$. In that case, we write $X_n \xrightarrow{\mathbf{P}} X$. $\{X_n\}$ is said to be stable if there exists a sequence of constants $\{b_n\}$ such that $X_n - b_n \xrightarrow{\mathbf{P}} 0$.

Consider a sequence of random variables $\{X_n\}$ and a sequence of non-zero real numbers $\{a_n\}$. Put $S_n = \sum_{k=1}^{n} X_k$. We say that the sequence $\{X_n\}$ satisfies the weak law of large numbers with the normalizing sequence $\{a_n\}$ if $\{S_n/a_n\}$ is a stable sequence.

The following theorem of Feller is one of the most important results on the weak law of large numbers for sequences of independent random variables. There are no assumptions about the existence of any moments in this theorem.

Let $\{X_n\}$ be a sequence of independent random variables and $\{a_n\}$ a sequence of positive numbers such that $a_n \uparrow \infty$. Put $V_n(x) = \mathbf{P}\{X_n < x\}$. In order that $\frac{1}{a_n}\sum_{k=1}^n X_k \xrightarrow{\mathbf{P}} 0$, it is necessary and sufficient that

$$\sum_{k=1}^{n} \mathbf{P}\{|X_k| \geq a_n\} \to 0,$$

$$\frac{1}{a_n^2}\sum_{k=1}^{n}\left\{\int_{|x|<a_n} x^2\, dV_k(x) - \left(\int_{|x|<a_n} x\, dV_k(x)\right)^2\right\} \to 0,$$

and

$$\frac{1}{a_n}\sum_{k=1}^{n}\int_{|x|<a_n} x\, dV_k(x) \to 0.$$

For the special case of identical distributions, Feller's theorem implies the following result. If $\{X_n\}$ is a sequence of independent random variables with a common distribution function $V(x)$, then the relation $\frac{1}{n}\sum_{k=1}^n X_k \xrightarrow{\mathbf{P}} 0$ is equivalent to the set of conditions $n\mathbf{P}\{|X_1| \geq n\} \to 0$ and $\int_{|x|<n} x\, dV(x) \to 0$. In turn, the last result implies the following theorem of Khinchine: If $\{X_n\}$ is a sequence of independent and identically distributed random variables with $\mathbf{E}X_1 = a$, then $\frac{1}{n}\sum_{k=1}^n X_k \xrightarrow{\mathbf{P}} a$.

There are many results concerning rates of convergence in the weak law of large numbers. We confine ourselves to stating the following theorems of Baum and Katz (1965). The relation $\mathbf{P}\{|S_n| \geq n\varepsilon\} = o(n^{-t})$ for every $\varepsilon > 0$ is equivalent to the conditions $\mathbf{P}\{|X_1| \geq n\} = o(n^{-t-1})$ and $\int_{|x|<n} x\, dV(x) = o(1)$. Here $t \geq 0$ and $V(x)$ is the distribution function of $X_1, X_2, \ldots$. Thus the conditions $\mathbf{E}X_1 = 0$ and $\mathbf{E}|X_1|^p < \infty$ for some $p \geq 1$ are sufficient for $\mathbf{P}\{|S_n| \geq n\varepsilon\} = o(n^{-p})$ for every $\varepsilon > 0$. If $pr > 1$ and $r > 1/2$, then the following conditions are equivalent:

(a) $\mathbf{E}|X_1|^p < \infty$ and (for $p \geq 1$) $\mathbf{E}X_1 = 0$;
(b) $\sum_{n=1}^{\infty} n^{pr-2}\mathbf{P}\{|S_n| \geq n^r\varepsilon\} < \infty$ for every $\varepsilon > 0$.

§2.2. The Strong Law of Large Numbers

Let $X, X_1, X_2, \ldots$ be a sequence of random variables defined on a probability space $(\Omega, \mathcal{A}, \mathbf{P})$. By definition, the sequence $\{X_n\}$ converges to X almost surely (a.s.) if $X_n(\omega) \to X(\omega)$ for all $\omega \in \Omega$ except for a set of ω's of $\mathbf{P}$-measure zero. $\{X_n\}$ is said to be strongly stable if there exists a sequence of constants $\{b_n\}$ such that $X_n - b_n \to 0$ a.s.

Let $\{X_n\}$ be a sequence of random variables, $S_n = \sum_{k=1}^{n} X_k$ and $\{a_n\}$ be a sequence of non-zero constants. We shall say that $\{X_n\}$ obeys the strong law of large numbers with the sequence of normalizing constants $\{a_n\}$ if $\{S_n/a_n\}$ is a strongly stable sequence.

The following Kolmogorov theorems are among the most important results about the strong law of large numbers. If $\{X_n\}$ is a sequence of independent random variable such that $\sum_{n=1}^{\infty} \operatorname{Var} X_n/n^2 < \infty$, then $(S_n - \mathbf{E}S_n)/n \to 0$ a.s. If $\{X_n\}$ is a sequence of independent and identically distributed random variables, then there exists a constant b such that $S_n/n \to b$ a.s. if and only if $\mathbf{E}|X_1| < \infty$. If the latter condition is satisfied, then $b = \mathbf{E}X_1$.

In the second of Kolmogorov's theorems, the independence assumption may be replaced by the weaker condition of pairwise independence (Etemadi (1981)). Let us state a generalization of the first theorem of Kolmogorov. If $\{X_n\}$ is a sequence of independent random variables, $a_n \uparrow \infty$ and $\sum_{n=1}^{\infty} a_n^{-p}\mathbf{E}|X_n|^p < \infty$ for some positive $p \le 2$, then $S_n/a_n \to 0$ a.s. if $p \le 1$; the proposition remains true for $1 < p \le 2$ under the additional assumption $\mathbf{E}X_n = 0$, $n = 1, 2, \dots$.

It is possible to generalize the last result. Let $\{X_n\}$ be a sequence of independent random variables and $g(x)$ an even function which is positive and non-decreasing for $x > 0$ and which satisfies at least one of the following conditions: (a) $x/g(x)$ does not decrease for $x > 0$; (b) $x/g(x)$ and $g(x)/x^2$ do not increase for $x > 0$. If (b) is satisfied, then we assume in addition that $\mathbf{E}X_n = 0$ for all n. Let $a_n \uparrow \infty$ and $\sum_{n=1}^{\infty} \mathbf{E}g(X_n)/g(a_n) < \infty$. Then $S_n/a_n \to 0$ a.s.

Note that the function $g(x) = |x|^p$ satisfies (a) if $0 < p \le 1$ and (b) if $1 \le p \le 2$. Kolmogorov's theorem corresponds to $g(x) = x^2$ and $a_n = n$.

The next theorem explains some connections between the weak and strong laws of large numbers for the traditional normalizing sequence $a_n = n$. Let $\{X_n\}$ be a sequence of independent random variables. Then the relation $S_n/n \to 0$ a.s. is equivalent to the set of conditions $S_n/n \xrightarrow{\mathbf{P}} 0$ and $\sum_{n=1}^{\infty} \mathbf{P}\{|S_{2^{n+1}} - S_{2^n}| \ge 2^n\varepsilon\} < \infty$ for every $\varepsilon > 0$.

Another result is closely connected with Feller's theorem on the weak law of large numbers. Let $\{X_n\}$ be a sequence of independent random variables with the distribution functions $\{V_n(x)\}$. Suppose that $a_n \uparrow \infty$ and $\sum_{n=1}^{\infty} \dfrac{1}{a_n^2} \int_{|x|<a_n} x^2\,dV_n(x) < \infty$. In order that $S_n/a_n \to 0$ a.s., it is necessary and sufficient that $\sum_{n=1}^{\infty} \mathbf{P}\{|X_n| \ge a\} < \infty$ and $\dfrac{1}{a_n} \sum_{k=1}^{n} \int_{|x|<a_n} x\,dV_k(x) \to 0$.

Necessary and sufficient conditions for the applicability of the strong law of large numbers with the traditional normalizing constants $a_n = n$ were obtained by Prokhorov (1959) in terms of the variances of the summands under the assumption that $X = O(n/\log\log n)$. This condition was shown to be essential by Prokhorov (1959). In the general case of unbounded random variables and $a_n = n$, other sets of necessary and sufficients conditions were found by Prokhorov (1950) and Nagaev (1972). Arbitrary (not necessarily monotone) sequences of normalizing constants are permitted in the theorems of Martikainen (1979c) and Buldygin (1975). We state a result of Martikainen.

Let $\{a_n\}$ be a sequence of non-zero real constants. Fix an arbitrary number $c > 1$. Let i_n be the greatest integer for which $|a_{i_n}| \leq c^n$ if there are finitely many integers with this property; otherwise put $i_n = 0$. A sequence of independent random variables $\{X_n\}$ satisfies the strong law of large numbers with the normalizing sequence $\{a_n\}$ if and only if

(a) $|a_n| \to \infty$ or all X_n are degenerate; and

(b) $\sum\limits_{n=1}^{\infty} \mathbf{P}\{|S_{i_n} - S_{i_{n-1}} - m(S_{i_n} - S_{i_{n-1}})| \geq \varepsilon c^n\} < \infty$

for every $\varepsilon > 0$. Here mX is a median of random variable X.

Martikainen (1979b) also found necessary and sufficient conditions for the strong law of large numbers to be applicable to a sequence of independent random variables with an arbitrary normalizing sequence of constants; they are expressed in terms of the distributions of the individual summands and not their sums.

Most studies on the applicability of the strong law of large numbers were done under the assumption that the normalizing sequence of constants is non-decreasing. This assumption does not lessen the generality of the considerations. Martikainen (1979c) proved that a sequence of independent random variables $\{X_n\}$ satisfies the strong law of large numbers with an arbitrary non-zero normalizing sequence of constants $\{a_n\}$ if and only if the variables satisfy the strong law of large numbers with the normalizing sequence $\{c_n\} = \{\inf_{k \geq n} |a_k|\}$.

Under certain moment conditions, one may derive estimates of the order of growth of the sums S_n in terms of the sums of moments of the summands (Petrov (1969)). We need some additional notation. Let Ψ_c (or, respectively, Ψ_d) denote the set of functions $\psi(x)$ such that each $\psi(x)$ is positive and non-decreasing for $x > x_0$ for some x_0 and $\sum \dfrac{1}{n\psi(n)}$ converges (respectively, diverges). For example, $x^\alpha \in \Psi_c$ for any $\alpha > 0$; $\log x \in \Psi_d$.

Let $g(x)$ be a continuous even function, which is positive and strictly increasing for $x > 0$ and such that $g(x) \to \infty$ as $x \to \infty$. Suppose that any of the two following conditions are satisfied:

(a) $x/g(x)$ is non-decreasing for $x > 0$;
(b) $x/g(x)$ and $g(x)/x^2$ do not increase for $x > 0$.

Suppose that $\{X_n\}$ is a sequence of independent random variables and $\mathbf{E}g(X_n) < \infty$ for all n. Suppose in addition that $\mathbf{E}X_n = 0$ for all n if condition (b) is satisfied. If $M_n = \sum_{k=1}^{n} \mathbf{E}g(X_k) \to \infty$ as $n \to \infty$, then

$$S_n = o(g^{-1}(M_n\psi(M_n))) \text{ a.s.}$$

for any $\psi \in \Psi_c$, where g^{-1} is the inverse of g.

The assertion may fail to be true if instead of $\psi \in \Psi_c$ we choose a more slowly increasing function $\psi \in \Psi_d$.

In the special case $g(x) = |x|^p$, where $0 < p \le 2$, the statements are much simpler. In particular, if $p = 2$, then we obtain the following proposition. If $\{X_n\}$ is a sequence of independent random variables with finite variances and $B_n = \sum_{k=1}^{n} \operatorname{Var} X_k \to \infty$, then $S_n - \mathbf{E}S_n = o\left(\sqrt{B_n\psi(B_n)}\right)$ a.s. for any function $\psi \in \Psi_c$. If $\psi \in \Psi_d$, this assertion may fail to be true.

Let us note a simple consequence. The sums S_n of independent random variables with finite variances whose variances $B_n = \operatorname{Var} S_n$ increase infinitely, admit the following growth estimates (in order of increasing severity):

$$S_n - \mathbf{E}S_n = o(B_n^{1/2+\varepsilon}) \text{ a.s.},$$
$$S_n - \mathbf{E}S_n = o(B_n^{1/2}(\log B_n)^{1/2+\varepsilon}) \text{ a.s.},$$
$$S_n - \mathbf{E}S_n = o(B_n^{1/2}(\log B_n)^{1/2}(\log\log B_n)^{1/2+\varepsilon}) \text{ a.s.}$$

and so on. We cannot replace ε by zero without introducing additional restrictions.

Much research has been devoted to the rates of convergence in the strong law of large numbers. We shall state some results of Baum and Katz (1965) concerning sequences of independent and identically distributed random variables $\{X_n\}$. If $t > 0$, then for every $\varepsilon > 0$ the relation

$$\mathbf{P}\left\{\sup_{k\ge n} \frac{1}{k}|S_k| \ge \varepsilon\right\} = o(n^{-t})$$

is equivalent to the conditions

$$\mathbf{P}\{|X_1| \ge n\} = o(n^{-t-1}) \quad \text{and} \quad \int_{|x|<n} x\, dV(x) = o(1),$$

where $V(x)$ is the distribution function of X_1. If $0 < t < 2$, then the following conditions are equivalent:

(A) $\mathbf{E}|X_1|^t \log(1 + |X_1|) < \infty$ and (for $t \ge 1$) $\mathbf{E}X_1 = 0$;

(B) $\displaystyle\sum_{n=1}^{\infty}(1/n)\mathbf{P}\left\{\sup_{k\ge n} k^{-1/t}|S_k| \ge \varepsilon\right\} < \infty$ for every $\varepsilon > 0$;

(C) $\displaystyle\sum_{n=1}^{\infty}(\log n/n)\mathbf{P}\left\{|S_n| \ge n^{1/t}\varepsilon\right\} < \infty$ for every $\varepsilon > 0$.

§2.3. Approximation of Sums of Independent Random Variables by Sums of Normally Distributed Independent Random Variables

Let $V(x)$ be a distribution function such that

$$\int_{-\infty}^{\infty} x\,dV(x) = 0, \qquad \int_{-\infty}^{\infty} x^2\,dV(x) = 1 \tag{1}$$

and

$$\int_{-\infty}^{\infty} e^{tx}\,dV(x) < \infty \quad \text{for } |t| < b \quad \text{and some } b > 0. \tag{2}$$

Then there exists a probability space on which it is possible to construct a sequence of independent random variables $\{X_n\}$ with a common distribution function $V(x)$ and a sequence of independent random variables $\{Y_n\}$ having the standard normal distribution such that the sums

$$S_n = \sum_{k=1}^{n} X_k, \qquad T_n = \sum_{k=1}^{n} Y_k \tag{3}$$

satisfy

$$S_n - T_n = O(\log n) \text{ a.s.} \tag{4}$$

If $V(x)$ satisfies the conditions (1) and if $\int_{-\infty}^{\infty} |x|^r\,dV(x) < \infty$ for some $r > 3$, then this proposition remains true with (4) replaced by the weaker relation

$$S_n - T_n = o(n^{1/r}) \text{ a.s.}$$

These results were obtained by Komlós, Major and Tusnády (1975). The last result is a refinement of a theorem of Strassen.

Suppose that $V(x)$ is a distribution function satisfying only (1) with no additional moment conditions. Then the above proposition holds with (4) replaced by the relation

$$S_n - T_n = o\left(\sqrt{n \log\log n}\right) \text{ a.s.}$$

(Strassen (1964)); another proof is due to Major (1976b). That this estimate is optimum is illustrated by the following theorem which belongs to Major (1976b).

Let $f(n)$ be an arbitrary positive function with $f(n) \to \infty$ as $n \to \infty$. Then there exists a distribution function $V(x)$ such that (1) is satisfied and any pair of sequences of independent random variables $\{X_n\}$ and $\{Y_n\}$ with the respective distribution functions $V(x)$ and $\Phi(x) = \dfrac{1}{\sqrt{2\pi}} \int_{-\infty}^{x} e^{-t^2/2}\,dt$ satisfy

$$\limsup_{n\to\infty} f(n) \frac{|S_n - T_n|}{\sqrt{n \log\log n}} = \infty \text{ a.s.}$$

Here S_n and T_n are defined in (3).

Generalizations and refinements of these results were proved for sequences of independent and non-identically distributed random variables by Sakhanenko (1984, 1985). For the case of identical distributions, we shall state two more results of Major (1976a, 1979).

Let $V(x)$ be a distribution function satisfying (1) and $\int_{-\infty}^{\infty} x^2 g(|x|)\,dV(x) < \infty$, where $g(x)$ is a function such that $x/g(x)$ and $g(x)/x^{\varepsilon}$ do not decrease for $x > 0$ for some $\varepsilon > 0$. Then it is possible to construct (on a probability space) a sequence of independent random variables $\{X_n\}$ with the common distribution function $V(x)$ and a sequence of independent random variables $\{Y_n\}$ having the standard normal distribution such that

$$\limsup_{n\to\infty} \frac{|S_n - T_n|}{h(n)} \le C \text{ a.s.},$$

where C is a constant and $h(x)$ is the inverse of $x^2 g(x)$. It follows that if $\int_{-\infty}^{\infty} |x|^r\,dV(x) < \infty$ for some r $(2 < r \le 3)$, then $S_n - T_n = o(n^{1/r})$ a.s.

Let $V(x)$ be a distribution function satisfying (1). Put

$$\sigma_k^2 = \int_{|x|<2^{n/2}} x^2\,dV(x) - \left(\int_{|x|<2^{n/2}} x\,dV(x)\right)^2$$

if $2^n \le k < 2^{k+1}$. Then it is possible to construct (on a probability space) a sequence of independent random variables with the common distribution function $V(x)$ and a sequence of normally distributed independent random variables $\{Y_n\}$ such that $\mathbf{E}Y_k = 0$, $\mathbf{E}Y_k^2 = \sigma_k^2$ for all k and $S_n - T_n = O(\sqrt{n})$ a.s.

Chapter 3
The Law of the Iterated Logarithm

Theorems on the law of the iterated logarithm for a sequence of random variables $\{X_n\}$ involve conditions under which $\limsup \dfrac{|S_n|}{a_n} = 1$ a.s. or $\limsup \dfrac{|S_n|}{a_n} \le 1$ a.s., where $S_n = \sum_{k=1}^{n} X_k$ and $\{a_n\}$ is a sequence of positive numbers. These relations strengthen the estimates provided by the strong law of large numbers.

§3.1. The Kolmogorov Theorem. The Hartman–Wintner Theorem

These theorems are the most important results concerning the law of the iterated logarithm.

Kolmogorov Theorem. *Let $\{X_n\}$ be a sequence of independent random variables with zero means and finite variances. Put $B_n = \sum_{k=1}^{n} \mathbf{E}X_k^2$. Let $B_n \to \infty$ as $n \to \infty$. Suppose that there exists a sequence of positive constants $\{M_n\}$ such that*

$$|X_n| \leq M_n, \tag{1}$$

and

$$M_n = o\left(\left(\frac{B_n}{\log\log B_n}\right)^{1/2}\right). \tag{2}$$

Then

$$\limsup_{n\to\infty} \frac{S_n}{(2B_n \log\log B_n)^{1/2}} = 1 \text{ a.s.} \tag{3}$$

If $\{X_n\}$ satisfies the conditions of this theorem, then so does the sequence $\{-X_n\}$. Thus Kolmogorov's theorem implies that

$$\liminf_{n\to\infty} \frac{S_n}{(2B_n \log\log B_n)^{1/2}} = -1 \text{ a.s.}$$

and therefore

$$\limsup_{n\to\infty} \frac{|S_n|}{(2B_n \log\log B_n)^{1/2}} = 1 \text{ a.s.} \tag{4}$$

Kolmogorov's theorem is the first general result about the law of the iterated logarithm. Earlier Khinchine proved the relation (4) for a sequence of two-valued independent random variables having identical distributions. Condition (2) cannot be weakened. Marcinkiewicz and Zygmund (1937) constructed a sequence of independent random variables having non-identical distributions with two values such that $B_n \to \infty$,

$$M_n = O\left(\left(\frac{B_n}{\log\log B_n}\right)^{1/2}\right),$$

but

$$\limsup_{n\to\infty} \frac{S_n}{(2B_n \log\log B_n)^{1/2}} < 1 \text{ a.s.}$$

Weiss (1959) and Egorov (1972) constructed sequences of independent random variables such that the first two of these relations hold but

$$\limsup_{n\to\infty} \frac{S_n}{(2B_n \log\log B_n)^{1/2}} > 1 \text{ a.s.}$$

Hartman–Wintner Theorem. *Let* $\{X_n\}$ *be a sequence of independent and identically distributed random variables with* $\mathbf{E}X_1 = 0$ *and* $\mathbf{E}X_1^2 = \sigma^2 < \infty$. *Then*

$$\limsup_{n\to\infty} \frac{S_n}{(2n\log\log n)^{1/2}} = \sigma \text{ a.s.},$$

Strassen (1966) proved that if $\{X_n\}$ is a sequence of independent and identically distributed random variables and $\mathbf{E}X_1^2 = \infty$, then

$$\limsup_{n\to\infty} \frac{S_n}{(n\log\log n)^{1/2}} = \infty \text{ a.s.}$$

That this result is optimal is illustrated by the following theorem of Berkes (1972). If $f(n)$ is an arbitrary function for which $\lim f(n) = \infty$, then there exists a sequence of independent and identically distributed random variables $\{X_n\}$ such that $\mathbf{E}X_1 = 0$, $\mathbf{E}X_1^2 = \infty$ and

$$\lim_{n\to\infty} \frac{S_n}{f(n)(n\log\log n)^{1/2}} = 0 \text{ a.s.}$$

Martikainen (1980), Rosalsky (1980) and Pruitt (1981) proved a converse to the law of the iterated logarithm. If $\{X_n\}$ is a sequence of independent and identically distributed random variables and

$$\limsup_{n\to\infty} \frac{S_n}{(2n\log\log n)^{1/2}} = 1 \text{ a.s.},$$

then $\mathbf{E}X_1 = 0$ and $\mathbf{E}X_1^2 = 1$.

§3.2. Connection between the Law of the Iterated Logarithm and the Central Limit Theorem

The conditions in the Hartman–Wintner theorem coincide with those of Lévy's theorem – a form of the central limit theorem. Thus a sequence of independent and identically distributed random variables with finite variances obeys the law of the iterated logarithm as well as the central limit theorem. A natural question arises about connections between the applicability of the central limit theorem and the applicability of the law of the iterated logarithm to sequences of non-identically distributed random variables having finite variances. It is possible to show that if such a sequence obeys the central limit theorem, it need not obey the law of the iterated logarithm (with traditional sequences of normalizing constants) and vice versa. Marcinkiewicz and Zygmund (1937) constructed (for other purposes) sequences of independent bounded random variables with different distributions that obey the central limit theorem but do not obey the law of the iterated logarithm. However the law of the iterated logarithm will be applicable if a sequence of independent random variables satisfies a stronger condition than under which the central limit theorem is applicable. The next result was proved by Petrov (1966).

Let $\{X_n\}$ be a sequence of independent random variables with zero means and finite variances. Put $B_n = \sum_{k=1}^{n} \mathbf{E}X_k^2$ and

$$R_n = \sup_x \left| \mathbf{P}\{S_n < x\sqrt{B_n}\} - \Phi(x) \right|.$$

If $B_n \to \infty$, $B_{n+1}/B_n \to 1$ and $R_n = O\left((\log B_n)^{-1-\delta}\right)$ for some positive δ, then relation (3) of §3.1 holds.

The assumptions of this theorem are satisfied if $\liminf B_n/n > 0$ and

$$\limsup_{n\to\infty} \frac{1}{n} \sum_{k=1}^{n} \mathbf{E}X_k^2 \left|\log|X_k|\right|^{1+\delta} < \infty$$

for some $\delta > 0$. It is possible to indicate other sufficient conditions avoiding assumptions about the existence of any moments of order greater than two.

Egorov (1969) proved that it is impossible to replace the positive number δ by zero in the above result. There exists a sequence of independent random variables $\{X_n\}$ with zero means and finite variances such that $\mathbf{E}X_n^2 \asymp 1$ and $R_n = O\left((\log B_n)^{-1}\right)$ but the relation (3) does not hold. It follows that if a sequence of independent random variables obeys the central limit theorem (i.e., if $R_n \to 0$ as $n \to \infty$), then the law of the iterated logarithm may fail.

Tomkins (1991) proved the following theorem. Let $\{X_n\}$ be a sequence of independent random variables with zero means and finite variances, $B_n = \sum_{k=1}^{n} \mathbf{E}X_k^2 \to \infty$, $\mathbf{E}X_n^2/B_n \to 0$ and $X_n = o\left((B_n/\log\log B_n)^{1/2}\right)$ a.s. If $\{X_n\}$ obeys the central limit theorem, then it obeys the law of the iterated logarithm. The converse is not always true.

§3.3. The Generalized Law of the Iterated Logarithm

One may avoid any assumptions about the existence of moments and deduce generalizations of the classical law of the iterated logarithm for sequences of random variables with finite variances.

Let $\{X_n;\ n = 1, 2, \ldots\}$ be a sequence of independent random variables and $\{a_n;\ 1, 2, \ldots\}$ a non-decreasing sequence of positive numbers such that $a_n \to \infty$ as $n \to \infty$. Put $S_n = \sum_{k=1}^{n} X_k$ for $n \geq 1$, $S_0 = 0$, $a_0 = 0$. Introduce the following condition (B): for every $\delta > 0$, there exists $\lambda > 0$ such that $\mathbf{P}\{(S_n - S_j) \geq -\delta a_n\} \geq \lambda$ for all non-negative $j \leq n$ and all n sufficiently large.

For every integer $n \geq 0$ and fixed $c > 1$, define $i_n = i_n(c)$ as the greatest integer for which $a_{i_n} \leq c^n$.

We point out that condition (B) is certainly satisfied if the random variables X_n have symmetric distributions or if $S_n/a_n \xrightarrow{\mathbf{P}} 0$.

If (B) is satisfied, then the relation

$$\limsup_{n\to\infty} \frac{S_n}{a_n} \le 1 \text{ a.s.}$$

is equivalent to any one of the following:

(D_1) for every $\varepsilon > 0$ and every non-decreasing sequence of integers $k_n \to \infty$,

$$\sum_{n=2}^{\infty} \mathbf{P}\left\{S_{k_n} - S_{k_{n-1}} > (1+\varepsilon)a_{k_n}\right\} < \infty; \tag{5}$$

(E_1) for every $\varepsilon > 0$, every integer $r \ge 1$ and every $c > 1$,

$$\sum_{n=r}^{\infty} \mathbf{P}\left\{S_{i_n} - S_{i_{n-r}} > (1+\varepsilon)c^n\right\} < \infty. \tag{6}$$

Under condition (B), the relation

$$\limsup_{n\to\infty} \frac{S_n}{a_n} = 1 \text{ a.s.}$$

is equivalent to any one of the following:

(D_2) for every $\varepsilon > 0$ and every non-decreasing sequence of integers $k_n \to \infty$, the relation (5) holds; for every $\varepsilon > 0$, there exists a non-decreasing sequence of integers $k_n \to \infty$ such that

$$\sum_{n=2}^{\infty} \mathbf{P}\left\{S_{k_n} - S_{k_{n-1}} > (1-\varepsilon)a_{k_n}\right\} = \infty;$$

(E_2) for every $\varepsilon > 0$, every integer $r \ge 1$ and every $c > 1$, the relation (6) holds; for every $\varepsilon > 0$, there exists $c > 1$ and $r \ge 1$ such that

$$\sum_{n=r}^{\infty} P\left\{S_{i_n} - S_{i_{n-r}} > (1-\varepsilon)c^n\right\} = \infty.$$

These results were obtained by Martikainen and Petrov (1977). Similar results were obtained later by Tomkins (1980) expressed in different terms. Martikainen (1979a) found generalizations of these results involving sequences of normalizing constants that are not necessarily monotone.

It is of interest to find necessary and sufficient conditions for the applicability of the generalized law of the iterated logarithm which are expresssed in terms of the distributions of the individual summands and not in terms of their sums. Such conditions were determined by Martikainen (1979b). Sufficient conditions of this kind were established by Klass and Tomkins (1984).

References*

Baum, L.E., and Katz. M. (1965): Convergence rates in the law of large numbers. Trans. Am. Math. Soc. **120**, No. 1, 108–123. Zbl. 142.14802

Berkes, I. (1972): A remark to the law of the iterated logarithm. Studia Sci. Math. Hungar. **7**, No. 1–2, 189–197. Zbl. 265.60025

Berry, A.C. (1941): The accuracy of the Gaussian approximation to the sum of independent variates. Trans. Amer. Math. Soc. **49**, No. 1, 122–136. Zbl. 025.34603

Bikelis, A. (1966): Estimates of the remainder term in the central limit theorem. Lit. Mat. Sb. 6, No. 3 (in Russian), 323–346. Zbl. 149.14002

Buldygin, V.V. (1978): The strong law of large numbers and convergence of Gaussian sequences to zero. Teor. Veroyatn. Mat. Stat., Kiev, No. 19, 33–41. Engl. transl.: Theory Probab. Math. Stat. **19**, 35–43 (1980; Zbl. 485.60029). Zbl 407.60021

Csörgő, M., and Révész, P. (1981): Strong Approximations in Probability and Statistics. Akadémiai Kiadó, Budapest. Zbl. 539.60029

Egorov, V.A. (1969): On the law of the iterated logarithm. Teor. Veroyatn. Primen. **14**, No. 4. 722–729. Engl. transl.: Theor. Probab. Appl. **14**, No. 4, 693–699. Zbl. 211.48903

Egorov, V.A. (1972): On Kolmogorov's theorem on the law of the iterated logarithm. Vestn. Leningr. Univ. Ser I **13**, 140–142 (Russian). Zbl. 244.6022

Egorov, V.A. (1973): On the rate of convergence to normal law which is equivalent to the existence of a second moment. Teor. Veroyatn. Primen. **18**, No. 1, 180–185. Engl. transl.: Theory Probab. Appl. **18**, No. 1, 175–180. Zbl. 307.60025

Esseen, C.G. (1942): On the Liapounoff limit of error in the theory of probability. Ark. Mat. Astron. Fys. **A28**, No. 2, 1–19. Zbl. 027.33902

Esseen, C.G. (1945): Fourier analysis of distribution functions. Acta Math. **77**, 1–125. Zbl. 060.28705

Etemadi, N. (1981): An elementary proof of the strong law of large numbers. Z. Wahrschlichkeitstheorie Verw. Geb. **55**, No. 1, 119–122. Zbl. 448.60024

Feller, W. (1968): On the Berry–Esseen theorem. Z. Wahrscheinlichkeitstheorie Verw. Geb. **10**, No. 3, 261–268. Zbl. 167.17304

Gnedenko, B.V., and Kolmogorov, A.N. (1949): Limit Distributions for Sums of Independent Random Variables. Gostekhizdat, Moscow-Leningrad. Engl. transl.: Limit Distributions for Sums of Independent Random Variables. Addison-Wesley, Reading, Mass. (1954). Zbl. 056.36001

Hall, P. (1980): Characterizing the rate of convergence in the central limit theorem. Ann. Probab. **8**, No. 6, 1037–1048. Zbl. 456.60018

Hall, P. (1982): Rates of Convergence in the Central Limit Theorem. Pitman, Boston, London and Melbourne. Zbl. 497.60001

Heyde, C.C. (1967): On the influence of moments on the rate of convergence to the normal distribution. Z. Wahrscheinlichkeitstheorie Verw. Geb. **8**, No. 1, 12–18. Zb. 149.14001

* For the convenience of the reader, references to reviews in Zentralblatt für Mathematik (Zbl.), compiled by means of the MATH database, and Jahrbuch über die Fortschritte der Mathematik (Jbuch) have, as far as possible, been included in this bibliography.

Heyde, C.C. (1969): Some properties of metrics in a study of convergence to normality. Z. Wahrscheinlichkeitstheorie Verw. Geb. **11**, No. 3, 181–192. Zbl. 169.20902

Heyde, C.C. (1973): On the uniform metric in the context of convergence to normality. Z. Wahrscheinlichkeitstheorie Verw. Geb. **25**, No. 2, 83–95. Zbl. 699.62095

Heyde, C.C., and Nakata, T. (1984): On the asymptotic equivalence of L_p metrics for the convergence to normality. Z. Wahrscheinlichkeitstheorie Verw. Geb. **68**, No. 1, 97–106. Zbl. 546.60035

Ibragimov, I.A. (1966): On the approximation of distribution functions of sums of independent variables. Teor. Veroyatn. Primen. **11**, No. 4, 632–655. Engl. transl.: Theor. Probab. Appl. **11**, No. 4, 559–580. Zbl. 161.15207

Ibragimov, I.A., and Linnik, Yu.V. (1965): Independent and Stationarily Connected Sequences of Variables. Nauka, Moscow. Zbl. 154.42201. Engl. transl.: Independent and Stationary Sequences of Random Variables, Wolters-Noordhoff, Groningen (1971). Zbl. 219.60027

Katz, M. (1963): Note on the Berry–Esseen theorem. Ann. Math. Stat. **34**, No. 3, 1107–1108. Zbl. 122.36704

Klass, M.J., and Tomkins, R.J. (1984): On the limiting behavior of normed sums of independent variables. Z. Wahrscheinlichkeitstheorie Verw. Geb. **68**, No. 1, 107–120, Zbl. 552.60026

Komlós, J., Major, P., and Tusnády, G. (1975): An approximation of partial sums of independent RV's, and the sample DF, I, II. Z. Wahrscheinlichkeitstheorie Verw. Geb. **32**, No. 1–2, 111–131; **34**, No. 1, 33–58. Zbl. 308.60029, Zbl. 315.60031

Kruglov, V.M. (1976): Global limit theorems. Zap. Nauch. Semin. Leningr. Otd. Mat. Inst. Steklova **61**, 84–191. Engl. transl.: J. Sov. Math. **16**, 1396–1409. Zbl. 358.60036

Loève, M. (1960): Probability Theory. 2nd ed., Van Nostrand, Pinceton, NJ. Zbl. 095.12201

Major, P. (1976a): The approximation of partial sums of independent RV's. Z. Wahrscheinlichkeitstheorie Verw. Geb. **35**, No. 3, 213–220. Zbl. 338.60031

Major, P. (1976b): Approximation of partial sums of i.i.d.r.v's when the summands have only two moments. Z. Wahrscheinlichkeitstheorie Verw. Geb. **35**, No. 3, 221–229. Zbl. 338.60032

Major, P. (1979): An improvement of Strassen's invariance principle. Ann. Probab. **7**, No. 1, 55–61. Zbl. 392.60034

Marcinkiewicz, J., and Zygmund, A. (1937): Remarque sur la loi du logarithme itéré. Fundam. Math. **29**, 215–222. Zbl. 018.03204

Martikainen, A.I. (1979a): Three theorems on the limit superior of sums of independent random variables. Vestn. Leningr. Univ., Mat. Meth. Astron., No. 1, 45–51. Engl. transl.: Vestn. Leningr. Univ. Math. **12**, 29–36. Zbl. 411.60048

Martikainen, A.I. (1979b): An exponential criterion for the law of the iterated logarithm. Zap. Nauchn. Semin. Leningr. Otd. Mat. Inst. Steklova **85**, 158–168. Engl. transl.: J. Sov. Math. **20**, 2214–2221 (1982). Zbl. 417.60042

Martikainen, A.I. (1979c): On necessary and sufficient conditions for the strong law of large numbers. Teor. Veroyatn. Primen. **24**, No. 4, 814–821. Engl. transl.: Theory Probab. Appl. **24**, No. 4, 813–820. Zbl. 432.60036

Martikainen, A.I. (1980): A converse to the law of the iterated logarithm for a random walk. Teor. Veroyatn. Primen. **25**, No. 2, 364–366. Engl. transl.: Theory Probab. Appl. **25**, No. 2, 361–362. Zbl. 432.60037

Martikainen, A.I., and Petrov, V.V. (1977): On necessary and sufficient conditions for the law of the iterated logarithm. Teor. Veroyatn. Primen. **22**, No. 1, 18–26; No. 2, 442. Engl. transl.: Theor. Probab. Appl. **22**, No. 1, 16–23; No. 2, 430. Zbl. 377.60036

Nagaev, S.V. (1965): Some limit theorems for large deviations. Teor. Veroyatn. Primen. **10**, No. 2, 231–254. Engl. transl.: Theor. Probab. Appl. **10**, No. 2, 214–235. Zbl. 144.18704

Nagaev, S.V. (1972): On necessary and sufficient conditions for the strong law of large numbers. Teor. Veroyatn. Primen. **17**, No. 4, 609–618. Engl. transl.: Theor. Probab. Appl. **17**, No. 4, 573–581. Zbl. 276.60035

Osipov, L.V. (1966): Refinement of Lindeberg's theorem. Teor. Veroyatn. Primen. **11**, No. 2, 339–342. Engl. transl.: Theor. Probab. Appl. **11**, No. 2, 299–302. Zbl. 147.37001

Osipov, L.V. (1967): Asymptotic expansions in the central limit theorem. Vestn. Leningr. Univ., Ser I **19**, 45–62 (in Russian). Zbl. 189.18003

Osipov, L.V. (1968): On the closeness with which the distribution of the sum of independent random variables approximates the normal distribution. Dokl. Akad. Nauk SSSR **178**, No. 5, 1013–1016. Engl. transl.: Sov. Math., Dokl. **9**, No. 1, 233–236. Zbl. 185.46803

Osipov, L.V. (1971): Asymptotic expansions for the distributions of sums of independent random variables. Teor. Veroyatn. Primen. **16**, No. 2, 328–338. Engl. transl.: Theor. Probab. App. **16**, No. 2, 333–343. Zbl. 248.60015

Osipov, L.V., and Petrov, V.V. (1967): On an estimate of the remainder term in the central limit theorem. Teor. Veroyatn. Primen. **12**, No. 2, 322–329. Engl. transl.: Theor. Probab. Appl. **12**, No. 2, 281–286. Zbl. 185.46801

Petrov, V.V. (1962): On some polynomials encountered in probability theory. Vestn. Leningr. Univ., Ser I **19**, 150–153 (in Russian). Zbl. 128.38003

Petrov, V.V. (1965): An estimate of the deviation of the distribution of a sum of independent random variables from the normal law. Dokl. Akad. Nauk **160**, No. 5, 1013–1015. Engl. transl.: Sov. Math., Dokl. **6**, No. 1, 242–244. Zbl. 135.19203

Petrov, V.V. (1966): On a relation between an estimate of the remainder in the central limit theorem and the law of the iterated logarithm. Teor. Veroyatn. Primen. **11**, No. 3, 514–518. Engl. transl.: Teor. Probab. Appl. **11**, No. 3, 454–458. Zbl. 203.19602

Petrov, V.V. (1969): On the strong law of large numbers. Teor. Veroyatn. Primen. **14**, No. 2, 193–202. Engl. transl.: Theor. Probab. Appl. **14**, No. 2, 183–192. Zbl. 196.20903

Petrov, V.V. (1972): Sums of Independent Random Variables. Nauka, Moscow. Engl. transl.: Springer-Verlag, Berlin-Heidelberg – New York (1975). Zbl. 267.60055 (Zbl. 322.60042)

Petrov, V.V. (1987): Limit Theorems for Sums of Independent Random Variables. Nauka, Moscow (in Russian). Zbl. 621.60022

Petrov, V.V. (1995): Limit Theorems of Probability Theory. Oxford University Press, Oxford. Zbl. 826-60357

Prokhorov, Yu.V. (1950): On the strong law of large numbers. Izv. Akad. Nauk SSSR, Ser. Mat. **14**, No. 6, 523–536 (in Russian). Zbl. 040.07301

Prokhorov, Yu.V. (1959): Some remarks on the strong law of large numbers. Teor. Veroyatn. Primen. **4**, No. 2, 215–220. Engl. transl.: Theor. Probab. Appl. **4**, No. 2, 204–208. Zbl. 089.13903

Pruitt, W.E. (1981): General one-sided laws of the iterated logarithm. Ann. Probab. **9**, No. 1, 1–48. Zbl. 462.60030

Révész, P. (1967): The Laws of Large Numbers. Akadémiai Kiadó, Budapest and NewYork, Academic Press, 1968. Zbl. 203.50403

Rosalsky, A. (1980): On the converse to the iterated logarithm law. Sankhya **A42**, No. 1–2, 103–108. Zbl. 486.60031

Rozovskii, L.V. (1975): Asymptotic expansions in the central limit theorem. Teor. Veroyatn. Primen. **20**, No. 4, 810–820. Engl. transl.: Theory Probabl. Appl. **20**, Mno. 4, 794–804. Zbl. 347.60020

Rozovskii, L.V. (1978a): On a lower bound for the remainder in the central limit theorem. Mat Zametki **24**, No. 3, 403–410. Engl. transl.: Math. Notes **24**, No. 3–4, 715–719. Zbl. 396.60026

Rozovskii, L.V. (1978b): On the exactness of an estimate of the remainder term in the central limit theorem. Teor. Veroyatn. Primen. **24**, No. 4, 744–761. Engl. transl.: Theory Probabl. Appl. **23**, No. 4, 712–730. Zbl. 388.60026

Sakhanenko, A.I. (1984): Rate of convergence in an invariance principle for non-identically distributed variables with exponential moments. Tru. Inst. Mat. **3**, 4–49. Engl. transl.: Advances in Pribability Theory, Transl. for Math. Engl. 2–73 (1986). Zbl. 541.60024

Sakhanenko, A.I. (1985): Estimates in an invariance principle, Tru. Inst. Mat. **5**, 27–44 (in Russian). Zbl. 585.60044

Stout, W.F. (1974): Almost Sure Convergence, Academic Press, New York. Zbl. 321.60022

Strassen, V. (1964): An invariance principle for the law of the iterated logarithm. Z. Wahrscheinlichkeitstheorie Verw. Geb. **3**, No. 3, 211–226. Zbl. 132.12903

Strassen, V. (1966): A converse to the law of the iterated logarithm. Z. Wahrscheinlichkeitstheorie Verw. Geb. **4**, 265–268. Zbl. 141.16501

Tomkins, R.J. (1980): Limit theorems without moment hypotheses for sums of independent random variables. Ann. Probab. **8**, No. 2, 314–324. Zbl. 432.60034

Tomkins, R.J. (1991): Refinements of Kolmogorov's law of the iterated logarithm. Stat. Probab. Lett. **14**, No. 4, 321–325 (1992), Zbl. 756.60033

Weiss, M. (1959): On the law of the iterated logarithm. J. Math. Mech. **8**, No. 1, 121–132. Zbl. 091.14206

Zolotarev, V.M. (1986): Modern Theory of Summation of Independent Random Variables. Nauka, Moscow (in Russian). Engl. transl.: VSP, Utrecht (1997). Zbl. 649.60016

II. The Accuracy of Gaussian Approximation in Banach Spaces

V. Bentkus, F. Götze, V. Paulauskas
and A. Račkauskas

Contents

Chapter 0
Introduction and Notation

Let B be a real separable Banach space with norm $||\cdot|| = ||\cdot||_B$. Suppose that $X, X_1, X_2, \ldots \in B$ are independent and identically distributed (i.i.d.) random elements (r.e.'s) taking values in B. Furthermore, assume that $EX = 0$ and that there exists a zero-mean Gaussian r.e. $Y \in B$ such that the covariances of X and Y coincide. Define

$$S_n = S_n(X) = n^{-1/2}(X_1 + \cdots + X_n).$$

Let $f : B \to \mathbb{R}$ (or more generally $f : B \to F$, where F is a real Banach spaçe) denote a function such that the expectations $EF(S_n)$, $Ef(Y)$ are well defined. The central limit theorem (CLT) in this context requires

$$\lim_{n\to\infty} Ef(S_n) = Ef(Y) \tag{0.1}$$

for all functions $f \in \mathcal{F}$, where $\mathcal{F}$ is some class of functions $f : B \to \mathbb{R}$. In the classical definition of the CLT, $\mathcal{F}$ coincides with the class $C_b(B)$ of bounded continuous functions $f : B \to \mathbb{R}$. In recent approaches $\mathcal{F}$ can differ from $C_b(B)$ and the expectation is understood not only as the traditional Lebesgue and Bochner integrals but also as other integrals.

Denote by ρ a semi–metric (i.e., $\rho(x, y) = 0$ does not necessary imply $x = y$) on a class of probability measures on B. Another formulation of the CLT is given by

$$\lim_{n\to\infty} \rho(\mathcal{L}(S_n), \mathcal{L}(Y)) = 0, \tag{0.2}$$

where $\mathcal{L}(X)$ denotes the distribution of the r.e. X. If ρ is the Prokhorov metric, then (0.2) is equivalent to the classical CLT. In what follows we shall consider the CLT in the classical sense.

The aim of the paper is to give a review of results and methods used to estimate the rate of convergence in the CLT in infinite-dimensional spaces. A review of results concerning asymptotic expansions is also presented. It is appropriate to remark that the earlier books by Sazonov (1981), Koroliuk and Borovskikh (1984) and Paulauskas and Račkauskas (1989) contain related reviews. In this paper we restrict ourselves to the case of i.i.d. summands and sums having a Gaussian limit. We do not try to give the most general or accurate formulations (for this we refer to original papers). Rather, we shall stress the differences between the finite-dimensional case and infinite-dimensional case, emphasizing new phenomena arising in connection with the large class of geometrical structures in Banach spaces. Thus the paper contains mainly sketches of proofs and many technical arguments are skipped. Our intention was also to reflect an increasing number of applications of Banach space results to certain asymptotic problems of mathematical statistics.

We shall use the following abbreviations:

$$\begin{aligned} \varepsilon &= \varepsilon(n) = n^{-1/2}, \ n = 1, 2, \ldots, \\ g(\varepsilon) &= g(\varepsilon, f) = Ef(S_n), \\ g(0) &= g(0, f) = Ef(Y), \\ h(\varepsilon) &= h(\varepsilon, \rho) = \rho(\mathcal{L}(S_n), \mathcal{L}(Y)). \end{aligned}$$

Then (0.1) means that the function $g(\varepsilon)$ of the discrete argument $\varepsilon = n^{-1/2}$, $n = 1, 2, \ldots$ is continuous at the point $\varepsilon = 0$:

$$\lim_{\varepsilon \to 0} g(\varepsilon) = g(0). \tag{0.3}$$

Similarly, (0.2) is equivalent to

$$\lim_{\varepsilon \to 0} h(\varepsilon) = 0. \tag{0.4}$$

A stronger assertion than (0.3) is to assume that g at the point $\varepsilon = 0$ satisfies a Hölder condition (say, with an exponent $\alpha > 0$ and a constant $C < \infty$)

$$|g(\varepsilon) - g(0)| \le C\varepsilon^\alpha, \tag{0.5}$$

(similarly,

$$h(\varepsilon) \le C\varepsilon^\alpha \tag{0.6}$$

in the case of (0.4)). The inequalities (0.5), (0.6) are usually called estimates of the convergence rate in the CLT. Clearly, one can rewrite (0.5) and (0.6) as

$$|Ef(S_n) - Ef(Y)| \le Cn^{-\alpha/2}, \tag{0.7}$$

$$\rho(\mathcal{L}(S_n), \mathcal{L}(Y)) \le Cn^{-\alpha/2}. \tag{0.8}$$

More precise approximations of $g(\varepsilon)$ than (0.5), (0.6) are given by the asymptotic expansion

$$g(\varepsilon) = g(0) + a_1\varepsilon + \cdots + a_k\varepsilon^k + R \tag{0.9}$$

with coefficients $a_1, \ldots, a_k$ and the remainder term $R = R_k(\varepsilon)$ such that $R_k(\varepsilon) = o(\varepsilon^k)$ as $\varepsilon \to 0$. Clearly, one can rewrite (0.9) as

$$Ef(S_n) = Ef(Y) + a_1 n^{-1/2} + \cdots + a_k n^{-k/2} + R. \tag{0.10}$$

Relations like (0.10) are usually called asymptotic expansions in the CLT.

In the first chapter of the paper we give a review of methods and results concerning estimate (0.7) when f is the indicator function of a set $A \subset B$: $f(x) = \chi_A(x)$. Then (0.7) yields an estimate for probabilities

$$|P\{S_n \in A\} - P\{Y \in A\}| \le Cn^{-\alpha/2}. \tag{0.11}$$

When the space B has finite dimension one can prove (0.11) for relatively large classes of sets A, for instance for the class of all convex sets (see, e.g.,

Bhattacharya and Rao (1976), Sazonov (1981)). The situation changes dramatically in the infinite–dimensional case. In this case there does not exist a uniform estimate of type (0.11) even for such relatively small classes as the class of all balls or the class of all halfspaces (see Sazonov (1981)). Therefore we have to consider special sets A. Let $F : B \to \mathbb{R}$ be a functional and let $A = A_r(F) = \{x \in B : F(x) < r\}$. Then (0.11) yields

$$|P\{F(S_n) < r\} - P\{F(Y) < r\}| \leq Cn^{-\alpha/2}. \tag{0.12}$$

It turns out that in the infinite-dimensional case the estimate (0.12) depends strongly on the smoothness properties of F. The Fourier method and the method of integration by parts discussed in §§1.1, 1.3 are adapted to the investigation of (0.12) for smooth functions F, while the Lindeberg method and the method of finite-dimensional approximation discussed in §§1.2, 1.4 are useful for the case of non–smooth functions F. The last section of the chapter discusses the results concerning the convergence rate in the CLT, estimated by means of the Prokhorov and bounded Lipschitz (BL) metrics.

The second chapter is devoted to asymptotic expansions. In §§2.1 and 2.2 we consider asymptotic expansions for the expectation $Ef(S_n)$ with $f : B \to \mathbb{R}$ a sufficiently smooth function or f a function having isolated points of non–differentiability. The expectation $E||S_n||^p$, $p > 0$, with a sufficiently smooth norm–function, is a typical example. The results discussed in these sections are obtained with no explicit condition like the classical one of Cramér. In §2.3 asymptotic expansions for $P\{F(S_n) < r\}$ are reviewed. In §2.4 we consider asymptotic expansions for the density function $(d/dr)P\{F(S_n) < r\}$.

In the third chapter we present examples applying the methods and results described in the previous chapters. We emphasize that our aim is to give illustrations of applications only. Here we distinguish the following situations:

i) cases where a limit theorem for a statistical test can be interpreted as a particular case of a general result in Banach spaces. The ω^2–statistic is a typical example of this kind (see §3.1). Every such statistic can be represented as the norm of the sum of i.i.d.r.e's in L_2 space;

ii) cases where a problem can be reduced to known facts from probability theory in Banach spaces. L–statistics (i.e. linear combinations of rank statistics) are a typical example (see §3.2). Such statistics can be represented as a sum of i.i.d. $\mathbb{R}$–valued r.v's with an additional remainder term. The remainder term can be majorized by the norm of a sum of Banach-space-valued r.e's;

iii) cases where methods and ideas from probability theory in Banach space can be used in statistical problems. The typical examples here are the so–called U–statistics;

iv) cases where it is necessary to use a combination of the previous approaches. The estimation of the convergence rate for empirical processes is an example (see §3.3 and §3.4).

If not stated otherwise, we suppose that all random elements and random variables under consideration are independent. We refer to Vakhaniya et al.

(1987) and Araujo and Giné (1980) for the general information on probability theory and the CLT in Banach spaces, respectively.

Let us introduce the notation we shall use throughout.

B, F real separable Banach spaces with norm $||\cdot|| = ||\cdot||_B$, $||\cdot|| = ||\cdot||_F$.

B^* the dual Banach space of B consisting of all linear continuous functionals with the standard *sup* norm.

H a real separable Hilbert space. We denote the scalar product by $(\cdot,\cdot)$ and the norm by $||\cdot|| = ||\cdot||_H$.

$\mathbb{R}$ the real line.

$\mathbb{R}^k$ the k–dimensional Euclidean space; $(x, y) = x_1y_1 + \cdots + x_ky_k$ denotes the scalar product.

ℓ_p, c_0 the classical Banach spaces of sequences.

$L_p(S, S, \nu)$, L_p, $C[0, 1]$, $D[0, 1]$ the classical spaces of functions.

a.s. almost surely.

i.i.d. independent and identically distributed.

r.e. random element.

B–r.e. r.e. with values in B.

r.v. random variable ($\mathbb{R}$–r.e.)

c.f. characteristic function

$X, X_1, X_2, \ldots \in B$ a sequence of i.i.d. B–r.e.'s such that $EX = 0$

$Y, Y_1, Y_2, \ldots \in B$ a sequence of i.i.d. Gaussian B–r.e.'s such that $EY = 0$ and the covariances of X and Y coincide, i.e. $Ef(X)g(X) = Ef(Y)g(Y)$ for all $f, g \in B^*$.

$S_n = S_n(X) = n^{-1/2}(X_1 + \cdots + X_n)$.

$\mu = \mathcal{L}(X)$ the distribution of X.

$\nu = \mathcal{L}(Y)$ the distribution of Y.

$cov\, X$ the covariance operator of X. By definition $cov\, X : B^* \to B$, $cov\, X(f) = Ef(X)X$ for all $f \in B^*$.

$\beta_s = E||X||^s$, $s \in \mathbb{R}$ the s–th moment of X.

$\nu_s = \nu_s(\mathcal{L}(X), \mathcal{L}(Y)) = \int_B ||x||^s |\mathcal{L}(X) - \mathcal{L}(Y)|(dx)$ the s–th pseudomoment.

$|\mu - \nu|$ the variation of the signed measure $\mu - \nu$.

$\chi_A(t)$ the indicator function of a set A.

$f^{(s)}(x)$ the s–th Fréchet derivative of the function f at the point x.

$f^{(s)}(x)h_1 \ldots h_s = f^{(s)}(x)[h_1, \ldots, h_s]$ the value of the derivative $f^{(s)}(x)$ as an s–linear continuous form at the point $(h_1, \ldots, h_s)$.

$f^{(s)}(x)h^s = f^{(s)}(x)[h^s] = f^{(s)}(x)h \ldots h$.

$||f^{(s)}(x)|| = \sup\{||f^{(s)}(x)h^s|| : ||h|| \le 1\}$ the norm of the s–linear form $f^{(s)}(x)$.

$C^s = C^s(B; F)$ the space of all s–times continuously Fréchet differentiable functions $f : B \to F$.

$C^s_b = C^s_b(B; F)$ the space of all bounded functions in C^s having bounded derivatives.

Chapter 1
Rates of Convergence

§1.1. The Fourier Method

The Fourier method has a long tradition in mathematics and particularly in probability. It is especially effective when used to investigate the distribution of sums of independent r.v.'s. In number theory it was employed by Gauss to obtain quadratic reprocity laws and to study the representation of integers by sums of squares or higher powers (the Waring problem); for the latter, use was made of the so-called Hardy–Littlewood circle method in analytic number theory (Hardy and Littlewood (1920)). In two papers, Lyapunov (1900, 1901) adopted the Fourier method to prove the CLT on the real line under the so-called Lyapunov moment conditions. This method relies on inequalities of the following type between the difference of probability measures and its Fourier transform (called the characteristic function or c.f. for short). Let μ be a probability measure and ν a measure of finite variation on $\mathbb{R}$. If ν is absolutely continuous with respect to Lebesgue measure and its density p is bounded: $m = \sup_{x\in\mathbb{R}} |p(x)| < \infty$, then for each $T > 0$ we have

$$\sup_{x\in\mathbb{R}} |\mu((-\infty, x)) - \nu((-\infty, x))| \le C_1 \int_{-T}^{T} \frac{|\hat{\mu}(t) - \hat{\nu}(t)|}{|t|}\, dt + C_2 \frac{m}{T}, \tag{1.1}$$

where C_1 and C_2 are absolute constants and the c.f.

$$\hat{\nu}(t) = \int_{\mathbb{R}} \exp(itx)\nu(dx).$$

This is the so-called Berry–Esseen lemma (Berry (1941), Esseen (1942)). It is used to prove convergence rates and higher order approximations to a measure μ by a measure ν with smooth density by means of their characteristic functions. Here T is chosen so that T^{-1} has a desired error size. The advantage of the Fourier method in Lyapunov's approach is apparent from the proof of the classical Berry–Esseen estimate

$$\sup_{x\in\mathbb{R}} |P\{S_n < x\} - P\{Y < x\}| = O(n^{-1/2}), \tag{1.2}$$

where $S_n = n^{-1/2}(X_1 + \cdots + X_n)$ and X, X_1, X_2, ... are i.i.d. real r.v.'s such that

$$EX = 0, \quad \beta_3 := E|X|^3 < \infty$$

and Y denotes a Gaussian r.v. such that $EY = 0$ and $\beta_2 := EX^2 = EY^2 > 0$. Setting in (1.1) $\mu = \mu_n = \mathcal{L}(S_n)$, $\nu = \mathcal{L}(Y)$ and $T = A\sqrt{n}$ for some $A = A(\mathcal{L}(Y)) > 0$ to be chosen later on, we can reduce (1.2) to the estimate

$$\int_{-T}^{T} \frac{|\hat{\mu}_n(t) - \hat{\nu}(t)|}{|t|} \, dt = O(n^{-1/2}). \tag{1.3}$$

Usually the estimation of the difference $\hat{\mu}_n(t) - \hat{\nu}(t)$ is based on different arguments for "small" and "large" values of $|t| \le T$. Clearly

$$\hat{\mu}_n(t) = E\exp(itS_n) = \Big(E\exp(itn^{-1/2}X)\Big)^n.$$

Expanding the exponential in a Taylor series we obtain

$$\Big|E\exp(itn^{-1/2}X)\Big| \le |1 - \beta_2 t^2/2n| + \beta_3|t|^3/6n^{3/2} \le \exp(-\beta_2 t^2/3n),$$

$$|\hat{\mu}_n(t)| \le \exp(-\beta_2 t^2/3) \tag{1.4}$$

for $|t| \le An^{1/2}$ with $A = \min(\beta_2/\beta_3, \sqrt{2\beta_2})$. Relation (1.4) and the equality $\hat{\nu}(t) = \exp(-\beta_2 t^2/2)$ allow one to show that on every interval $n^\varepsilon \le |t| \le A\sqrt{n}$ the estimate (1.3) holds for each fixed $\varepsilon > 0$. For "small" values of $|t| \le n^\varepsilon$ one should additionally take into account the coincidence of the moments of X and Y up to second order, that is

$$\frac{d^s}{dt^s}\left(\hat{\mu}_n(t) - \hat{\nu}(t)\right)\Big|_{t=0} = 0, \quad s = 0, 1, 2.$$

For example, one can expand

$$\ln \hat{\mu}_n(t) = n \ln E\exp(itn^{-1/2}X)$$

in a Taylor series in a neighborhood of zero,

$$n \ln E\exp(itn^{-1/2}X) = -\beta_2 t^2/2 + O(\beta_3|t|^3 n^{-1/2}),$$

and obtain

$$\frac{|\hat{\mu}_n(t) - \hat{\nu}(t)|}{|t|} = O\Big(t^2\exp(-t^2\beta_2/2)n^{-1/2}\Big).$$

This guarantees the necessary estimate $O(n^{-1/2})$ of the integral (1.3) on the interval $|t| \le n^\varepsilon$. We refer the reader to Lyapunov (1900), Esseen (1945), Ibragimov and Linnik (1971) and Petrov (1975) for more information on the Fourier method in the one-dimensional case.

Similar argumentation is possible in the finite-dimensional case. Extensions of the Berry–Esseen Lemma are obtainable there (see, e.g., Bhattacharya and Rao (1976), Bhattacharya (1977), Sweeting (1977) and Sazonov (1981)). The c.f.

$$\hat{\mu}_n(t) = E\exp(i(t, S_n)) = \Big(E\exp(in^{-1/2}(t, X))\Big)^n,$$

where $(x, y) = x_1y_1 + \cdots + x_ky_k$, $x, y \in \mathbb{R}^k$, is the standard scalar product and r.e. $X \in \mathbb{R}^k$, still has the multiplicative structure and, generally speaking, one can repeat the one-dimensional arguments with certain, sometimes very complicated, technical changes.

Now let X, X_1, X_2, ... denote r.e.'s taking values in a measurable space $(\mathcal{X}, \mathcal{B})$. Instead of sums one can study more general statistics

$$T = t_n(X_1, \dots, X_n), \tag{1.5}$$

which are symmetric in $X_1, \dots, X_n$ and such that the influence of each X_j is asymptotically negligible. An example of (1.5) is provided by U-statistics (see Götze (1979), Bickel, van Zwet and Götze (1986), Götze (1987)). Another example is

$$T = F(S_n),$$

where the summands X, X_1, X_2, ... of S_n take values in a separable Banach space B and are such that $EX = 0$, $E||X||^s < \infty$, $s \geq 2$. Here $F : B \to \mathbb{R}$ denotes a function that has enough Fréchet derivatives. Therefore T is no longer a sum of i.i.d. r.v.'s and the c.f. of T does not have a multiplicative structure. Thus a new technique for analyzing the c.f. of T has to be developed to still be able to use the Berry–Esseen lemma. Let us start with the simplest infinite-dimensional case where B denotes a Hilbert space H and $F(x) = ||x||^2$. The following extremely useful symmetrization inequality allows one to reduce the analysis of $E \exp(it||S_n||^2)$ to that of a product of certain characteristic functions (see Lemma 1.8 for generalizations).

The symmetrization $\bar{U}$ of a r.e. U is defined as $\bar{U} = U_1 - U_2$, where U_1 and U_2 are independent copies of r.e. U.

Lemma 1.1 (Götze (1979)). *For arbitrary independent r.e.'s* U, V, $W \in H$ *and* $t \in \mathbb{R}$, *the inequality*

$$|E \exp(it||U + V + W||^2)|^4 \leq E \exp(2it(\bar{U}, \bar{V}))$$

holds.

Proof. Note that for any real-valued function $f(u, v)$,

$$|E \exp(itf(U, V))|^2 \leq E \exp(it\Delta_2(V_1 - V_2)f(U, V_2)), \tag{1.6}$$

where V_1, V_2 are independent copies of V and the difference operator $\Delta_2(h)f(u, v) = f(u, v+h) - f(u, v)$ acts on the second argument of f. Indeed, applying Hölder's inequality and using the fact that V, V_1, V_2 are i.i.d., we have

$$\begin{aligned} |E \exp(itf(U, V))|^2 &\leq E|E(\exp(itf(U, V))|U)|^2 \\ &= E \exp(itf(U, V_1)) \exp(-itf(U, V_2)), \end{aligned}$$

which coincides with (1.6). Applying (1.6) twice with $f(U, V) = ||U+V+W||^2$ and using in this case that

$$\Delta_1(h_1)\Delta_2(h_2)f(u, v) = 2(h_1, h_2),$$

where the difference operator Δ_1 acts on the first argument of f, we complete the proof.

In order to illustrate how the symmetrization inequality works, we sketch the proof of the following theorem.

Theorem 1.2. *Let $X, Y \in H$. Suppose that X is a bounded r.e., $P\{||X|| \leq M\} = 1$, and X is not concentrated in a finite-dimensional subspace of H. Then*

$$\Delta_n := \sup_{x>0} |P\{||S_n||^2 < x\} - P\{||Y||^2 < x\}| = O(n^{-1/2}).$$

In the proof of the theorem we assume that X is not concentrated in a subspace of H, otherwise we can replace H by this subspace. Thus, there exists an orthonormal basis $\{e_k, k \in \mathbb{N}\}$ of H such that

$$Y = \sum_{k=1}^{\infty} \sigma_k \eta_k e_k, \quad \sum_{k=1}^{\infty} \sigma_k^2 < \infty,$$

where $\eta, \eta_1, \eta_2, \ldots \in \mathbb{R}$ are i.i.d. standard normal r.v.'s, $E\eta = 0$, $E\eta^2 = 1$ and $\sigma_1 \geq \sigma_2 \geq \cdots \geq 0$.

Lemma 1.3. *For all $t \in \mathbb{R}$ and $l = 1, 2, \ldots,$*

$$|E \exp(it||Y||^2)| \leq (1 + 4t^2\sigma_l^4)^{-l/4}. \tag{1.7}$$

Moreover,

$$|E \exp(it(Y_1, Y_2))| \leq (1 + t^2\sigma_l^4)^{-l/2}. \tag{1.8}$$

Proof. It is easy to verify that

$$E \exp(it\eta^2) = 1/\sqrt{1 - 2it}.$$

Therefore

$$E \exp(it||Y||^2) = \prod_{k=1}^{\infty} (1 - 2it\sigma_k^2)^{-1/2}$$

and this clearly implies (1.7). Furthermore, since

$$E \exp(it(x, Y)) = \exp(-(Dx, x)/2),$$

where $D = cov\, Y$, we have

$$E \exp(it(Y_1, Y_2)) = E \exp(-t^2(DY, Y)/2) \leq (E \exp(-t^2\sigma_l^4\eta^2))^l$$

which obviously implies (1.8).

Lemma 1.4. *If $P\{||X|| \leq M\} = 1$, then for all $p > 0$ and m, l such that $m + l \leq n$ the following inequality holds:*

$$E||n^{-1/2}(X_1 + \cdots + X_m + Y_1 + \cdots + Y_l)||^p \leq C(p)M^p.$$

Proof. It is sufficient to prove that

$$E||n^{-1/2}(Y_1+\cdots+Y_l)||^p \le C(p)M^p, \tag{1.9}$$

$$E||n^{-1/2}(X_1+\cdots+X_m)||^p \le C(p)M^p. \tag{1.10}$$

The inequality (1.9) follows from $\mathcal{L}(Y_1+\cdots+Y_l)=\mathcal{L}(\sqrt{l}Y)$ and

$$E||Y||^p \le C(p)(E||Y||^2)^{p/2} = C(p)(E||X||^2)^{p/2} \le C(p)M^p.$$

To prove (1.10) we apply the Zygmund–Marcinkiewicz inequality (see e.g. Vakhaniya et al. (1987))

$$E||Z_1+\cdots+Z_m||^p \le C(p)E(||Z_1||^2+\cdots+||Z_m||^2)^{p/2}$$

which holds for arbitrary i.i.d. r.e.'s $Z_1,\ldots,Z_m \in H$ with zero mean. Therefore (1.10) reduces to the obvious inequality $n^{-p/2}E(||X_1||^2+\cdots+||X_m||^2)^{p/2} \le M^p$.

Lemma 1.5. *Let $P\{||X||\le M\}=1$. Define $\bar{U}:=n^{-1/2}(\bar{X}_1+\cdots+\bar{X}_m)$, $m\le n$. Suppose that the r.e.'s Z, $\bar{U}\in H$ are independent. Then for all $s\in\mathbb{R}$ and $L>0$ such that*

$$|s|LM\le\sqrt{n}, \tag{1.11}$$

we have

$$0\le E\exp(is(Z,\bar{U}))\le P\{||Z||>L\}+E\exp\left(-\frac{s^2m}{2n}(DZ,Z)\right), \tag{1.12}$$

where $D=cov\,X=cov\,Y$.

Proof. Since the r.e. $\bar{U}$ is the symmetrization of $U:=n^{-1/2}(X_1+\cdots+X_m)$, the inequality on the left-hand side of (1.12) holds. To prove the inequality on the right-hand side of (1.12), it suffices to show that for each (non-random) $z\in H$ we have

$$E\exp(is(z,\bar{U}))\le\chi\{||z||>L\}+\exp\left(-\frac{s^2m}{2n}(Dz,z)\right). \tag{1.13}$$

If $||z||>L$, then (1.13) is obvious. Therefore we assume that $||z||\le L$. Then (1.13) reduces to

$$E\exp(is(z,\bar{U}))\le\exp\left(-\frac{s^2m}{2n}(Dz,z)\right). \tag{1.14}$$

We have

$$E\exp(is(z,\bar{U}))=\left[E\cos\left(\frac{s}{\sqrt{n}}(s,\bar{X})\right)\right]^m \tag{1.15}$$

since $\bar{X}$, $\bar{X}_1$, $\bar{X}_2$, ... are i.i.d. and $\bar{X}$ is symmetric. Put $x=s(z,\bar{X})/\sqrt{n}$. Then (1.11) together with $||z||<L$, $||\bar{X}||\le 2M$ ensure that $|x|\le 2$. Therefore in

(1.15) we can apply the obvious inequality $\cos x \le 1 - x^2/4$ valid for $|x| \le 2$. Noting that $E(z, \bar{X})^2 = 2(Dz, z)$ we have

$$E\exp(is(z,\bar{U})) \le \left[1 - \frac{s^2}{2n}(Dz,z)\right]^m \le \exp\left(-\frac{s^2 m}{2n}(Dz,z)\right),$$

which completes the proof.

Proof of Theorem 1.2. The distribution function $r \to P\{||Y||^2 < r\}$ has a bounded density provided $\sigma_3 > 0$. Indeed, it follows from Lemma 1.3 that the c.f.

$$|E\exp(it||Y||^2)| \le \min(1; \sigma_3^{-3}|t|^{-3/2})$$

is integrable. This allows one to apply the Berry–Esseen lemma with $T = \sqrt{n}$. Therefore the estimate $\Delta_n = O(n^{-1/2})$ will result from the following four bounds:

$$\begin{aligned}
I_1 &:= \int_{n^\varepsilon < |t| < \sqrt{n}} \frac{1}{|t|}|g(t)|\,dt = O(n^{-1/2});\\
I_2 &:= \int_{n^\varepsilon < |t| < \sqrt{n}} \frac{1}{|t|}|f_n(t)|\,dt = O(n^{-1/2});\\
I_3 &:= \int_{|t|\le 1} \frac{1}{|t|}|f_n(t) - g(t)|\,dt = O(n^{-1/2});\\
I_4 &:= \int_{1\le|t|\le n^\varepsilon} \frac{1}{|t|}|f_n(t) - g(t)|\,dt = O(n^{-1/2}),
\end{aligned}$$

where $0 < \varepsilon \le \frac{1}{2}$ is a number to be chosen later and

$$g(t) := E\exp(it||Y||^2), \quad f_n(t) := E\exp(it||S_n||^2).$$

Let us estimate I_1 first. It follows from Lemma 1.3 that

$$|g(t)| \le 1/(\sigma_l^l |t|^{l/2}).$$

Therefore $I_1 \le 2/(l\sigma_l^l n^{\varepsilon l/2}) = O(n^{-1/2})$ if $\varepsilon l \ge 1$. But such an $l = l(\varepsilon)$ exists since $\sigma_l > 0$ for all l.

Next we shall estimate I_2. It is sufficient to show for $1 \le |t| \le \sqrt{n}$ and a sufficiently large fixed constant $A > 0$ that

$$|f_n(t)| = O(|t|^{-A} + n^{-A}). \tag{1.16}$$

Write $S_n = U + V + W$, where

$$U = n^{-1/2}(X_1 + \cdots + X_m), \quad V = n^{-1/2}(X_{m+1} + \cdots + X_{m+k}), \quad W = S_n - U - V$$

and $k + m \le n$. Then by the symmetrization inequality (see Lemma 1.1), (1.16) follows from

$$E\exp(it(\bar{V},\bar{U})) = O(|t|^{-A} + n^{-A}). \tag{1.17}$$

Using Lemma 1.5 we obtain for all $L > 0$

$$E\exp(it(\bar{V},\bar{U})) \le P\{\|\bar{V}\| > L\} + E\exp\left(-\frac{t^2 m}{2n}(D\bar{V},\bar{V})\right), \tag{1.18}$$

provided $|t|LM \le \sqrt{n}$. Choose (noting that $1 \le |t| \le \sqrt{n}$)

$$L = 1/(M|t|^{1/4}), \quad k \sim n/(2|t|), \quad m = n - k \sim n(1 - 1/(2|t|)).$$

Using Chebyshev's inequality and Lemma 1.4, we have

$$P\{\|\bar{V}\| > L\} = P\{\|\bar{S}_k\| > L\sqrt{n/k}\} \le C(M,A)(k/(nL^2))^{2A} = O(|t|^{-A}). \tag{1.19}$$

Similarly,

$$\begin{aligned} E\exp\left(-\frac{t^2 m}{2n}(D\bar{V},\bar{V})\right) &\le E\exp\left(-\frac{|t|}{8}(D\bar{S}_k,\bar{S}_k)\right) \\ &\le P\{(D\bar{S}_k,\bar{S}_k) \le 8/\sqrt{|t|}\} + \exp(-\sqrt{|t|}). \end{aligned} \tag{1.20}$$

Estimates (1.18)–(1.20) reduce (1.17) to

$$P\{(D\bar{S}_k,\bar{S}_k) \le 8/\sqrt{|t|}\} = O(|t|^{-A} + n^{-A}), \tag{1.21}$$

where $k \sim n/(2|t|)$, and $1 \le |t| \le \sqrt{n}$. But (1.21) is a consequence of the following concentration inequality (note that $k \ge \sqrt{n}/2$):

$$P\{(D\bar{S}_k,\bar{S}_k) < \varepsilon^2\} = O(\varepsilon^l + k^{-l/2}), \tag{1.22}$$

valid for all $\varepsilon > 0$ and $l = 1, 2, \ldots$. Such a concentration inequality for balls in Hilbert space can be reduced to the analogous statement in $\mathbb{R}^l$ via $\|x\|^2 \ge \sum_{s=1}^{l}(x,e_s)^2$. In $\mathbb{R}^l$ concentration inequalities of this kind have been proved by Esseen (1968), Paulauskas (1973), Götze (1979), etc. For the infinite-dimensional case, see, e.g., Siegel (1981), Bentkus (1985b), etc. The proof of (1.22) in $\mathbb{R}^l$ can be obtained as follows. For $s \ge 0$ we have

$$P\{\|S_n\|^2 \le \varepsilon^2\} \le \exp(\varepsilon^2 s^2/2) E\exp\left(-\frac{s^2}{2}\|S_n\|^2\right).$$

If we denote by $Z \in \mathbb{R}^l$ a r.e. with the standard normal distribution, then

$$E\exp\left(-\frac{s^2}{2}\|S_n\|^2\right) = E\exp(is(Z,S_n)).$$

Now one can apply Lemma 1.5.

Let us estimate I_3. Put $G(x) = \exp(it\|x\|^2)$. Then $G \in C^\infty$ and

$$G'''(x)h^3 = -8t^2 G(x)[it(x,h)^3 + (x,h)(h,h)]. \tag{1.23}$$

The estimate $I_3 = O(n^{-1/2})$ clearly follows from

$$|EG(S_n) - EG(Y)| = O(|t|n^{-1/2}). \tag{1.24}$$

We shall prove (1.24) using Lindeberg's method (see the beginning of § 1.2). Obviously

$$|EG(S_n) - EG(Y)| \le J_1 + \cdots + J_n,$$

where

$$J_k = |EG(W_k + n^{-1/2}X) - EG(W_k + n^{-1/2}Y)|,$$
$$W_k = n^{-1/2}(X_1 + \cdots + X_{k-1} + Y_{k+1} + \cdots + Y_n).$$

To estimate J_k, we use the Taylor formula

$$G(x+h) = G(x) + G'(x)h + \frac{1}{2}G''(x)h^2 + \frac{1}{2}E(1-\tau)^2G'''(x+\tau h)h^3$$

with $x = W_k$, $h = n^{-1/2}X$ and $h = n^{-1/2}Y$, respectively. Here the r.v. τ is uniformly distributed on [0, 1]. The terms containing derivatives of G up to the second order vanish since the means and covariances of X and Y coincide and we have

$$J_k \le n^{-3/2}(J_k' + J_k''),$$

where

$$\begin{aligned} J_k' &= |E(1-\tau)^2G'''(W_k + \tau n^{-1/2}X)X^3|, \\ J_k'' &= |E(1-\tau)^2G'''(W_k + \tau n^{-1/2}Y)Y^3|. \end{aligned} \tag{1.25}$$

It follows from (1.23) for $|t| \le 1$ that

$$|G'''(x)h^3| = O(|t| \cdot ||h||^3(1 + ||x||^3)).$$

Therefore

$$J_k' = O\left(|t|E||X||^3(1 + ||W_k + \tau n^{-1/2}X||^3)\right) = O(|t|)$$

since $||X|| \le M$ and, according to Lemma 1.4, we have $E||W_k||^3 = O(1)$. Similarly, $J_k'' = O(|t|)$ since all of the moments of the Gaussian r.e. Y are finite. Therefore $J_k = O(n^{-3/2}|t|)$, which yields (1.24).

It remains to estimate I_4. Here we shall combine the methods employed in estimating I_2 and I_3. Repeating the estimation of I_3, we arrive at, for example, the integral J_k' (see (1.25)) and it suffices to show for $|t| \ge 1$ and each $A > 0$ that

$$J_k' = O(|t|^3(|t|^{-A} + n^{-A})). \tag{1.26}$$

It is clear from the explicit formula (1.23) for G''' that

$$G'''(W_k + \tau n^{-1/2}X)X^3 = EP(t, X, W_k, \tau n^{-1/2}X)\exp(it||W_k + n^{-1/2}\tau X||^2),$$

where $P(\cdot,\cdot,\cdot,\cdot)$ is a polynomial not exceeding third degree in each argument. Applying the triangle inequality, we can reduce the estimation of J_k' to that of a certain sum of quantities like $|t|^3\gamma$, where

$$\gamma := |E(W_k, X)^3 \exp(it||W_k + \tau n^{-1/2} X||^2)|,$$

and so (1.26) will follow from

$$\gamma = O(|t|^{-A} + n^{-A}). \tag{1.27}$$

Let us split $W_k = T_1 + T_2 + T_3 + T_4$ into four sums in such a way, that each sum T_1, T_2, T_3, T_4 has approximately the same number (equivalent to $n/4$) of terms X_j or Y_j. Then

$$(W_k, X)^3 = \sum_{1 \le l_1, l_2, l_3 \le 4} (T_{l_1}, X)(T_{l_2}, X)(T_{l_3}, X)$$

and the estimation of γ is reduced to estimating

$$\gamma_1 := |E(T_{l_1}, X)(T_{l_2}, X)(T_{l_3}, X) \exp(it||W_k + \tau n^{-1/2} X||^2)|.$$

But among T_{l_1}, T_{l_2}, T_{l_3} (with fixed l_1, l_2, l_3) at least one of T_1, T_2, T_3, T_4 is absent, say T_1. Therefore

$$\gamma_1 \le E||X||^3 \cdot ||T_{l_1}|| \cdot ||T_{l_2}|| \cdot ||T_{l_3}|| \gamma_2,$$

where

$$\gamma_2 = |E_{T_1} \exp(it||T_1 + W||^2)|, \quad W = W_k + \tau n^{-1/2} X - T_1,$$

and E_{T_1} denotes the expectation taken only with respect to the r.e. T_1. We can estimate the moments of the r.e.'s T_{l_1}, T_{l_2}, T_{l_3} by Lemma 1.4. To estimate γ_2, we repeat the arguments of the estimation of $E \exp(it||S_n||^2)$ used in the proof of the relation $I_2 = O(n^{-1/2})$. Here we replace S_n by T_1 and use the fact that T_1 contains at least $n/4$ summands, and that T_1 and W are independent. In this way we arrive at $\gamma_2 = O(|t|^{-A} + n^{-A})$, which implies (1.27). This completes the proof of the theorem.

Write

$$\Delta_n(a, r) = |P\{||S_n - a||^2 < r\} - P\{||Y - a||^2 < r\}|,$$
$$\Delta_n(a) = \sup_{r \ge 0} \Delta_n(a, r).$$

Estimates of $\Delta_n(a, r)$ in Hilbert space were obtained in a number of papers by various methods. For $\beta_3 = E||X||^3 < \infty$ and fixed $a \in H$, the bounds on $\Delta_n(a)$ were improved from logarithmic order in Kandelaki (1965) to $O(n^{-1/6})$ by Paulauskas (1976b) and to $O(n^{-1/4})$ by Yurinskii (see Sazonov (1981)). Assuming additionally that the coordinates of r.e. X are independent, Nagaev and Chebotarev (1978) showed that $\Delta_n(0) = O(n^{-1/2})$. To prove the same bound, Borovskikh and Račkauskas (1979) and Račkauskas (1981) needed only seven coordinates independent of the rest. The essential step is due to Götze (1979) who introduced the symmetrization inequality (Lemma 1.1) for investigations of $\Delta_n(a, r)$ by the Fourier method. Götze (1979) proved the bound $\Delta_n(a) = O(n^{-1/2})$ for a fixed $a \in H$ for $\beta_6 < \infty$ and the bound $\Delta_n(0) = O(n^{-1+\varepsilon})$, $\varepsilon > 0$, for $\beta_8 < \infty$. Yurinskii (1982) introduced in addition a truncation technique and obtained the following result.

Theorem 1.6. *There exists a constant $C = C(\mathcal{L}(Y)) > 0$ such that for each $a \in H$ the following inequality holds:*

$$\Delta_n(a) \leq C(1 + ||a||^3)\beta_3 n^{-1/2}. \tag{1.28}$$

The Fourier method was refined and various improvements and generalizations were obtained by Zalesskii (1982, 1985), Nagaev (1983, 1985, 1988, 1989a), Bentkus (1984d, e), Sazonov and Zalesskii (1985), Nagaev and Chebotarev (1986), Senatov (1986), Sazonov, Ul'yanov and Zalesskii (1987a, b, 1988, 1989a), Aliev (1989) and Sazonov and Ul'yanov (1990, 1991). It is interesting to note than in infinite-dimensional Hilbert space, one can obtain better rates than $O(n^{-1/2})$ without conditions like the classical Cramér condition for the c.f. Namely, if $E||X||^{2+2\delta} < \infty$, $0 \leq \delta < 1$, then $\Delta_n(a) = o(n^{-\delta})$ if $a = 0$ or X is symmetric (see §2.3 for details). This goes back to a result in $\mathbb{R}^k$ due to Esseen (1945) and has been proved by Zalesskii (1982) when $a = 0$ and by Bentkus (1984e) in the general case. However the following problem remains open: *is it true that for infinite-dimensional r.e. $X \in H$ the condition $E||X||^4 < \infty$ implies $\Delta_n(0) = O(n^{-1})$?*

The structure of the constant C in (1.28) has been investigated by various authors. The dependence of the estimate on the eigenvalues σ_i^2 of $cov\, Y$ was studied in particular. Here things have progressed from thirteen eigenvalues in Yurinskii (1982) to seven in Nagaev (1983), Sazonov, Ul'yanov and Zalesskii (1987) and to six in Senatov (1989b). Recently Nagaev (1989b) and Zalesskii, Sazonov and Ul'yanov (1988) gave the estimate

$$\Delta_n(a) \leq \frac{C_1\beta_3}{\sigma_1 \cdots \sigma_6}(\sigma^3 + ||a||^3)n^{-1/2}, \tag{1.29}$$

where $\sigma^2 = E||X||^2$, whereas Senatov (1989a) gave the result

$$\Delta_n(a, r) \leq \frac{C_1\beta_3}{\sigma_1 \cdots \sigma_6}(r^3 + \beta_3 n^{-1/2})n^{-1/2}. \tag{1.30}$$

In both estimates C_1 is an absolute constant. The proof of (1.29) it is now 1999–2000 appeared in Sazonov, Ul'yanov and Zalesskii (1989b).

Lower bounds of the convergence rate in Hilbert space were found by Senatov (1985a, b, 1986), Barsov (1987), Aliev (1987) and Bloznelis (1989). Senatov (1986) proved the necessity of the first six eigenvalues of $cov\, Y$ in (1.29) and (1.30). Barsov has shown that $\Delta_n(0) = O(n^{-(s-2)/2})$ for some $s > 2$ implies $E||X||^{s-\varepsilon} < \infty$ for each $\varepsilon > 0$, $\varepsilon \leq s$.

A truncation lemma adapted for nonuniform estimates for $\Delta_n(a, r)$ was introduced by Sazonov and Zalesskii (1985). Nonuniform estimates of $\Delta_n(a, r)$ were obtained by Sazonov and Zalesskii (1985), Bentkus and Zalesskii (1985) and Sazonov, Ul'yanov and Zalesskii (1987b, 1988). The following bound provides an example of a nonuniform estimate

$$\Delta_n(a, r) = O\left(\frac{1 + ||a||^3}{(1 + \rho)^s\sqrt{n}}\right),$$

where $E||X||^s < \infty$ for some $s \geq 3$ and $\rho = |\sqrt{r} - ||a|||$ denotes the distance between 0 and the boundary of the ball $\{x \in H : ||x - a|| < \sqrt{r}\}$.

The case of non-identically distributed summands was considered in Bentkus (1984e), Ul'yanov (1987) and Borisov (1989). For results concerning the local CLT in H see §2.4.

We end the review of results obtained in Hilbert space with the following one on large deviations.

Theorem 1.7 (Yurinskii (1988)). *Suppose that*

$$E\exp(c||X||) < \infty$$

for some constant $c > 0$. Then there exist constants $A_i = A_i(\mathcal{L}(X)) > 0$ such that

$$P\{||S_n|| > r\} = P\{||Y|| > r\}I(r, n, \mathcal{L}(X))(1 + \theta A_1 r n^{-1/2})$$

for $A_2 \leq r \leq A_3 n^{1/2}$. Here $|\theta| \leq 1$ and $I(r, n, \mathcal{L}(X))$ denotes a certain analog of the Cramér series in large deviations theorems. Furthermore,

$$P\{||S_n|| > r\} = P\{||Y|| > r\}(1 + \theta A_4 r^3 n^{-1/2})$$

if $A_2 \leq r \leq A_5 n^{1/6}$.

For details concerning the construction of $I(r, n, \mathcal{L}(X))$, see Yurinskii (1988). In the case of Hilbert space this improves results of Osipov (1978a, b), Bentkus (1986c), Bentkus and Račkauskas (1990), Zalesskii (1989) and Račkauskas (1988).

Consider now Banach spaces. Put

$$\Delta_{n,F}(r) = |P\{F(S_n) < r\} - P\{F(Y) < r\}|,$$

where X and Y assume values in the Banach space B and a function $F : B \to \mathbb{R}$. As in the Hilbert space case the estimation of $\Delta_{n,F}$ strongly depends on the symmetrization inequalities. An example of such an inequality is provided by the following lemma.

Lemma 1.8. *If a function $f(u_1, \ldots, u_k)$ of arguments $u_1, \ldots, u_k \in B$ is real-valued and the r.e.'s $U_1, \cdots, U_k \in B$ are independent, then*

$$\begin{aligned} &|E\exp(itf(U_1, \ldots, U_k))|^{2^k} \\ &\leq E\exp(i\Delta_1(U_1' - U_1)\cdots\Delta_k(U_k' - U_k)f(U_1, \ldots, U_k)), \end{aligned}$$

where U_s' denotes an independent copy of U_s and the difference operator

$$\Delta_s(h)f(\ldots, u_s, \ldots) = f(\ldots, u_s + h, \ldots) - f(\ldots, u_s, \ldots)$$

acts on the s-th argument of f.

In particular, for all independent $U_1, \ldots, U_{k+1} \in B$ and every polynomial $\pi(x) = \pi_k(x) + \cdots + \pi_0(x)$, $x \in B$, where $\pi_s(x) = \pi_s(x, \ldots, x)$, $s = 0, \ldots, k$, denote symmetric continuous s-linear forms on B, the following holds:

$$|E\exp(i\pi(U_1+\cdots+U_{k+1}))|^{2^k} \leq E\exp(ik!\pi_k(\bar{U}_1,\ldots,\bar{U}_k)), \tag{1.31}$$

where $\bar{U}$ denotes the symmetrization of U.

Lemma 1.8 is an easy generalization of Lemma 1.1 if one uses an estimate of the type (1.6) k times. Inequality (1.31) was given in a somewhat different form by Weyl (1916) for polynomials $\pi(x)$ where x has a uniform distribution on the discrete set $\{1,\ldots,N\}$. Here it was necessary to split x in the form $x = x_1+\cdots+x_k+x_{k+1}$, where $x_1,\ldots,x_k$ are uniformly distributed and x_{k+1} is a function of these variables, so as to yield a uniformly distributed sum. The generality of Weyl's inequality made possible numerous applications in analytic number theory (see, e.g., Schmidt (1984)). For polynomials in one variable, there are sharper estimates due to Vinogradov (1934). These seem not easily extendable to general probability distributions. The inequality (1.31) for $k=2$ was derived for the probability context independently in Götze (1979). The immediate extension to $k>2$ was used by Yurinskii (1983) to prove the following result.

Theorem 1.9 (Yurinskii (1983)). *Let $B=\ell_p$, $p=2, 4, \ldots$. Then*

$$|P\{||S_n-a||<r\}-P\{||Y-a||<r\}| \leq C\beta_3\sigma^{-3}(1+||a/\sigma||^{3p-3})n^{-1/2},$$

where $\sigma^2 = E||X||^2$ and $C=C(\mathcal{L}(Y))$ is a constant.

To formulate the next result we need the following conditions.
Differentiability condition (D_5). There exists $p \geq 0$ such that

$$\sup_{x\in B} P(1+||x||)^{-p}||F^{(s)}(x)|| < \infty, \quad s=0,\ldots,5.$$

Variance condition (V). For sufficiently large $M>0$ and for each fixed c

$$\sup_{||a||\leq c} \{\sigma(Y+a)<\delta\} = O(\delta^M) \text{ as } \delta\downarrow 0,$$

where $\sigma^2(x) = E(F'(x)Y)^2$.

Theorem 1.10 (Götze (1983)). *If $E||X||^3 < \infty$ and (D_5) and (V) are satisfied, then*

$$\sup_{r\in\mathbb{R}} |P\{F(S_n)<r\}-P\{F(Y)<r\}| = O(n^{-1/2}). \tag{1.32}$$

Actually Götze (1983) proved a slightly more exact result. Under certain natural conditions Zalesskii (1985) made the estimate (1.32) nonuniform in the Hilbert space $B=H$. The proof of Theorem 1.9 uses the Fourier method and exploits a slightly more general symmetrization inequality as Lemma 1.8. Götze (1986) introduced the method of integration by parts (see § 1.3) which allows one to prove more precise results than Theorem 1.10. Nevertheless, generalizations and extensions of the Fourier method, used to prove Theorems like 1.9 and 1.10, are useful for the construction of asymptotic expansions (see § 2.3).

Remark 1.11. It seems that verification of variance condition (V) is not simple in general. For example, it is still unknown whether (V) (or some other suitable condition) is satisfied for each infinite-dimensional r.e. $Y \in B = L_p$, $1 < p < \infty$, with $F(x) = ||x||$ or $F(x) = ||x||^s$ with an appropriate $s > 0$.

§1.2. The Lindeberg Method

Lindeberg's proof of the central limit theorem, which appeared in 1920 and 1922, is very simple and can be easily extended to investigate the rate of convergence even for B-r.e.'s. A similar remark applies to Trotter's proof given in 1959. Actually Trotter's proof differs from Lindeberg's only in terminology, but the method is presented in an intuitively more understandable manner. Besides the original papers of Lindeberg (1920, 1922) and Trotter (1959), we mention the books of Billingsley (1968), Thomasian (1969) and Feller (1971).

In this section we shall concentrate our attention on some extensions of Lindeberg's method that have been used to investigate the rate of convergence in the CLT in Banach spaces. Recall that X, Y are B-r.e.'s such that $EX = EY = 0$, $cov\, X = cov\, Y$ and Y is Gaussian. Furthermore, $S_n = n^{-1/2}(X_1 + \cdots + X_n)$, where $X, X_1, X_2, \ldots$ are i.i.d.

Briefly, Lindeberg's approach is as follows. Suppose we have to estimate the difference of the measures $\mu_n = \mathcal{L}(S_n)$ and $\nu = \mathcal{L}(Y)$ on a certain measurable set $A \subset B$, i.e., we have to estimate the quantity

$$\mu_n(A) - \nu(A) = \int_B \chi_A(x)(\mu_n - \nu)(dx).$$

The first step is to replace the discontinuous indicator function χ_A by a sufficiently smooth function say $g = g_{A,\varepsilon}$, which coincides with χ_A everywhere with the exception of an ε-neighborhood $(\partial A)_\varepsilon = \{x \in B : \inf_{y\in\partial A} ||x - y|| < \varepsilon\}$ of the boundary ∂A of the set A. After this substitution (usually called the smoothing lemma) two terms are to be estimated: the integral $I = \int_B g(x)(\mu_n - \nu)(dx)$ and the quantity $\nu((\partial A)_\varepsilon)$. The estimation of I is based on the Taylor formula and the following identity:

$$\begin{aligned}\mu_n - \nu &= \mathcal{L}(S_n(X)) - \mathcal{L}(S_n(Y)) \\ &= \sum_{k=1}^{n} \Big[\mathcal{L}(W_{n,k} + n^{-1/2}X_k) - \mathcal{L}(W_{n,k} + n^{-1/2}Y_k)\Big], \qquad (2.1)\end{aligned}$$

where $W_{n,k} = n^{-1/2}\left(\sum_{i=1}^{k-1} X_i + \sum_{i=k+1}^{n} Y_i\right)$. If g can be chosen to be three-times Fréchet differentiable such that $\sup_{x\in B} ||g'''(x)|| \le C\varepsilon^{-3}$, we can expand each $g(W_{n,k} + n^{-1/2}X_k)$ and $g(W_{n,k} + n^{-1/2}Y_k)$ around $W_{n,k}$ and obtain

$$g(W_{n,k}+n^{-1/2}X_k) = g(W_{n,k})+g'(W_{n,k})[X_k]n^{-1/2}+\frac{1}{2}g''(W_{n,k})[X_k^2]n^{-1}$$
$$+\frac{1}{6}g'''(W_{n,k}+\theta X_k n^{-1/2})[X_k^3]n^{-3/2},$$

where $|\theta| \leq 1$. Due to the equality of the means and covariances of X_k and Y_k (see Lemma 2.2), the difference $Eg(W_{n,k}+n^{-1/2}X_k) - Eg(W_{n,k}+n^{-1/2}Y_k)$ will contain only terms involving third derivatives and we easily arrive at the bound

$$I \leq Cn^{-1/2}\varepsilon^{-3}\nu_3,$$

where

$$\nu_3 = \int_B ||x||^3|\mathcal{L}(X)-\mathcal{L}(Y)|(dx).$$

The quantity $\nu((\partial A)_\varepsilon)$ is usually of order $C\varepsilon$. Therefore, if $\varepsilon^4 = n^{-1/2}\nu_3$, we obtain the bound $C\nu_3^{1/4}n^{-1/8}$ for the difference $\mu_n(A)-\nu(A)$. Such a rate was found for the first time using Lindeberg's method in Hilbert space by Kuelbs and Kurtz (1974). It is clear, however, that in order to derive better bounds than $O(n^{-1/8})$, one needs some additional arguments.

Firstly, there are several ways to construct a smooth approximation to an indicator function such that the smoothing error is as small as possible. In finite-dimensional space, one can use the convolution of the indicator function with a Gaussian distribution. This type of smoothing leads to the convolution method, which yields Berry–Esseen-type convergence rate results. This method was applied explicitly for the first time by Bergström (1944). We refer the reader to Sazonov's book (1981) for more details on the convolution method. Bentkus (1986a) constructed directly the approximation of an indicator function by a once-differentiable function whose derivative satisfies Hölder's condition with a constant independent of the dimension and exploited the smoothing properties of the Gaussian terms in the identity (2.1). This resulted in a better dependence of the remainder term on the dimension.

However, neither the arguments of the convolution method nor those used by Bentkus have a satisfactory straightforward extension to the infinite-dimensional case. This is due to the limited smoothness properties of Gaussian measures and to the fact that there is no analogue of Lebesgue measure in infinite-dimensional Banach spaces. Efforts to provide such an extension to Hilbert space were made by Osipov and Rotar' (1985). They considered the rate of convergence on balls and obtained Berry–Esseen-type bounds with a logarithmic factor. Optimal bounds had already been found by the methods of characteristic functions (see the previous section). Nevertheless the arguments used by Osipov and Rotar' prove to be useful in the case of dependent r.e.'s (see, for example, Račkauskas (1990) for the martingale case).

If g is a smoothed indicator function such that $g(x) = 1$ if $x \in A$ and $g(x) = 0$ if $x \in B - A_\varepsilon$, then one may use the simple fact that the derivatives of g vanish outside an ε-neighborhood of the boundary ∂A. This idea has been exploited in several papers using in addition either iteration or induction arguments. It yields bounds of at most order $O(n^{-1/6})$ by assuming a finite third moment. Despite the great difference as compared to the finite-dimensional Berry–Esseen-type results, such orders are in general unimprovable (see Theorem 2.6 below). Iteration arguments appeared in Paulauskas (1976b) and under certain assumptions, he gave bounds on balls of order $O(n^{-1/6})$. Refinements and generalizations were given by Butzer et al. (1979), Ul'yanov (1981), Bernotas (1980) and Bentkus and Račkauskas (1982), (1983). Bentkus and Račkauskas (1982) applied different notions of smoothness. This allowed them to prove rates for sets with very unsmooth boundaries such as balls in the spaces $C[0,1]$, c_0, etc. These papers mostly considered the convergence rate on balls under the assumption that for $a \in B$ the function $r \to (d/dr)P\{||Y+a|| < r\}$ is bounded and admits certain estimates. Such an assumption can be verified in Hilbert space and some other "good" Banach spaces but it may fail in general Banach spaces (especially in $C[0,1]$ and c_0). Račkauskas proved (see Paulauskas and Račkauskas (1989)) that $O(n^{-1/6})$ remains valid under the natural assumption that the density $r \to (d/dr)P\{||Y|| < r\}$ is bounded and the third moment is finite.

We shall consider the estimation of the quantity

$$\Delta_{n,q} := \sup_{r \geq 0} |P\{q(S_n(X)) < r\} - P\{q(Y) < r\}|$$

in detail, where $q : B \to \mathbb{R}$ is a contuous semi-norm. Without loss of generality, we may assume that $q(x) \leq ||x||$ for all $x \in B$. First we introduce some conditions needed to formulate results obtained using Lindeberg's method with induction.

Smoothness condition (A_3). For each $r \geq 0$ and $\varepsilon > 0$ there exists a function $g_{r,\varepsilon} : B \to \mathbb{R}$ such that

(a) for all $x \in B$,

$$\chi(q(x) < r) \leq g_{r,\varepsilon}(x) \leq \chi(q(x) < r + \varepsilon);$$

(b) the function $g_{r,\varepsilon}$ is three times continuously Frechét differentiable and there is a constant $C > 0$ such that for all $r \geq 0$, $\varepsilon > 0$ and $i = 1, 2, 3$,

$$\sup_{x \in B} ||g_{r,\varepsilon}^{(i)}(x)|| \leq C\varepsilon^{-i}.$$

Density condition (D). There exists a constant $C = C(q, \mathcal{L}(Y)) > 0$ such that for all $\varepsilon > 0$

$$\sup_{r \geq 0} P\{r - \varepsilon \leq q(Y) \leq r + \varepsilon\} \leq C\varepsilon.$$

Actually condition (D) is equivalent to the existence of a bounded density of the distribution function $P\{q(Y) < r\}$.

Theorem 2.1. *Suppose that conditions (A_3), (D) hold and that the Gaussian r.e. Y is infinite-dimensional. Then there exists a constant $C = C(q, \mathcal{L}(Y)) > 0$ such that*

$$\Delta_{n,q} \leq C\nu_3^{1/3} n^{-1/6}.$$

Proof. Define for $k = 1, \ldots, n$ the quantity

$$\Delta_{n,k,q} = \sup_{r \geq 0} |P\{q(U_{k,n}) < r\} - P\{q(Y) < r\}|,$$

where $U_{k,n} = n^{-1/2}(\sum_{i=1}^{k} X_i + \sum_{i=k+1}^{n} Y_i)$, with $\sum_{i \in \emptyset} \equiv 0$. We now proceed to prove inductively the bound

$$\delta_n := \max_{1 \leq k \leq n} \Delta_{n,k,q} \leq C_0 \nu_3^{1/3} n^{-1/6},$$

which of course yields the desired result since $\Delta_{n,q} \leq \delta_n$.

Since Y is infinite-dimensional, we have $\delta_1 \leq C_0 \nu_3^{1/3}$ (see, e.g., Theorem 5.1.11 in Paulauskas and Račkauskas (1989)). So let $n > 1$ and suppose that according to the induction assumption

$$\delta_{n-1} \leq C_0 \nu_3^{1/3} (n-1)^{-1/6}. \tag{2.2}$$

Let $\varepsilon > 0$ denote a parameter which will be specified at the end of the proof. If $r \leq 2\varepsilon$, we have $\Delta_{n,k,q}(r) \leq \Delta_{n,k,q}(2\varepsilon) + C\varepsilon$, where

$$\Delta_{n,k,q}(r) := |P\{q(U_{n,k}) < r\} - P\{q(Y) < r\}|.$$

Therefore it is sufficient to estimate $\Delta_{n,k,q}(r)$ for $r \geq 2\varepsilon$ only. Define $g_1(x) = g_{r,\varepsilon}(x)$, $g_2(x) = g_{r-\varepsilon,\varepsilon}(x)$, where $g_{r,\varepsilon}$ is the same function as in condition (A_3). Define

$$G_{n,k} = \mathcal{L}(U_{n,k}), \quad P_{n,k} = \mathcal{L}(W_{n,k}), \quad H = \mathcal{L}(X) - \mathcal{L}(Y).$$

One easily shows the following smoothing inequality (see Kuelbs and Kurtz (1974)):

$$\Delta_{n,k,q}(r) \leq \max_{i=1,2} \left| \int_B g_i(x)(G_{n,k} - G)(dx) \right| + P\{r - \varepsilon \leq q(Y) \leq r + \varepsilon\}. \tag{2.3}$$

The last probability does not exceed $C\varepsilon$ due to condition (D). To estimate the integrals in (2.3) we proceed as in Lindeberg's method. Using (2.1), we obtain

$$\left| \int_B g_i(x)(G_{n,k} - G)(dx) \right| \leq \sum_{j=1}^{k} \left| \int_B \int_B g_i(y + xn^{-1/2}) P_{n,j}(dy) H(dx) \right|.$$

We now expand $g_i(y + xn^{-1/2})$ in a Taylor series in a neighborhood of y obtaining

$$g_i(y+xn^{-1/2}) = g_i(y)+g_i'(y)[x]n^{-1/2}+\frac{1}{2}n^{-1}g_i''(y)[x^2] + \frac{1}{2}n^{-3/2}\int_0^1(1-\theta)^2 g_i'''(y+\theta n^{-1/2}x)[x^3]\,d\theta. \quad (2.4)$$

Note that $\int_B g_i'(y)[x]H(dx)=0$ since $EX=EY=0$. Moreover,

$$\int_B g_i''(y)[x^2]H(dx)=0$$

since the covariances of X and Y coincide. This is not trivial because $cov\,X = cov\,Y$ implies $Ef(X)g(X)=Ef(Y)g(Y)$ for all $f,g\in B^*$ only. The second derivative $g_i''(y)$, which is continuous bilinear form on B, in general may not be generated by linear functionals. But we have the following result due to Borisov (1989).

Lemma 2.2. *Let B be a separable Banach space. Let r.e.'s X, $Y\in B$, $E||X||^2<\infty$ and Y be Gaussian. If $EX=EY=0$ and $cov\,X=cov\,Y$, then for each continuous bilinear form T on B the following holds:*

$$ET(X,X)=ET(Y,Y).$$

We continue with the proof by integrating both sides of (2.4) with respect to $P_{n,j}(dy)$ and $H(dx)$. We obtain

$$\begin{aligned} I_j &:= \int_B\int_B g_i(y+n^{-1/2}x)P_{n,j}(dy)H(dx)\\ &= \frac{1}{2}n^{-3/2}\int_0^1(1-\theta)^2\\ &\quad\times\int_B\int_B g_i'''(y+\theta n^{-1/2}x)[x^3]P_{n,j}(dy)H(dx)d\theta. \end{aligned} \quad (2.5)$$

Split the integral with respect to $H(dx)$ into two parts over $\{||x||\le\varepsilon\sqrt{n}\}$ and $\{||x||>\varepsilon\sqrt{n}\}$. Denote the first integral by $I_{j,1}$ and the second one by $I_{j,2}$. In order to estimate $I_{j,1}$ we use the following bound on the third derivative of the function g_i, which follows from the property (a) of condition (A_3) and the assumption $q(x)\le||x||\le\varepsilon n^{1/2}$:

$$\begin{aligned} |g_i'''(y+\theta n^{-1/2}x)[x^3]| &\le C||x||^3\varepsilon^{-3}\chi(r-\varepsilon\le q(y+\theta n^{-1/2}x)\le r+\varepsilon)\\ &\le C||x||^3\varepsilon^{-3}\chi(r-2\varepsilon\le q(y)\le r+2\varepsilon). \end{aligned}$$

Therefore we have

$$|I_{j,1}|\le C\varepsilon^{-3}n^{-3/2}\nu_3P_{n,j}\{r-2\varepsilon\le q(y)\le r+2\varepsilon\}. \quad (2.6)$$

Due to condition (D),

$$\begin{aligned} &P_{n,j}\{r-2\varepsilon\le q(y)\le r+2\varepsilon\}\le 2\delta_{n-1}+P\{(r-2\varepsilon)(n/n-1)^{1/2}\\ &\le q(Y)\le (r+2\varepsilon)(n/n-1)^{1/2}\}\le 2\delta_{n-1}+C\varepsilon, \end{aligned} \quad (2.7)$$

and we arrive at the estimate

$$|I_{j,1}| \le C\varepsilon^{-3} n^{-3/2} \nu_3(\delta_{n-1} + \varepsilon). \tag{2.8}$$

For $\{||x|| > \varepsilon\sqrt{n}\}$ we have

$$\begin{aligned}&\chi(r - \varepsilon \le q(y + \theta n^{-1/2}x) \le r + \varepsilon)\\ &\le \chi(r - 2||x||n^{-1/2} \le q(y) \le r + 2||x||n^{-1/2}),\end{aligned}$$

and, as above, we obtain an estimate of the integral of this indicator function with respect to $P_{n,j}(dy)$ of order $C(\delta_{n-1} + n^{-1/2}||x||)$. Since we can use the third pseudomoment only, one needs to reduce the third derivative of g_i to the second one. Integrating by parts, we can rewrite the remainder term in (2.4) as

$$-\frac{1}{2} g_i''(y + n^{-1/2}x)[x^2]n^{-1} + \int_0^1 (1-\theta) g_i''(y + \theta n^{-1/2}x)[x^2]\, d\theta$$

and proceed as above. In this way we can prove bound (2.8) for $I_{j,2}$ and finally we arrive at the estimate

$$\Delta_{n,k,q}(r) \le C\varepsilon^{-3} n^{-1/2} \nu_3(\delta_{n-1} + \varepsilon) + C\varepsilon,$$

valid for all $r \ge 0$; thus, we have the following recursive inequality:

$$\delta_n \le C\varepsilon^{-3} n^{-1/2} \nu_3(\delta_{n-1} + \varepsilon).$$

In order to finish the proof we choose $\varepsilon = C_1 \nu_3^{1/3} n^{-1/6}$ with an appropriate constant C_1 and use the induction assumption (2.2).

Theorem 2.1 is taken from Paulauskas and Račkauskas (1989). We refer the reader to this book for more information about the results obtained by Lindeberg's method combined with induction. Some generalizations of this approach are described there in detail.

The same method can be used to obtain nonuniform bounds on the convergence rate and to investigate probabilities of large deviations as well. This is possible due to the "good" estimates of the Gaussian measure of the set $\{r - \varepsilon \le q(x) \le r + \varepsilon\}$ for any continuous semi-norm q on B. For instance, when r.v. $q(Y)$ has a bounded density, a well-known result of Fernique yields for all $\lambda > 0$, $\varepsilon > 0$, $r \ge 0$

$$P\{r - \varepsilon \le q(Y) \le r + \varepsilon\} \le C\varepsilon \exp(-\lambda r),$$

where $C = C(\lambda, \mathcal{L}(Y))$. More precise results of this type are needed when considering large deviations. To this end the following lemma proved by Lifshits is quite useful (see Bentkus and Račkauskas (1990)).

Lemma 2.3. *Let q be a continuous semi-norm on B. Then*
(a) *there exists a constant $C = C(\mathcal{L}(Y)) > 0$ such that for all $\varepsilon > 0$, $r \geq 0$*

$$P\{q(Y) > r - \varepsilon\} \leq C \exp(Cr\varepsilon) P\{q(Y) > r\};$$

(b) *for any $r_0 > 0$ there exists a constant $C = C(r_0, \mathcal{L}(Y)) > 0$ such that for all $\varepsilon > 0$, $r \geq r_0$*

$$P\{r - \varepsilon \leq q(Y) \leq r + \varepsilon\} \leq C\varepsilon(r+1) P\{q(Y) > r - \varepsilon\}. \tag{2.9}$$

Moreover, (2.9) *holds for all $r \geq 0$, $\varepsilon > 0$ with $C = C(\mathcal{L}(Y)) > 0$ if and only if the distribution of the r.e. $q(Y)$ has a bounded density.*

Now, if we examine the formulas (2.6), (2.7) and make use of Lemma 2.3, we easily see the possibility of obtaining nonuniformity of power type in the bounds on $|I_{j,1}|$ and $|I_{j,2}|$. With some additional technical arguments the following result can be proved (for details, see Paulauskas and Račkauskas (1991)).

Theorem 2.4. *Let $m \geq 3$. Suppose that the conditions (A_3) and (D) are fulfilled. Then there exists a constant $C = C(m, \mathcal{L}(Y)) > 0$ such that*

$$\sup_{r \geq 0} r^m |P\{q(S_n(X)) < r\} - P\{q(Y) < r\}| \leq C \max(\nu_m n^{-m/2+1}, \nu_m^{1/m} n^{-1/6}).$$

Exponential nonuniformity is also available (see Paulauskas and Račkauskas (1991)). The validity of Lemma 2.3 in Hilbert space was pointed out by Bentkus (1986c), who investigated $P\{||S_n(X)|| > r\}$ for $0 \leq r \leq r_n$, where $r_n = O(n^{1/6})$. This result was generalized to Banach space and improved in Bentkus and Račkauskas (1990) by Lindeberg's method with induction and iteration arguments. A typical result is the following one.

Theorem 2.5 (Bentkus and Račkauskas (1990)). *Suppose that $q : B \to \mathbb{R}$ is continuous semi-norm and $E||X||^3 < \infty$. Suppose that the conditions (A_3), (D) are fulfilled. Then the following statements are equivalent:*
(a) *there exists $h > 0$ such that*

$$E \exp(hq^{1/2}(X)) < \infty;$$

(b) *there exist constants C_1, $C_2 < 0$ such that*

$$P\{q(S_n) > r\} \leq C_1 P\{q(Y) > r\}$$

for $0 \leq r \leq C_2 n^{1/6}$;
(c) *there exist constants C_1, $C_2 > 0$ such that*

$$P\{q(S_n) > r\} = P\{q(Y) > r\}(1 + \theta C_1 n^{-1/6}(1+r))$$

for $0 \leq r \leq C_2 n^{1/6}$, where $|\theta| \leq 1$;

(d) *for every function* $f : \mathbb{R} \to \mathbb{R}$ *such that* $f(x) \to 0$ *as* $x \to \infty$,

$$P\{q(S_n) > r\}/P\{q(Y) > r\} \to 1, \quad \text{as } n \to \infty$$

uniformly in r, $0 \le r \le f(n)n^{1/6}$.

Remarks on the proof. Assuming (a) to be true and employing Lemma 2.3 (instead of (2.7)) to help estimate $|I_{j,1}|$ and $|I_{j,2}|$, we obtain

$$|P\{q(S_n) > r\} - P\{q(Y) > r\}| \le CP\{q(Y) > r\}n^{-1/8}(1+r)^{3/4}$$

when $0 \le r \le C_2 n^{1/6}$. This implies (b) which can be viewed as a first iteration for proving (c). The proof that (d) $\Rightarrow$ (b) proceeds by investigating the sequence $\sup[P\{q(S_n) > r\}/P\{q(Y) > r\} : r \le f(n)n^{1/6}]$, $n \in \mathbb{N}$. One shows that if (b) is false, then this sequence fails to be bounded. To prove (b) $\Rightarrow$ (a), one applies Lévy's inequality.

As mentioned above, the bound given by Theorem 2.1 is in general sharp. The following result presents an example in Hilbert space.

Theorem 2.6 (Bentkus (1986b)). *Let a sequence* $b_n \downarrow 0$. *There exist a zero-mean Gaussian* l_2*-r.e.* Y, *a symmetric* l_2*-r.e.* X *and a continuous semi-norm* q *on* l_2 *such that*

(a) $\operatorname{cov} X = \operatorname{cov} Y$, $P\{\|X\| \le 1\} = 1$;
(b) *both conditions* (A_3) *and* (D) *are fulfilled;*
(c) $\Delta_{n,q} \ge b_n n^{-1/6}$ *for infinitely many* n

Sketch of the proof. Let I_s, $s \ge 1$, be blocks of natural numbers of length k_s, $s \ge 1$. Let $\lambda_i = 2^{-3i/2}k_i^{-2/3}$, $i \in \mathbb{N}$. Define

$$q(x) = \sum_{s=1}^{\infty} \lambda_s \sum_{k \in I_s} |x_k|.$$

The condition (A_3) for the semi-norm q can be deduced by means of the arguments of the proof of Theorem 2.2.31 in Paulauskas and Računskas (1989). To determine X and Y, we first define r.e.'s $X^{(j)}$, $j \in \mathbb{N}$, by setting

$$P\{X^{(j)} = a_j \lambda_j^{-1} e_i\} = P\{X^{(j)} = -a_j \lambda_j^{-1} e_i\} = (2k_j)^{-1}, \ i \in I_j,$$

where $\{e_i, i \in \mathbb{N}\}$ is the standard basis in l_2 and $a_j = 2^{-5j/3}k_j^{-2/3}$, $j \in \mathbb{N}$. Now let $X = \sum_{j=1}^{\infty} X^{(j)}$. It is easy to check that X is an l_2-r.e. The corresponding Gaussian l_2-r.e. is defined by $Y = \sum_{i=1}^{\infty} \tilde{\sigma}_i \gamma_i e_i$, where γ_i, $i \in \mathbb{N}$, is a sequence of i.i.d. standard normal r.v.'s and $\tilde{\sigma}_i = a_j \lambda_j^{-1} k_j^{-1/2}$ for $i \in I_j$, $j \in \mathbb{N}$. Since Y has independent coordinates, one can easily check that the distribution of r.v. $q(Y)$ has a bounded density. Therefore, the condition (D) follows. To prove (c) note, that

$$P\{q(S_n(X)) < r\} \ge P\{q(S_n(\tilde{X}^{(s)})) < r - \alpha_s\},$$

where $\tilde{X}^{(s)} = \sum_{i=1}^{s-1} X^{(i)}$, $\alpha_s = n^{1/2}\sum_{i=s}^{\infty} a_i$. Using the estimate of the rate of convergence in the CLT in $(k_1 + \cdots + k_{s-1})$-dimensional space (e.g., in Bentkus (1986a)), one deduces that

$$\Delta_{n,q} \geq \sup_{r\geq 0}[P\{q(\tilde{Y}^{(s)}) < r - \alpha_s\} - P\{q(Y) < r\}]$$
$$-C(k_1 + \cdots + k_{s-1})^2 n^{-1/2},$$

where $\tilde{Y}^{(s)} = \sum_{i=1}^{s-1}\sum_{k\in I_i} \tilde{\sigma}_k \gamma_k e_k$. To finish the proof one has to use the properties of the Gaussian r.e. Y and choose an appropriate sequence k_i, $i \in \mathbb{N}$.

In the case where one of the conditions (A_3), (D) fails to hold, the rate of convergence may be arbitrarily slow. Appropriate examples have been constructed by Bentkus (1984a), Rhee and Talagrand (1984) and Borisov (1985).

The verification of conditions (A_3), (D) is a separate problem. For details on the properties of the density of the r.v. $q(Y)$ for various functions $q : B \to \mathbb{R}$ we refer to the survey paper by Davydov and Lifshits (1985) and to Lifshits (1983) and Rhee and Talagrand (1986). Smooth approximation of an indicator function with various notions of smoothness are investigated in Bentkus and Račkauskas (1983) and Paulauskas and Račkauskas (1989), (1990).

We close this section with two results connected with conditions (A_3), (D). The modulus of convexity of a continuous semi-norm q on B is defined by

$$\tau_q(\varepsilon) = \inf\{1 - q(x - y/2) : q(x) = q(y) = 1,\ q(x - y) \geq \varepsilon\}.$$

Theorem 2.7 (Rhee and Talagrand (1986)). *If there exist constants $C > 0$, $\beta \geq 2$ such that $\tau_q(\varepsilon) \geq C\varepsilon^{\beta}$ for each $0 < \varepsilon \leq 2$, then the condition (D) holds.*

Note that the norm of the space L_p, $p > 1$, satisfies Theorem 2.7. The next result is usually used to construct a smooth approximation of an indicator function.

Theorem 2.8 (Bentkus and Račkauskas (1984)). *Condition (A_3) is equivalent to the following one. For each $\varepsilon > 0$ there exists a function $f_\varepsilon : B \to \mathbb{R}$ such that*

(a) *for all $x \in B$*
$$|q(x) - f_\varepsilon(x)| \leq \varepsilon;$$

(b) *the function f_ε is three times continuously Fréchet differentiable and*
$$\sup_{x\in B} ||f_\varepsilon^{(i)}(x)|| \leq C\varepsilon^{-i+1},\ i = 1, 2, 3.$$

§1.3. The Method of Integration by Parts

The method of integration by parts was introduced by Götze (1986). It allows one to obtain the Berry–Esseen-type bound for the speed of convergence in the CLT for sets $\{x \in B : F(x) < r\}$, $r \in \mathbb{R}$, where $F : B \to \mathbb{R}$ is a smooth function. When F is a linear function, this method reduces to Stein's method of differential equations. Further developments of Götze's approach are due to Zalesskii (1988). In order to describe the method we introduce the following conditions.

Differentiability condition (D_m). The function $F : B \to \mathbb{R}$ is m times Fréchet differentiable and there exist constants $C_F > 0$, $p \geq 0$ such that

$$||F^{(i)}(x)|| \leq C_F(1 + ||x||^p), \ i = 1, \dots, m.$$

To formulate the next condition, let

$$\sigma^2(x) = E(F'(x)(Y))^2.$$

Variance condition (V). For any $c > 0$ and for sufficiently large $M > 0$ there exists a constant $C = C(c, M, \mathcal{L}(Y))$ such that for all $t \in \mathbb{R}$

$$\sup_{||a|| \leq c} E \exp(-t^2 \sigma^2(Y + a)) \leq C(1 + |t|)^{-M}.$$

Theorem 3.1. *If conditions* (D_3), (V) *are fulfilled and* $E||X||^3 < \infty$, *then*

$$\Delta_n := \sup_{r \in \mathbb{R}} |P\{F(S_n) < r\} - P\{F(Y) < r\}| = O(n^{-1/2}).$$

Götze (1986) proved Theorem 3.1 under the additional condition that the third derivative of F satisfies

$$||F'''(x + y) - F'''(y)|| \leq C_F(1 + ||x||^p + ||y||^p)||x||^\varepsilon$$

for some $\varepsilon > 0$, $p \geq 0$, and under a more complicated variance condition. Theorem 3.1 was proved by Zalesskii (1988). Actually Zalesskii proves a nonuniform result. To state it, let ρ denote the distance from $0 \in B$ to the boundary of the set $\{x \in B : F(x) < r\}$.

Theorem 3.2 (Zalesskii (1988)). *Suppose that conditions* (D_3) *and* (V) *hold and* $E||X||^3 < \infty$. *Then for any* $s \geq 0$ *there exists a constant* $C = C(\mathcal{L}(X), s)$ *such that*

$$|P\{F(S_n) < r\} - P\{F(Y) < r\}| \leq Cn^{-1/2}[(1 + \rho)^{-3} + (1 + |r|)^{-s}].$$

Now we give the main ideas of the proof of Theorem 3.1. For details, see the papers of Götze (1986) and Zalesskii (1988). The first step is to replace X_j by its truncation at the level $n^{1/2}$. One easily deduces that $\Delta_n \le \bar{\Delta}_n + Cn^{-1/2}$, where

$$\bar{\Delta}_n = \sup_{r\in\mathbb{R}} |P\{F(\bar{S}_n) < r\} - P\{F(Y) < r\}|,$$

$\bar{S}_n = n^{-1/2}(\bar{X}_1 + \cdots + \bar{X}_n)$; $\bar{X}_j = X_j$ if $||X_j|| \le n^{1/2}$, and $\bar{X}_j = 0$ otherwise. Then one smoothes the indicator function of the set $\{F(x) < r\}$. To this end, let $f = f_{r,n} : \mathbb{R} \to \mathbb{R}$ be a monotone non-decreasing function such that $f(t) = 0$ if $t < r - n^{-1/2}$, $f(t) = 1$ if $t \ge r$, f is three times differentiable and $|f^{(i)}(t)| \le Cn^{i/2}$, $i = 1, 2, 3$. Define $f_1(x) = f(F(x))$, $f_2(x) = f(F(x) - n^{-1/2})$. Then we have

$$\bar{\Delta} \le \max_{i=1,2} \sup_{r\in\mathbb{R}} |I_i| + \sup_{r\in\mathbb{R}} P\{r \le F(Y) \le r + n^{-1/2}\},$$

where

$$I_i = I_i(r) = Ef_i(\bar{S}_n) - Ef_i(Y),\ i = 1, 2.$$

The next lemma proves the boundedness of the density of the r.v. $F(Y)$.

Lemma 3.3. *Let $m \ge 2$. If the function $F : B \to \mathbb{R}$ satisfies (D_{m+1}) and (V) and a function $\varphi : B \to \mathbb{R}$ satisfies (D_m), then for each $\varepsilon > 0$ and all $a \in \mathbb{R}$ there exists a constant $C = C(a, \varepsilon, \mathcal{L}(Y)) > 0$ such that for all $t \in \mathbb{R}$*

$$\sup_{||x||\le a} |E \exp(itF(Y + x))\varphi(Y + x)| \le C(1 + |t|)^{-m+\varepsilon}.$$

From Lemma 3.3 we easily obtain the estimate $\sup_{r\in\mathbb{R}} P\{r \le F(Y) \le r + \varepsilon\} \le C\varepsilon$. Therefore it remains to estimate I_i, $i = 1, 2$.

If identity (2.1) and a Taylor series are employed, then the third derivatives of the smoothed indicator function f would occur which are of order $O(n^{3/2})$. This is much too large to obtain an error bound of order $O(n^{-1/2})$. The main idea in integrating by parts is to insert factors in the integrals in I_i in such a way that integrating by parts helps to replace the derivatives $f^{(k)}$, $k = 1, 2, 3$, by the function f. To determine such factors, we need some preparation. Let X_{n+1}, X_{n+2}, X_{n+3} (respectively Y_{n+1}, Y_{n+2}, Y_{n+3}) denote independent copies of X (respectively Y). Let θ be a r.v. uniformly distributed on [0, 1]. Define random functions

$$g_{1,j}(x) = F'(x)[X_{n+j}],\ g_{2,j}(x) = F'(x)[Y_{n+j}],\ j = 1, 2, 3,$$

and functions

$$\sigma_1^2(x) = E(F'(x + \theta n^{-1/2}X)[X])^2,\ \sigma_2^2(x) = E(F'(x + \theta n^{-1/2}Y)[Y])^2.$$

Then $h_i(x) = \prod_{j=1}^3 g_{i,j}^2(x)(\sigma_i^2(x) + n^{-a})^{-3}$, $i = 1, 2$, $a > 1/2$, are the factors mentioned above. Put

$$\bar{U}_{k,n} = n^{-1/2}(\bar{X}_1 + \cdots + \bar{X}_k + Y_{k+1} + \cdots + Y_n);$$

and

$$Z_1 = t_1 X_{n+1} + t_2 X_{n+2} + t_3 X_{n+3}, \; Z_2 = t_1 Y_{n+1} + t_2 Y_{n+2} + t_3 Y_{n+3}.$$

Lemma 3.4. *For any* $[n/2] \le k \le n$,

$$|Ef(F(\bar{U}_{k,n})) - Ef(F(\bar{U}_{k,n} + Z_1))h_1(\bar{U}_{k,n} + Z_1)| = O(n^{-1/2}).$$

Lemma 3.5. *For any* $1 \le k \le [n/2]$,

$$|Ef(F(\bar{U}_{k,n})) - Ef(F(\bar{U}_{k,n} + Z_2))h_2(\bar{U}_{k,n} + Z_2)| = O(n^{-1/2}).$$

Now let us continue with the consequences of these results. We have

$$I_1 \le J_1 + J_2 + Cn^{-1/2},$$

where

$$\begin{aligned} J_1 &= |Eh(\bar{U}_{n,n_0} + Z_1) - Eh(\bar{U}_{n,n} + Z_1)|, \\ J_2 &= |Eh(\bar{U}_{n,n_0} + Z_2) - Eh(\bar{U}_{n,0} + Z_2)|, \end{aligned}$$

$h(z) = f(F(z))h_1(z)$ and n_0 is the integer $[n/2]$.

Now one proceeds as usual. One uses the identity (2.1) and then expands the function h in a Taylor series up to terms of second order obtaining, e.g.,

$$J_1 \le \sum_{k=n_0}^{n} [|J_{1,k}^{(1)}| + |J_{1,k}^{(2)}| + |J_{1,k}^{(3)}|],$$

where

$$\begin{aligned} J_{1,k}^{(1)} &= \int_B \int_B \chi(||x|| \ge \sqrt{n}) h(y) Q_{k,n}(dy) H(dx); \\ J_{1,k}^{(2)} &= \int_B \int_B \chi(||x|| \ge \sqrt{n}) h'(y)[x] n^{-1/2} Q_{k,n}(dy) H(dx); \\ J_{1,k}^{(3)} &= \int_B \int_B h''(y + \lambda n^{-1/2} x)[x^2] n^{-1} Q_{k,n}(dy) \bar{H}(dx), \; \lambda \in [0,1], \end{aligned}$$

$$H = \mathcal{L}(X) - \mathcal{L}(Y), \; \bar{H} = \mathcal{L}(\bar{X}) - \mathcal{L}(Y),$$

$$Q_{n,k} = \mathcal{L}(\bar{W}_{k,n}), \; \bar{W}_{n,k} = n^{-1/2} \left(\sum_{i=1}^{k-1} \bar{X}_i + \sum_{i=k+1}^{n} Y_i \right).$$

The following lemma is important for the further estimations. Put

$$\sigma_0^2(x) = E(F'(x)[\bar{X}_1])^2.$$

Lemma 3.6. *For any $c > 1/2$, $b > 0$, $a \in \mathbb{R}$ there exists a positive constant C such that for any $k = 1, 2, \ldots, n$*

$$\sup_{||x|| \le a} E(\sigma_0^2(\bar{W}_{n,k} + x) + n^{-c})^{-b} \le C.$$

Integration by parts is applied after rewriting

$$g_{1,1}(y + Z_1) f'(F(y + Z_1)) = \frac{\partial}{\partial t_1} f(F(y + Z_1));$$

$$g_{1,1}(y + Z_1) g_{1,2}(y + Z_1) f''(F(y + Z_1)) = \frac{\partial}{\partial t_1}\frac{\partial}{\partial t_2} f(F(y + Z_1)),$$

etc., and using the fact that the factors $g_{1,j}(y+Z_1)$ are always attached to the derivatives of the function f. Now one considers $J_{1,k}^{(i)} = \int_0^1 \int_0^1 \int_0^1 J_{1,k}^{(i)} \, dt_1 \, dt_2 \, dt_3$ and integrates by parts to estimate these integrals.

Integration by parts is also useful when investigating the probabilities of large deviations. As an illustration, we state the following result (cf. Theorem 2.5 and Theorem 1.7).

Theorem 3.7 (Račkauskas (1988)). *Let X, Y be H-valued r.e.'s. The following statements are equivalent:*

(a) *there exists $\lambda > 0$ such that*

$$E \exp(\lambda ||X||^{1/2}) < \infty;$$

(b) *there exist constants $C_i = C(\mathcal{L}(Y)) > 0$, $i = 1, 2$, such that*

$$P\{||S_n|| > r\} = P\{||Y|| > r\}(1 + C_1 \theta (1 + r) n^{-1/2}),$$

when $0 \le r \le C_2 n^{1/6}$. Here $|\theta| \le 1$.

Earlier Zalesskii (1989) proved (b) under the slightly stronger moment condition $E \exp(\lambda ||X||) < \infty$.

§1.4. The Method of Finite-Dimensional Approximation

This section discusses the method of finite-dimensional approximation in the context of estimating the convergence rates in limit theorems in infinite-dimensional spaces. At the end of the seventies, it seemed that this method yields less accurate results compared with other methods. The first estimate in the CLT in infinite-dimensional Hilbert space, due to Kandelaki (1965) (see also Vakhaniya and Kandelaki (1969)), by means of finite-dimensional approximation, had only an inverse logarithmic order of decay. Sazonov (1968, 1969) used finite-dimensional approximation in the CLT in Hilbert space for summands having a special structure (for the ω^2-test) and he achieved the rate

$O(n^{-1/6+\varepsilon})$, $\varepsilon > 0$. Giné (1976) and Paulauskas (1976a) used this method to find the convergence rate in the CLT in $C(S)$, where S is a compact metric space. For $S = [0,1]$ and the limiting Wiener process, they obtained the rate $n^{-1/20}$ or even less. These papers studied the special compact set $S = \{0, 1/2, 1/3, \ldots\}$ under very strict conditions on the terms: sub-Gaussian increments and a very slow rate of growth of the metric entropy, etc. The rate found in these papers was $n^{-1/6+\varepsilon}$, $\varepsilon > 0$. In a later paper, Paulauskas (1984) came up with $n^{-1/2+\varepsilon}$, $\varepsilon > 0$.

There are more papers where the method of finite-dimensional approximation has been applied under different settings. But a general feature of all is the rather slow rate of the bounds obtained. This can be explained by the fact that the estimates such as

$$\sup_{A \in \mathcal{A}} |P\{S_n(X) \in A\} - P\{Y \in A\}| \le C(k)C(\mathcal{L}(X), \mathcal{L}(Y))n^{-1/2}$$

on some class $\mathcal{A}$ of Borel sets in $\mathbb{R}^k$ (for example, the class of rectangles or the class of convex sets) depend rather heavily on the dimension k. The present estimates of the constant $C(k)$ are of the form $C(k) \le Ck^\beta$ for some absolute constant C and some exponent β. They depend on the form of the constant $C(\mathcal{L}(X), \mathcal{L}(Y))$ and on the class of sets $\mathcal{A}$ under consideration. For the class of convex sets and the identity covariance operator of X, Nagaev (1976) and Senatov (1983) proved the estimate with $\beta = 1$ and $C(\mathcal{L}(X), \mathcal{L}(Y)) = E||X||^3$. Bentkus (1986a) improved this result to $\beta = 1/2$. We note that for the method of finite-dimensional approximation usually one needs estimates in $\mathbb{R}^k$ over the class of rectangles only without assumptions about the covariance structure of random vectors. To this end, there is a useful result due to Bentkus (1984c, 1990). Namely, he constructed smooth functions approximating well the indicator functions of balls in l^k_∞ with precise estimates of the derivatives of these functions (see Lemma 4.6 below). This allowed him to obtain new bounds on the remainder term in the CLT in l^k_∞ for the class of balls with worse dependence on n but with much better dependence on k, namely, logarithmic. This resulted in a better convergence rate in the CLT in the space $C(S)$, S a compact metric space (see Bentkus (1982) and Paulauskas and Račkauskas (1989) for full information on this topic) and even in other infinite-dimensional spaces (see Paulauskas and Juknevičienė (1988) for a generalization to the Skorokhod space $D[0,1]$ and Norvaiša and Paulauskas (1990) for the case of general empirical processes).

Now we describe this method in detail. Suppose that our r.e.'s X, X_1, X_2, ... are random processes, defined on a probability space $(\Omega, \mathcal{A}, P)$ and indexed by some parameter set T, which we assume to be a compact metric set, i.e., $X : \Omega \times T \to \mathbb{R}$. We shall view X as a map from Ω to $l_\infty(T)$, the space of all bounded functions on T with supremum norm or to some smaller subspace such as $C(T)$. For $x \in l_\infty(T)$, $||x||_\infty \equiv ||x||_{T,\infty} := \sup_{t \in T} |x(t)|$; let $EX(t) = 0$ and $EX^2(t) < \infty$ for all $t \in T$. Suppose that the corresponding Gaussian random element $Y = \{Y(t), t \in T\}$ with $EY(t) = 0$ and

$EY(t)Y(s) = EX(t)X(s)$ for all $s, t \in T$, is in $l_\infty(T)$ (or in a smaller subspace if X is concentrated in it). Finally, (in order not to use outer measure and integrals) suppose that both $||S_n(X)||_\infty$ and $||Y||_\infty$ are measurable. Therefore the quantities

$$\Delta_n(r) = \Delta_n(T, r) = |P\{||S_n(X)||_\infty < r\} - P\{||Y||_\infty < r\}|, \tag{4.1}$$

$$\Delta_n = \sup_{r \geq 0} \Delta_n(T, r)$$

are well defined. Before proceeding to estimate (4.1), we give the main examples which fit into this framework.

(a) $T = [a, b]$, X and Y, as processes on $[a, b]$, are a.s. continuous. Here an estimate of $\Delta_n(r)$ yields the rate of convergence in the CLT in $C[a, b]$.

(b) $T = [a, b]$, X and Y have no discontinuities of the second kind a.s. Here $\Delta_n(r)$ measures the rate of convergence in the CLT in $D[a, b]$ on balls with respect to the supremum norm. Later on we shall consider this example in detail (see §3.3).

(c) It is possible to generalize (b) and to consider the space $D(\mathbb{A})$ of functions $x : \mathbb{A} \to \mathbb{R}$, indexed by some class $\mathbb{A}$ of closed Borel subsets of $I^d (\equiv [0, 1]^d)$, which are outer continuous and have inner limits. This space was introduced by Bass and Pyke (1985, 1987) where general limit theorems for a triangular array were proved. If one takes $T = \mathbb{A}$ in (4.1), it should be interesting to find estimates of the convergence rate in these general theorems.

(d) Let $Z, Z_1, Z_2, \ldots$ be i.i.d. r.v.'s with values in a measurable space $(\chi, \mathcal{A})$ and with a distribution m. The empirical process associated with m is given by

$$E_n = \sqrt{n} \left(\frac{1}{n} \sum_{i=1}^{n} \delta_{Z_i} - m \right).$$

We consider the process $E_n(f)$, $f \in \mathcal{F}$, indexed by some class $\mathcal{F} \subset L_2(\chi, \mathcal{A}, m)$ of real-valued measurable functions $f : \chi \to \mathbb{R}$. Here and in what follows we put $\mu(f) = \int_\chi f(x)\mu(dx)$. In particular, the class $\mathcal{F}$ can consist of the indicator functions of some class of sets. Let $\{G_m(f), f \in \mathcal{F}\}$ denote the limiting Gaussian process (see Giné and Zinn (1984) for details). For simplicity, assume that $\mathcal{F}$ is countably generated (for m), that is, there exists a countable subclass $\mathcal{F}_0 \subset \mathcal{F}$ such that $\sup_{f \in \mathcal{F}} |E_n(f)| = \sup_{f \in \mathcal{F}_0} |E_n(f)|$ a.s. for all $n \geq 1$. A possible way to estimate the convergence rate in the CLT for empirical processes $E_n(f)$, $f \in \mathcal{F}$, is to bound the quantity

$$\delta_n(\mathcal{F}, r) = |P\{\sup_{f \in \mathcal{F}} |E_n(f)| < r\} - P\{\sup_{f \in \mathcal{F}} |G_m(f)| < r\}|.$$

This coincides with $\Delta_n(T, r)$ in (4.1) if we choose $X = \delta_Z - m$ and $T = \mathcal{F}$. Rates of convergence for empirical processes will be described in the last chapter. This important example has an additional theoretical aspect. It is known that in any Banach space an estimate in the CLT on balls can be obtained by estimating the remainder term in the CLT for the empirical process with class $\mathcal{F}$ being the unit sphere of the dual space B^* (due to the fact $||x|| = \sup\{|f(x)| : f \in B^*,\ ||f|| = 1\}$).

We now return to estimating the quantity $\Delta_n(T, r)$ in (4.1). Suppose that ϱ is some pseudometric on T under which T is totally bounded. Usually ϱ is connected with the process X. The pseudometric $\tau(s,t) = E^{1/2}(X(s)-X(t))^2$, $s, t \in T$, provides such an example. Let $\delta > 0$ be arbitrary and $N = N_\varrho(\delta)$ be the number of elements of a minimal δ-net $T(\delta) = \{t_1, t_2, \dots, t_N\}$ of the totally bounded set (T, ϱ). For $x \in l_\infty(T)$ and $\delta > 0$, put

$$w_\varrho(x, \delta) = \sup\{|x(t) - x(s)| : t, s \in T,\ \varrho(t, s) < \delta\};$$

$$||x||_N := \max_{t \in T(\delta)} |x(t)| = \max_{1 \le i \le N} |x(t_i)|;$$

$$\Delta_n^N(r) = \Delta_n(T(\delta), r) = |P\{||S_n||_N > r\} - P\{||Y||_N > r\}|.$$

Obviously, for each n and any $0 < \varepsilon < r$, we have

$$\begin{aligned}
P\{||S_n||_\infty > r\} - P\{||Y||_\infty > r\} &= P\{||S_n||_\infty > r\} - P\{||S_n||_N > r - \varepsilon\} \\
&+P\{||S_n||_N > r - \varepsilon\} - P\{||Y||_N > r - \varepsilon\} \\
&+P\{||Y||_N > r - \varepsilon\} - P\{||Y||_\infty > r\} \\
&\le P\{||S_n||_\infty > r, ||S_n||_N \le r - \varepsilon\} \\
&+\Delta_n^N(r - \varepsilon) + P\{||Y||_N > r - \varepsilon, ||Y||_\infty \le r\} \\
&\le P\{w_\varrho(S_n, \delta) > \varepsilon, ||S_n||_\infty > r\} \\
&+\Delta_n^N(r - \varepsilon) + P\{r - \varepsilon \le ||Y||_\infty \le r\}.
\end{aligned}$$

Analogously we have the lower bound

$$\begin{aligned}
&P\{||S_n||_\infty > r\} - P\{||Y||_\infty > r\} \\
&\ge -\Delta_n^N(r) - P\{r \le ||Y||_\infty \le r + \varepsilon\} - P\{w_\varrho(Y, \delta) > \varepsilon, ||Y||_\infty > r\}.
\end{aligned}$$

These estimates lead to the following lemma which presents the core of the method of finite-dimensional approximation.

Lemma 4.1. *For all n and all $\varepsilon > 0$, $r > \varepsilon$, $\delta > 0$,*

$$\begin{aligned}
&\Delta_n(T, r) \le \Delta_n^N(r - \varepsilon) + \Delta_n^N(r) \\
&+P\{w_\varrho(S_n, \delta) > \varepsilon, ||S_n||_\infty > r\} + P\{w_\varrho(Y, \delta) > \varepsilon, ||Y||_\infty > r\} \\
&+P\{r - \varepsilon \le ||Y||_\infty \le r + \varepsilon\}.
\end{aligned} \tag{4.2}$$

Remark 4.2. The last term in (4.2) is the Gaussian measure of an ε-strip straddling the boundary of a ball in $l_\infty(T)$. Such terms appear in all methods used to estimate the convergence rate (see §§1.1, 1.2). In what follows we assume additionally that for each $m > 0$ there exists a constant $C = C(\mathcal{L}(Y), m)$ such that for all $\varepsilon > 0$, $r > 0$

$$P\{r \le ||Y||_\infty \le r + \varepsilon\} \le C\varepsilon(1 + r)^{-m}. \tag{4.3}$$

Remark 4.3. The bound (4.2) has been proved for $r > \varepsilon$. If we want to obtain a general estimate of $\Delta_n(T,r)$ involving uniform and nonuniform bounds simultaneously we need to estimate $\sup_{r\geq 0}(1+r)^m\Delta_n(T,r)$. It is easy to see that this can be reduced to estimating over $r > \varepsilon$ only. Indeed, utilizing (4.3), for any $0 < r \leq \varepsilon$, we have

$$(1+r)^m\Delta_n(T,r) \leq (1+\varepsilon)^m\Delta_n(T,\varepsilon) + 2P\{||Y||_\infty \leq \varepsilon\}$$
$$\leq \sup_{r>\varepsilon}(1+r)^m\Delta_n(T,r) + C\varepsilon.$$

Remark 4.4. Sometimes it is better to bound $P\{||S_n||_\infty > r, ||S_n||_N \leq r-\varepsilon\}$ by

$$P\{\sup_{1\leq i\leq N}\sup_{\varrho(t,t_i)<\delta}|S_n(t) - S_n(t_i)| > \varepsilon, ||S_n||_\infty > r\}.$$

We now discuss briefly the estimation of the terms in (4.2), except for the last one, for which we require (4.3). The terms

$$P\{w_\varrho(S_n,\delta) > \varepsilon, ||S_n||_\infty > r\},\ P\{w_\varrho(Y,\delta) > \varepsilon, ||Y||_\infty > r\}$$

in (4.2) control the oscillation of the processes S_n and Y, respectively. Using the elementary inequality

$$P\{A\cap B\} \leq (P\{A\})^\gamma(P\{B\})^{1-\gamma},$$

valid for all $0 \leq \gamma \leq 1$, we reduce the estimation of these quantities to that of

$$P\{w_\varrho(S_n,\delta) > \varepsilon\},\ P\{w_\varrho(Y,\delta) > \varepsilon\}, \tag{4.4}$$

$$P\{||S_n||_\infty > r\},\ P\{||Y||_\infty > r\}. \tag{4.5}$$

To bound the terms in (4.5) the Chebyshev inequality suffices. Since we are assuming that the Gaussian process Y, as a process on the parameter set T, is bounded, we can use an exponential bound for the tail of the distribution of the Gaussian process (see Fernique (1971) and Marcus and Shepp (1972)): there exists a finite positive constant C such that for all $\varepsilon,\delta \in \mathbb{R}^+$

$$P\{w_\varrho(Y,\delta) > \varepsilon\} \leq C\exp(-\varepsilon^2/4\sigma_\varrho^2(\delta)),$$

where $\sigma_\varrho^2(\delta) = \sup\{E(Y(t)-Y(s))^2 : \varrho(s,t) \leq \delta\}$. The only term for which we are unable to recommend a general approach is $P\{w_\varrho(S_n,\delta) > \varepsilon\}$ since under different assumptions on $\mathcal{L}(X)$ different bounds for this term are available.

The estimation of $\Delta_n^N(r)$ and $\Delta_n^N(r-\varepsilon)$ is equivalent to estimating the convergence speed in the CLT in the finite-dimensional Banach space l_∞^N. In estimating $\Delta_n^N(r)$, one must keep in mind that N (through its dependence on $\delta = \delta_n \downarrow 0$) will grow with n. Another feature in estimating $\Delta_n^N(r)$ is the appearance of the probability

$$P\{r \leq \max_{1\leq i\leq N}|Y(t_i)| \leq r+\varepsilon\}. \tag{4.6}$$

We can bound the quantity (4.6) by $C\varepsilon$ using (4.3) and the term controlling the oscillation of the process Y.

We shall state an estimate for $\Delta_n^N(r)$ when $E\|X\|_\infty^3 < \infty$. Write

$$\begin{aligned}
&W_\varrho(\mathcal{L}(Y), \delta, t) := \sup_{r\geq 0}(1+r)^3 P\{w_\varrho(Y,\delta) > t, \|Y\|_\infty > r\},\\
&M_3(n) = n^{-1/6}(1 \vee (E\|X\|_\infty^3 + E\|Y\|_\infty^3)^{1/3}),\\
&D_n(\delta) = M_3(n) \vee W_\varrho(\mathcal{L}(Y), \delta, M_3(n)),\\
&H_\varrho(u) = \log N_\varrho(u),\\
&a \vee b = \max(a,b), \ a \wedge b = \min(a,b).
\end{aligned}$$

Theorem 4.5. *There exists an absolute constant $C < \infty$ such that for all $\delta > 0$, $n \in \mathbb{N}$, $r \geq 0$ the following estimate holds:*

$$\Delta_n^{N_n}(r) \leq C(1+r)^{-3} H_\varrho^2(\delta)\big(D_n(\delta) \vee D_n^3(\delta)\big), \tag{4.7}$$

where $N_n = N_\varrho(\delta)$.

This result is taken from Norvaiša and Paulauskas (1990). Earlier results of this kind (roughly speaking, giving an order $n^{-1/6}$ for Δ_n) were proved by Paulauskas and Račkauskas (1989) (see Theorem 5.2.6), and Paulauskas and Juknevičienė (1988). In the formulation of (4.7) we did not seek the best possible accuracy. It is easy to see that instead of the moments $E\|X\|_\infty^3 + E\|Y\|_\infty^3$ in (4.7) one can use

$$\int \sup_{t\in T} |x(t)|^3 |\mathcal{L}(X) - \mathcal{L}(Y)|(dx).$$

But in estimating the oscillation of the processes S_n and Y, at present one cannot avoid using the moments $E\|X\|_\infty^p$ and $E\|Y\|_\infty^p$, $p > 3$. Therefore we had to restrict ourselves to (4.7). A more important question concerns the power with which the metric entropy $H_\varrho(\delta_n)$ occurs in (4.7). It is possible to show that this power can be lowered from 2 to 2/3 in two cases: if we confine ourselves to uniform estimates of $\Delta_n^N = \sup_{r\geq 0}\Delta_n^N(r)$ or if we allow $\sup_n E\|S_n\|_\infty^2$ to appear in the final estimate of $\Delta_n(T, r)$.

The proof of Theorem 4.5 runs along the lines of Lindeberg's method, described in the previous section. Therefore we shall give only a sketch of the proof. Define

$$\xi_j = (X_j(t_1), \ldots, X_j(t_N)), \ \eta_j = (Y_j(t_1), \ldots, Y_j(t_N)),$$

where Y, Y_1, Y_2, ... are i.i.d. r.e.'s. Instead of $\Delta_n^N(r)$ it is more convenient to estimate the quantity

$$\begin{aligned}
\delta_n^N \;=\; & \sup_{1\leq j\leq n}\sup_{r\geq 0} r^3 |P\{n^{-1/}\|\xi_1 + \cdots + \xi_j + \eta_{j+1} + \cdots + \eta_n\|_N \leq r\}\\
& -P\{\|\eta\|_N \leq r\}|,
\end{aligned}$$

which majorizes $\Delta_n^N(r) r^3$. The first step is usually called smoothing and uses the following result due to Bentkus (1990) (see also Bentkus (1984c)).

Lemma 4.6. *For all $r \geq 0$, $\varepsilon > 0$ there exists a function $f_{r,\varepsilon} : l_\infty^N \to [0,1]$, $f_{r,\varepsilon} \in C^\infty$, such that*

$$\chi\{||x||_N \leq r\} \leq f_{r,\varepsilon}(x) \leq \chi\{||x||_N \leq r + \varepsilon\},$$

$$||f_{r,\varepsilon}^{(m)}(x)|| \leq C(m)\varepsilon^{-m} \ln^{m-1}(N+1), \; m = 1, 2, \ldots .$$

Furthermore, the constant $C(m)$ depends on m only.

The further steps in the proof are standard: one uses the identity (2.1) and the Taylor expansion up to the third order. The only change is that one has to apply the estimate

$$P\{r - \varepsilon \leq ||\eta||_N \leq r + \varepsilon\} \leq$$
$$P\{r - \varepsilon - t \leq ||Y||_\infty \leq r + \varepsilon + t\} + P\{w_\varrho(Y, \delta) > t, ||Y||_\infty > r - \varepsilon\}$$

for any $t \geq 0$. This results in the following recursive inequality:

$$\delta_n^N \leq M_3^3(n)\left\{C + CH_\varrho^2(\delta)t^{-3}\left[\delta_{n-1}^{N_n} + t + W_\varrho(Y,\delta,t)\right]\right\}$$
$$+C[t + t^3 + W_\varrho(Y, \delta, t)].$$

A standard induction argument completes the estimation of $\delta_n^N \geq r^3\Delta_n^N(r)$. The bound for $\sup_{r\geq 0}\Delta_n^N(r)$ is obtained in an analogous way.

It is worth mentioning that the method of finite-dimensional approximation can be used not only for the supremum norm. For example, if we want to estimate the remainder term in the CLT in the space l_p, $1 \leq p < \infty$, on balls, then as a finite-dimensional approximation it is natural to take the first N coordinates and to estimate the remainder term in the CLT in l_p^N for balls, combining this with estimates of the tails of the coordinates. Such an approach has been used by Paulauskas (1981) and in l_2 by Sazonov (1968, 1969).

In the paper due to Asriev and Rotar' (1985) an estimate in l_∞^k for parallelepipeds was obtained (having the order $n^{-1/2}$ with a logarithmic factor). The estimate is expressed in such a form that it allows one to pass to the limit as $k \to \infty$ and to get a bound (of the same order) in $\mathbb{R}^N$. Unfortunately we are unable to apply this result in the context of finite-dimensional approximation since it is obtained under the assumption that the covariance matrix of the random vector X under consideration is diagonal. This assumption is not restrictive at all in $\mathbb{R}^k$ in the case of the class of convex sets since by means of an orthogonal transformation, the covariance matrix can be diagonalized. But this is impossible for the class of parallelepipeds with sides parallel to the coordinate axes.

We shall now demonstrate how the general scheme of finite-dimensional approximation applies to the CLT in the Skorokhod space $D[0,1]$. But before we do this we shall give a short (and therefore not full) review of what is known about the CLT in the space $D[0,1]$ itself, since this topic is not covered in the literature so thoroughly as the CLT in Banach spaces (see, for example, Araujo and Giné (1980) and Paulauskas and Račkauskas (1989)).

Let X, X_1, X_2, ... be i.i.d. r.e.'s with values in $D[0,1]$. We assume that $D[0,1]$ is equipped with the Skorokhod topology and metric under which it is a separable and complete metric space. Let us assume that $EX(t) = 0$, $EX^2(t) < \infty$, for all $t \in [0,1]$. We say that X satisfies the CLT in $D[0,1]$ ($X \in CLT(D)$ for short) if there exists a Gaussian zero-mean $D[0,1]$-valued r.e. Y such that $EX(t)X(s) = EY(t)Y(s)$ for all $s, t \in [0,1]$ and $S_n = n^{-1/2}\sum_{i=1}^n X_i$ converges in distribution to Y (see, e.g., Billingsley (1968)). The CLT in D was considered by Fisz (1959), Hahn (1978), Bass and Pyke (1985), Juknevičienė (1985), Paulauskas and Stieve (1990), and Bézandry and Fernique (1990). Many applied problems lead directly to the CLT in $D[0,1]$. One such example is in a paper of Phoenix and Taylor (1973) where the asymptotic strength distribution of a general fibre bundle was investigated. As a matter of fact this investigation goes back to an early paper of Daniels (1945). Pheonix and Taylor (1973) did prove the CLT in $D[0,1]$ (without stating it explicitly for i.i.d. random processes with a special structure. Influenced by this result, Hahn (1978) proved the following theorem.

Theorem 4.7. *Let X be a r.e. in $D[0,1]$, $EX(t) = 0$ and $EX^2(t) < \infty$ for all $t \in [0,1]$. Assume that there exist nondecreasing continuous functions F_1 and F_2 on $[0,1]$ and numbers $\alpha_1 > 1/2$, $\alpha_2 > 1$ such that for all $0 \le s \le t \le u \le 1$ the following two inequalities hold:*

$$E(X(t) - X(s))^2 \le (F_1(t) - F_1(s))^{\alpha_1}, \tag{4.8}$$

$$E(X(u) - X(t))^2(X(t) - X(s))^2 \le (F_2(u) - F_2(s))^{\alpha_2}. \tag{4.9}$$

Then $X \in CLT(D)$ and $\mathcal{L}(Y)(C[0,1]) = 1$, where Y is the limiting Gaussian process for $S_n(X)$.

Condition (4.9) requires the finiteness of the fourth moment $EX^4(s)$ for all $s \in [0,1]$. This shortcoming is eliminated in the following theorems.

Theorem 4.8 (Paulauskas and Stieve (1990)). *Suppose that X is a r.e. in $D[0,1]$ and $EX(t) = 0$, $EX^2(t) < \infty$ for all $t \in [0,1]$. Suppose that (4.8) is satisfied for some $\alpha_1 > 2/3$ and the following inequality holds for some $\alpha_2 > 1$:*

$$E\left((X(t) - X(s))^2 \wedge 1\right)\left(X(u) - X(t)\right)^2 \le \left(F_2(u) - F_2(s)\right)^{\alpha_2}. \tag{4.10}$$

Then $X \in CLT(D)$ and $\mathcal{L}(Y)(C[0,1]) = 1$.

Theorem 4.9 (Bézandry and Fernique (1990)). *Let X be a real-valued random function on $[0,1]$, defined on some probability space $(\Omega, \mathcal{A}, P)$. Suppose that there exist continuous increasing functions δ, η, and θ from $[0,1]$ to $\mathbb{R}$ such that θ is concave and $\delta(0) = \eta(0) = \theta(0) = 0$. Moreover, let the following conditions hold for all $0 \le s \le t \le u \le 1$ and all $A \in \mathcal{A}$:*

$$E|X(0)|^2 < \infty, \ E|X(t) - X(s)|^2 \le \delta^2(t - s), \tag{4.11}$$

$$\int_0^1 u^{-5/4}(\log(1+1/u))^{1/4}\delta(u)\,du < \infty, \tag{4.12}$$

$$E(|X(s)-X(t)|^2 \wedge |X(t)-X(u)|^2)\chi_A \le \eta^2(u-s)\theta(P(A)), \tag{4.13}$$

$$\int_0^1 u^{-3/2}\theta^{1/2}(u\log_2(1+1/u))\eta(u)\,du < \infty. \tag{4.14}$$

Then $X \in CLT(D)$ *and* $\mathcal{L}(Y)(C[0,1]) = 1$.

It seems that Theorems 4.7–4.9 are not comparable. It is worth mentioning that if we require the boundedness of a random process X, then a very mild condition on the increments of the process is sufficient for the CLT. This is demonstrated in the following result.

Theorem 4.10 (Giné and Zinn (1984)). *Let* X *be a centered stochastically continuous and uniformly bounded process with sample functions in* $D[0,1]$. *Assume that there exist a positive* C *and a nondecreasing function* $F \in D[0,1]$ *such that for all* $s,t \in [0,1]$

$$E|X(t)-X(s)| \le C|F(t)-F(s)|.$$

Then $X \in CLT(D)$. *Furthermore, if* $\mathcal{L}(X)(C[0,1]) = 1$, *then*

$$X \in CLT(C[0,1]).$$

In fact, this theorem is a consequence of a more general result which, in turn, is derived from the general CLT for empirical processes (see Giné and Zinn (1984)).

We mention that another result (also apparently uncomparable with others) on the CLT in $D[0,1]$ can be derived from Bass and Pyke (1987).

Remark 4.11. A final result on the CLT in $D[0,1]$ for i.i.d. summands, formulated in terms of moments of increments of a process, was obtained independently by Bloznelis and Paulauskas (1994) and Fernique (1994).

Next we shall state and give a sketch of the proof of a result due to Paulauskas and Stieve (1990), which provides an estimate of the convergence rate in Theorem 4.8. As usual, the modulus of continuity of a function f is defined by $\omega_f(\delta) = \sup\{|f(t)-f(s)| : |s-t| < \delta\}$.

Theorem 4.12. *Let* X *satisfy the conditions of Theorem 4.8. Additionally, assume that there exist a nondecreasing continuous function* F_3, *numbers* $\alpha_3 \ge 1$, β_1, β_2, $\beta_3 > 0$ *and* $0 \le \kappa \le 1$ *such that for all* $0 \le s \le t \le 1$

$$E|X(t)-X(s)|^{3+\kappa} \le (F_3(t)-F_3(s))^{\alpha_3}, \tag{4.15}$$

$$\omega_{F_i}(\delta) \le C\delta^{\beta_i},\ i = 1,2,3. \tag{4.16}$$

Suppose that $E\sup_{t\in[0,1]}|X(t)|^3 < \infty$ *and condition (4.3) holds with* $T = [0,1]$. *Then there exists a finite constant* $C = C(m, \mathcal{L}(X), \mathcal{L}(Y))$ *such that*

$$\Delta_n := |P\{||S_n(X)||_\infty < \lambda\} - P\{||Y||_\infty < \lambda\}| \\ \leq Cn^{-\varphi(\kappa)}(1+\lambda)^{-3}(\ln n + \ln(1+\lambda))^2, \tag{4.17}$$

where

$$\varphi(\kappa) = \begin{cases} 1/6 & \text{if } \alpha_3 > 1; \\ 1/6 \vee (1+\kappa)/10 & \text{if } \alpha_3 = 1. \end{cases}$$

Sketch of the proof. To avoid certain technicalities, we shall sketch the proof under stronger conditions then those stated in the theorem. Namely, we shall assume instead of (4.10) the conditions (4.9) and (4.15) with $\kappa = 1$ (therefore in (4.8) we can assume that $\alpha_1 > 1/2$). Under such conditions an estimate of the remainder term was given in Paulauskas and Juknevičienė (1988). Moreover, we assume that $\alpha_3 > 1$ and $E\sup_{t\in[0,1]}|X(t)|^p < \infty$ for some $p > 3$. Let $0 < \delta < 1$ be a number to be specified later on and let $N = [1/\delta] + 1$ and $0 = t_1 < t_2 < \cdots < t_N = 1$ be such that $t_k - t_{k-1} \leq \delta$. As earlier $||x||_N = \sup_{1\leq i\leq N}|x(t_i)|$ and subscript N refers to the corresponding quantities in the space l_∞^N. Applying Lemma 4.1 and Remark 4.4, for any $0 \leq \varepsilon \leq \lambda$ and $0 \leq \delta < 1$, we have the estimate

$$\Delta_n(\lambda) \leq I_1 + I_2 + I_3 + \Delta_n^N(\lambda - \varepsilon) + \Delta_n^N(\lambda), \tag{4.18}$$

where

$$I_1 = P\{\sup_{1\leq i\leq N-1}\sup_{t_i\leq t\leq t_{i+1}}|S_n(t) - S_n(t_i)| > \varepsilon,\ ||S_n||_\infty > \lambda\},$$
$$I_2 = P\{\sup_{1\leq i\leq N-1}\sup_{t_i\leq t\leq t_{i+1}}|Y(t) - Y(t_i)| > \varepsilon, ||Y||_\infty > \lambda\},$$
$$I_3 = P\{\lambda - \varepsilon \leq ||Y||_\infty \leq \lambda + \varepsilon\}.$$

By some straightforward computations, we arrive at having to estimate the quantities

$$I_{1,1} = P\{\sup_{1\leq i\leq N-1}\sup_{t_i\leq t\leq t_{i+1}}|S_n(t) - S_n(t_i)| > \varepsilon\},$$
$$I_{2,1} = P\{\sup_{|t-s|<\delta}|Y(t) - Y(s)| > \varepsilon\}.$$

To estimate $I_{2,1}$ it is sufficient to use the following result (see, for example, Marcus and Pisier (1981)).

Lemma 4.13. *Let η be a zero-mean Gaussian $C[0,1]$-valued r.e. and let $r^2(s,t) = E(\eta(t) - \eta(s))^2$. Then for any $\varepsilon > 0$ and $0 < \delta < 1$ the following inequality holds:*

$$P\{\sup_{\tau(s,t)<\delta}|\eta(s) - \eta(t)| > \varepsilon\}$$
$$\leq C\varepsilon^{-1}\left(\int_0^\delta (H([0,1],\tau,x))^{1/2}\,dx + \delta\ln^+\ln(4d\delta^{-1})^{1/2}\right),$$

where $d = \sup_{s,t\in[0,1]}\tau(s,t)$ and $\ln^+ u = \ln(1 \vee u)$.

In the case under consideration

$$\tau(s,t) = (E|Y(t) - Y(s)|^2)^{1/2} \le (F_1(t) - F_1(s))^{\alpha_1/2} \le C|t-s|^{\beta_1\alpha_1/2}.$$

After some rather rough estimates we obtain

$$I_{2,1} \le C\varepsilon^{-1}\delta^{\beta_1\alpha_1/4}.$$

To estimate $I_{1,1}$ one needs the following lemmas. The first two may be found in Billingsley (1968) and the third in Hahn (1978).

Lemma 4.14. *Let $[a,b] \subset [0,1]$. For $x \in D[0,1]$ the following inequality holds:*

$$\sup_{t\in[a,b]} |x(t) - x(a)| \le \sup_{t\in[a,b]} \min\{|x(t) - x(a)|,\ |x(b) - x(t)|\} + |x(b) - x(a)|.$$

Lemma 4.15. *Let $\xi_1, \dots, \xi_m$ be r.v.'s and let $S_0 = 0$ and $S_k = \sum_{i=1}^k \xi_i$, $k = 1, 2, \dots, m$. Suppose that there exist non-negative numbers $u_1, \dots, u_m$ such that for any $\lambda > 0$, $0 \le i \le j \le k \le m$ the inequality*

$$P\{|S_j - S_i| \ge \lambda,\ |S_k - S_j| \ge \lambda\} \le \lambda^{-\gamma}\left(\sum_{n=i}^{k} u_n\right)^{\alpha}$$

holds for some $\gamma \ge 0$ and $\alpha > 1$. Then for all $\lambda > 0$

$$P\{\max_{0\le j\le m} \min\{|S_j|,\ |S_m - S_j|\} \ge \lambda\} \le K_{\gamma,\alpha}\lambda^{-\gamma}\left(\sum_{i=1}^{m} u_i\right)^{\alpha},$$

where $K_{\gamma,\alpha}$ is a constant depending on γ and α only.

Lemma 4.16. *Suppose that X satisfies the conditions (4.8) and (4.9). Then for all $\lambda > 0$, $0 \le s \le t \le u \le 1$*

$$\lambda^4 P\{|S_n(t) - S_n(s)| \ge \lambda,\ |S_n(u) - S_n(t)| > \lambda\} \le (G(u) - G(s))^{\mu_1},$$

where $\mu_1 = \alpha_2 \wedge 2\alpha_1 > 1$ and $G(u) = 2^{1/\mu_1}(n^{-1/\mu_1}F_2(u) + 3^{1/\mu_1}F_1(u))$.

Applying Lemmas 4.14–4.16, we deduce that

$$I_{1,1} \le \sum_{i=1}^{N-1} \left(P\{R_i > \varepsilon/2\} + P\{|S_n(t_{i+1}) - S_n(t_i)| > \varepsilon/2\}\right), \tag{4.19}$$

$$P\{R_i > \varepsilon/2\} \le C\varepsilon^{-4}(G(t_{i+1}) - G(t_i))^{\mu_1},$$

where $R_i = \sup_{t_i\le t\le t_{i+1}} \min\{|S_n(t) - S_n(t_i)|,\ |S_n(t_{i+1}) - S_n(t)|\}$. To estimate the second term on the right-hand side of (4.19) we use (4.15) with $k = 1$ and obtain

$$P\{|S_n(t_{i+1}) - S_n(t_i)| > \varepsilon/2\} \leq C\varepsilon^{-4} E(S_n(t_{i+1}) - S_n(t_i))^4$$
$$\leq C\varepsilon^{-4}\big(n^{-1}(F_3(t_{i+1}) - F_3(t_i))^{\alpha_3} + 3(F_1(t_{i+1} - (F_1(t_i))^{2\alpha_1}\big).$$

Finally, applying (4.16) for all F_i, $i = 1, 2, 3$, we have

$$I_{1,1} \leq C\varepsilon^{-4}\big((\delta^{\beta_2} n^{-1/\mu_1} + \delta^{\beta_1})^{\mu_1 - 1} + \delta^{\beta_3(\alpha_3 - 1)} n^{-1}\big).$$

In order to apply Theorem 4.5 to estimate $\Delta_n^N(\lambda)$ and $\Delta_n^N(\lambda - \varepsilon)$ we can use the estimate for I_2, obtained above to bound $D_n(\delta)$. It is not difficult to deduce that

$$D_n(\delta) \leq C \max(n^{-1/6},\ n^{1/12}\delta^{\beta_1\alpha_1/8}).$$

It remains to collect all estimates which we have for the quantities entering inequality (4.18) and to choose the parameters ε and δ in an appropriate way. It is easy to see that if we put $\varepsilon \sim n^{-1/6}$ and $\delta \sim n^{-\rho}$ with some positive constant ρ depending on β_i, α_i, $i = 1, 2, 3$, and p, then all terms will not exceed $n^{-1/6}\lambda^{-3}$. It is worth mentioning that $\rho \to \infty$ in any of the following cases: if $\min_i \beta_i \to 0$, or if the α_i tend to their least possible values, or $p \to 3$. Also one can notice that under these stronger conditions which we had assumed for the final estimate of the remainder term there will be no term $\ln(1 + \lambda)$

§1.5. Rates of Convergence in Prokhorov and BL Metrics

Let $\mathcal{P}(B)$ denote the set of probability measures on B. We recall that the Prokhorov metric π and the bounded Lipschitz (BL) metric ϱ_{BL} on $\mathcal{P}(B)$ are defined as follows: for $\mu, \nu \in \mathcal{P}(B)$

$$\pi(\mu, \nu) := \inf\{\varepsilon > 0 : \mu(F) \leq \nu(F_\varepsilon) + \varepsilon \text{ for all closed sets } F \subset B\};$$

$$\varrho_{BL}(\mu, \nu) := \sup\left\{\left|\int_B f(x)(\mu - \nu)(dx)\right| : ||f||_{BL} \leq 1\right\},$$

where F_ε denotes an ε-neighborhood of the set $F \subset B$ and

$$||f||_{BL} := \sup_{x \in B} |f(x)| + \sup_{x \neq y} |f(x) - f(y)|/||x - y||$$

for $f : B \to \mathbb{R}$. For the sake of brevity we shall write $\pi(\xi, \eta)$ and $\varrho_{BL}(\xi, \eta)$ instead of $\pi(\mathcal{L}(\xi), \mathcal{L}(\eta))$ and $\rho_{BL}(\mathcal{L}(\xi), \mathcal{L}(\eta))$, respectively. Recall that both metrics π and ϱ_{BL} metrize the weak convergence on $\mathcal{P}(B)$. Therefore the rate of convergence, estimated by means of these metrics, is of great interest. Results on the rate of convergence for the Prokhorov metric in finite-dimensional spaces may be found, for example, in Yurinskii (1977), Dehling (1983), Zaitsev (1987), and Bentkus (1984f). In the case of infinite-dimensional spaces we have the following negative result.

Theorem 5.1 (Senatov (1981)). *For any monotone sequence $b_n \downarrow 0$ there exist a Gaussian ℓ_2-r.e. Y and ℓ_2-r.e. X such that*

(i) $EX = EY = 0$, $cov\, X = cov\, Y$;
(ii) $P\{||X|| \le 1\} = 1$;
(iii) $\liminf_{n\to\infty} \pi(S_n(X), Y) b_n^{-1} > 0$;
(iv) $\liminf_{n\to\infty} \varrho_{BL}(S_n(X), Y) b_n^{-1} > 0$.

Note that conditions (i), (ii) guarantee the CLT; therefore $\pi(S_n(X), Y) \to 0$ and $\varrho_{BL}(S_n(X), Y) \to 0$ as $n \to \infty$. Theorem 5.1 tells us that in order to have a convergence rate either in the Prokhorov metric or in the BL metric in infinite-dimensional spaces one needs stronger assumptions than the finiteness of the usual moments. The following results provide such conditions. Recall that the r.e.'s X, $Y \in B$ are such that $EX = EY = 0$, $cov\, X = cov\, Y$ and that Y is Gaussian.

Theorem 5.2. *Let r.e.'s $X, Y \in \ell_2$. Suppose that*

$$E||X||_\lambda^3 < \infty,$$

where $||x||_\lambda^2 = \sum_{i=1}^\infty (x_i/\lambda_i)^2$ and the sequence $\lambda = (\lambda_1, \lambda_2, \ldots)$, $\lambda_i > 0$, is such that

$$\sum_{i=1}^\infty \lambda_i^2 < \infty. \tag{5.1}$$

Then

$$\pi(S_n(X), Y) = O(n^{-1/8})$$

and

$$\varrho_{BL}(S_n(X), Y) = O(n^{-1/6}).$$

Theorem 5.3. *Let r.e.'s X, $Y \in c_0$. Suppose that*

$$E||X||_2^3 < \infty,$$

where $||x||_2^2 = \sum_{i=1}^\infty x_i^2$. Then

$$\pi(S_n(X), Y) = O(n^{-1/8})$$

and

$$\varrho_{BL}(S_n(X), Y) = O(n^{-1/6}).$$

Therefore the finiteness of the third moment under an appropriate seminorm, stronger than the original one, is sufficient to derive a convergence rate in the Prokhorov and BL metrics. Theorem 5.2 is due to Bentkus and Račkauskas (1984). Theorem 5.3 was proved by Bentkus (the proof is contained in Paulauskas and Račkauskas (1989)) and, in fact, is stronger than Theorem 5.2. Now we shall state more general results than Theorem 5.2. To this end, let a Hilbert space $H \subset B$ be linearly and continuously imbedded in the Banach space B (therefore, without loss of generality, one may assume that $||x||_H \ge ||x||_B$ for all $x \in H$). The imbedding operator $H \hookrightarrow B$ is said to be γ-radonifying if the Gaussian measure ν of the cylinder sets in B with characteristic functional

$$\int_B \exp(ix(y))\nu(dy) = \exp(-||x||_H^2/2)$$

for $x \in B^* \subset H^* = H$ is a σ-additive measure (see, e.g., Badrikian and Chevet (1974)). Put

$$\nu_{H,m} = \int_B ||x||_H^m |\mathcal{L}(X) - \mathcal{L}(Y)|(dx).$$

Theorem 5.4 (Bentkus and Račkauskas (1984)). *Suppose that the imbedding $H \hookrightarrow B$ is γ-radonifying. Then there exists a finite constant C such that*

$$\pi(S_n(X), Y) \le C\nu_{H,3}^{1/4} n^{-1/8}; \tag{5.2}$$

$$\varrho_{BL}(S_n(X), Y) \le C \max(\nu_{H,3} n^{-1/2},\ \nu_{H,3}^{1/3} n^{-1/6}). \tag{5.3}$$

We note that the corresponding Hilbert space H for the space $B = \ell_2$ can be taken as

$$H = \ell_2(\lambda) := \left\{ x \in \ell_2 : |x|_H^2 := \sum_{i=1}^{\infty} (x_i/\lambda_i)^2 < \infty \right\},$$

where the sequence $\lambda = (\lambda_1, \lambda_2, \ldots)$ of positive numbers is such that (5.1) holds.

In Hilbert space, Yurinskii (1977) obtained

$$\pi(S_n(X), Y) = O(n^{-\alpha/(6+8\alpha)} \log n)$$

under the assumption of a finite third moment, where α is a positive parameter characterizing the behaviour of the eigenvalues of the covariance operator $cov\, Y$. Conditions, guaranteeing some logarithmic order of $\pi(S_n(X), Y)$ in the case X, $Y \in \ell_2$, may be found in Kukuš (1981, 1982). Lapinskas (1978) investigated the convergence rate in the CLT in the Prokhorov metric in Banach spaces with a Schauder basis. He imposed certain assumptions on the coordinates of X, Y and obtained the rate $O(n^{-1/21})$. For X, $Y \in c_0$ and $E||X||_1^3 < \infty$, where $||\cdot||_1$ is the ℓ_1-norm, a result of Senatov (1981) yields the estimate $\pi(S_n(X), Y) = O(n^{-1/8} \log^{3/4} n)$ (compare with Theorem 5.2).

Note that both estimates (5.2) and (5.3) are in general unimprovable. Appropriate examples are constructed by Bentkus (1987) (see also Senatov (1981) and Theorem 26 in Bentkus and Račkauskas (1984)). It should be noted that the condition $\nu_{H,3} < \infty$ is rather restrictive. For example, it can be shown that when the limiting Gaussian element Y is a Wiener process in $B = C[0,1]$, we automatically have $\nu_{H,3} = \infty$ if $\mathcal{L}(X) \neq \mathcal{L}(Y)$. Bentkus and Račkauskas (1984) proposed a method which enables one to obtain estimates under less restrictive assumptions. However, the order in n obtained by this method is somewhat worse. We shall not go into detail but confine ourselves to a particular example concerning the space $B = C[0,1]$. If ω is the modulus of continuity, then we denote by H_ω the space of all functions x on $[0,1]$ such that

$$||x||_\omega := \sup_{t\in[0,1]} |x(t)| + \sup_{t\neq s} |x(t) - x(s)|/\omega(|t-s|) < \infty.$$

In the case $\omega(s) = s^\alpha$, we write $||\cdot||_\alpha$, H_α instead of $||\cdot||_\omega$, H_ω, respectively. Theorems 5.5 and 5.6 below assume instead of $EX = EY = 0$, $cov\, X = cov\, Y$ that $EX(s) = EY(s) = 0$, $EX(t)X(s) = EY(t)Y(s)$ for all $s, t \in [0,1]$. The next result is a slightly improved version of Bentkus and Račkauskas (1984) (see Paulauskas and Račkauskas (1989)).

Theorem 5.5. *Let $X, Y \in C[0,1]$. Suppose that the following conditions are fulfilled:*

$$P\{X \in H_\alpha\} = P\{Y \in H_\alpha\} = 1$$

and

$$\nu_{\alpha,3} := \int_{C[0,1]} ||x||_\alpha^3 |\mathcal{L}(X) - \mathcal{L}(Y)|(dx) < \infty.$$

If $1/2 < \alpha \le 1$, then there exists a constant $C(\alpha)$ such that

$$\pi(S_n(X), Y) \le C(\alpha)\nu_{\alpha,3}^{1/4} n^{-1/8},$$

and

$$\varrho_{BL}(S_n(X), Y) \le C(\alpha)\nu_{\alpha,3}^{1/3} n^{-1/6}.$$

If $0 < \alpha \le 1/2$, then for each $\varepsilon > 0$ there exists a constant $C(\alpha, \varepsilon)$ such that

$$\pi(S_n(X), Y) \le C(\alpha, \varepsilon) \max\left(\nu_{\alpha,3}^{1/3} n^{-1/6},\ (\nu_{\alpha,3} n^{-1/2})^{4\alpha/(9-2\alpha)-\varepsilon}\right), \qquad (5.4)$$

and

$$\varrho_{BL}(S_n(X), Y) \le C(\alpha, \varepsilon) \max\left(\nu_{\alpha,3}^{1/3} n^{-1/6}, (\nu_{\alpha,3} n^{-1/2})^{2\alpha/3-\varepsilon}\right). \qquad (5.5)$$

In the case where Y is a Wiener process on $[0,1]$ and $E||X||_\alpha^3 < \infty$ with $\alpha = 1/2 + \delta$, $\delta < 0$, consequences of Theorem 5.5 are

$$\pi(S_n(X), Y) = o(n^{-1/8+\varepsilon}),$$

and

$$\varrho_{BL}(S_n(X), Y) = o(n^{-1/6+\varepsilon}),$$

where $\varepsilon = \varepsilon(\delta) \downarrow 0$ when $\delta \uparrow 0$.

Combining the method of Bentkus and Račkauskas (1984) with the results due to Bogachev (1988), one can strengthen (5.4) and (5.5) for $\alpha = 1/2$. Namely, the following result holds.

Theorem 5.6. *Let* X, $Y \in C[0,1]$ *and* $\omega(t) = t^{1/2}(\log(1/t))^{-\beta}$, $\beta > 3/2$. *Suppose that the following conditions are fulfilled:*

$$P\{X \in H_\omega\} = P\{Y \in H_\omega\} = 1$$

and

$$\nu_{\omega,3} := \int_{C[0,1]} ||x||_\omega^3 |\mathcal{L}(X) - \mathcal{L}(Y)|(dx) < \infty.$$

Then there exists a constant $C = C(\beta)$ *such that*

$$\pi(S_n(X), Y) \le C\nu_{\omega,3}^{1/4} n^{-1/8};$$

and

$$\varrho_{BL}(S_n(X), Y) \le C\nu_{\omega,3}^{1/3} n^{-1/6}.$$

The convergence rate in the CLT estimated by means of other distances than the Prokhorov metric or BL metric was considered by Zolotarev (1976a, b, 1977), Bentkus and Račkauskas (1984), Sakalauskas (1983), Liubinskas (1987), Rachev and Yukich (1989), and Rachev and Rüschendorf (1990). Bounds of the convergence rate in the infinite-dimensional invariance principle in the Prokhorov metric and other metrics were studied by Borovkov and Sakhanenko (1980), Borovkov (1984), Sakhanenko (1988) and Bentkus and Liubinskas (1989).

Chapter 2
Asymptotic Expansions

§2.1. Short Expansion

Throughout this chapter we shall use the notation $\varepsilon = \varepsilon(n) = n^{-1/2}$. Let $f : B \to F$ denote a function on a Banach space B such that the expectations $Ef(S_n)$, $Ef(Y)$ are well defined. We set

$$g(\varepsilon) = g(\varepsilon; f) = Ef(S_n),$$
$$g(0) = g(0; f) = Ef(Y)$$

We shall describe the general idea for constructing asymptotic expansions of the type

$$g(\varepsilon) = g(0) + a_1\varepsilon + \cdots + a_k\varepsilon^k + R \tag{1.1}$$

or, equivalently,

$$Ef(S_n) = Ef(Y) + a_1 n^{-1/2} + \cdots + a_k n^{-k/2} + R \tag{1.2}$$

(the so-called asymptotic power series). Here $a_1, \ldots, a_k$ are "known" coefficients and the remainder term $R = R_k(\varepsilon)$ usually satisfies $\varepsilon^{-k} R_k(\varepsilon) \to 0$ as $\varepsilon \to 0$. By choosing various functions f in (1.2), one can obtain asymptotic expansions for moments, probabilities, etc. We remark that the structure of our asymptotic expansions depends neither on the structure of f and Y nor on the dimension and the structure of Banach spaces B and F.

The following lemma is obvious.

Lemma 1.1. *If the remainder term $R = R_k(\varepsilon)$ satisfies $\varepsilon^{-k} R_k(\varepsilon) \to 0$ as $\varepsilon \to 0$ then the coefficients $a_1, \ldots, a_k$ in the expansion (1.1) are unique and for $s = 1, \ldots, k$*

$$a_s = \lim_{\varepsilon \to 0} \varepsilon^{-s} (g(\varepsilon) - g(0) - a_1 \varepsilon - \cdots - a_{s-1} \varepsilon^{s-1}).$$

We shall demonstrate in detail the complete construction only in the case of the short asymptotic expansion

$$Ef(S_n) = Ef(Y) + a_1 \varepsilon + R.$$

The general case differs from this only in cumbersome technical details and more complicated notation. Also we restrict ourselves to the case of functions $f : B \to F$ with bounded derivatives. This restriction is essential and considerably simplifies the estimation of the remainder term. However, we point out that, roughly speaking, to estimate the remainder term for non-smooth f, one has to apply additionally the methods developed for estimating the convergence rates in the CLT.

Theorem 1.2. *If $f : B \to F$ is a function of the class C_b^6 and $E\|X\|^4 < \infty$, then*

$$Ef(S_n) = Ef(Y) + \frac{1}{6} \varepsilon E f'''(Y) X^3 + R, \tag{1.3}$$

where

$$\|R\| \le \varepsilon^2 C(f) \left(E\|X\|^4 + E\|Y\|^4 + (E\|X\|^3 + E\|Y\|^3)^2 + E\|X\|^3 E\|Y\| \right), \tag{1.4}$$

$$C(f) = \|f^{(4)}\|_\infty + \|f^{(6)}\|_\infty, \ \|f\|_\infty = \sup_{x \in B} \|f(x)\|.$$

If B is a Hilbert space, then

$$\|R\| \le C \varepsilon^2 (\|f^{(4)}\|_\infty + \|f^{(6)}\|_\infty)(1 + E\|Y\|^2) E\|X\|^4,$$

where C is an absolute constant.

The proof of Theorem 1.2 involves several steps (see Lemmas 1.3–1.6). To estimate the convergence rate, the following obvious algebraic identity is very useful (see §2.1):

$$\mu_1 \cdots \mu_n = \nu_1 \cdots \nu_n + \sum_{i=1}^{n} \mu_1 \cdots \mu_{i-1} (\mu_i - \nu_i) \nu_{i+1} \cdots \nu_n, \tag{1.5}$$

where $\mu_1, \ldots, \mu_n, \nu_1, \ldots, \nu_n$ are arbitrary measures and the multiplication is understood to be convolution of measures. For asymptotic expansions one needs to apply (1.5) iteratively several times. The number of iterations depends on the desired estimate of the remainder term. In the case of the short asymptotic expansions it is enough to use (1.5) twice, i.e. the identity

$$\mu_1 \cdots \mu_n = \nu_1 \cdots \nu_n + \sum_{i=1}^{n} \nu_1 \cdots \nu_{i-1}(\mu_i - \nu_i)\nu_{i+1} \cdots \nu_n + R, \tag{1.6}$$

where the remainder

$$R = \sum_{i=1}^{n} \sum_{j=1}^{i-1} \mu_1 \cdots \mu_{j-1}(\mu_j - \nu_j)\nu_{j+1} \cdots \nu_{i-1}(\mu_i - \nu_i)\nu_{i+1} \cdots \nu_n.$$

In the case of identically distributed summands, $\mu = \mu_1 = \cdots = \mu_n$, $\nu = \nu_1 = \cdots = \nu_n$ and (1.6) reduces to

$$\mu^n = \nu^n + n\nu^{n-1}(\mu - \nu) + R, \tag{1.7}$$

with

$$R = \sum_{i=1}^{n-1} (n-i)\mu^{i-1}(\mu - \nu)^2 \nu^{n-i-1}.$$

Integrating f with respect to the measure μ^n, choosing $\mu = \mathcal{L}(\varepsilon X)$, $\nu = \mathcal{L}(\varepsilon Y)$ and applying (1.7), we obtain

$$Ef(S_n) = Ef(Y) + n \int_B Ef(\varepsilon\sqrt{n-1}\,Y + \varepsilon x)H(dx) + R, \tag{1.8}$$

where

$$H = \mathcal{L}(X) - \mathcal{L}(Y)$$

and

$$R = \sum_{i=1}^{n-1} (n-i) \int_B \int_B f(a_i + \varepsilon x + \varepsilon y)H(dx)H(dy),$$

and where r.e. a_i has the distribution of $\varepsilon X_1 + \cdots + \varepsilon X_{i-1} + \varepsilon\sqrt{n-i-1}\,Y$.

Lemma 1.3. *For all $a \in B$ and $\varepsilon > 0$ the following estimates hold:*

$$\left\| \int_B f(a + \varepsilon x)H(dx) - \frac{1}{6}\varepsilon^3 \int_B f'''(a)x^3 H(dx) \right\|$$
$$\leq \frac{1}{24}\varepsilon^4 \|f^{(4)}\|_\infty (E\|X\|^4 + E\|Y\|^4), \tag{1.9}$$

$$\left\| \int_B \int_B f(a + \varepsilon x + \varepsilon y)H(dx)H(dy) \right\|$$
$$\leq \frac{1}{36}\varepsilon^6 \|f^{(6)}\|_\infty (E\|X\|^3 + E\|Y\|^3)^2. \tag{1.10}$$

Proof. To prove (1.9) it is sufficient to apply Taylor's formula (see, e.g., Cartan (1971))

$$h(u+v) = \sum_{j=0}^{s} h^{(j)}(u)v^j/j! + \int_0^1 (1-\tau)^s h^{(s+1)}(u+\tau v)v^{s+1}d\tau/s!$$

with $h = f$, $u = a$, $v = \varepsilon x$, $s = 3$ and to note that the means and covariances of X and Y coincide. For the proof of (1.10) it is sufficient to apply Taylor's formula twice with $s = 2$ and $v = \varepsilon x$ and $v = \varepsilon y$, respectively.

Lemma 1.4. *The following relation holds:*

$$Ef(S_n) = Ef(Y) + \varepsilon Ef'''(\varepsilon\sqrt{n-1}\,Y)X^3/6 + R,$$

where

$$||R|| \le \varepsilon^2(||f^{(4)}||_\infty + ||f^{(6)}||_\infty)(E||X||^4 + E||Y||^4 + (E||X||^3 + E||Y||^3)^2/3)/24.$$

Proof. The lemma is a consequence of the representation (1.8) and Lemma 1.3. We apply the estimate (1.9) to the integral in (1.8) with $a = \varepsilon\sqrt{n-1}\,Y$. The remainder term R in (1.8) is estimated with the help of (1.10) with $a = a_i$. Finally, we have

$$\int_B f'''(a)x^3 H(dx) = Ef'''(a)X^3$$

due to the symmetry of Y.

Lemma 1.5.

$$||Ef'''(Y)X^3 - Ef'''(\varepsilon\sqrt{n-1}\,Y)X^3|| \le \varepsilon||f^{(4)}||_\infty E||X||^3||E||Y||.$$

Proof. The r.e. Y has the same distribution as $\varepsilon\sqrt{n-1}\,Y + \varepsilon Y_1$, where Y_1 is an independent copy of Y. Therefore it is sufficient to apply Taylor's formula to $Ef'''(Y)X^3$ with $h = f'''$, $u = \varepsilon\sqrt{n-1}\,Y$, $v = \varepsilon Y_1$ and $s = 0$.

Lemma 1.6. *If B is a Hilbert space, then $E||X||^2 = E||Y||^2$. Furthermore $E||Y||^p \le C(p)E||X||^p$, $p > 0$.*

Proof. The equality $E||X||^2 = E||Y||^2$ is obvious since $||x||^2$ in a Hilbert space is a quadratic form and the covariances of X and Y coincide. Therefore the well-known inequality $E||Y||^p \le C(p)(E||Y||^2)^{p/2}$ and the Hölder inequality help to complete the proof of the lemma.

§2.2. The Smooth Case

In this section we shall consider the results obtained without (explicit) conditions similar to the classical Cramér condition

$$\limsup_{|t|\to\infty} |E\exp(itX)| < 1$$

for the characteristic function of a r.v. $X \in \mathbb{R}$. In the "smooth case" the formulations of results are not that overloaded by conditions and technical details. Therefore we shall consider different forms of expansions and certain technical aspects. We begin with the description of the coefficients $a_1, \ldots, a_k$ in the general asymptotic expansion. We shall show that

$$a_s = EP_s f(Y), \; s = 0, 1, 2, \ldots, \tag{2.1}$$

where the P_s are certain random differential operators. Let us repeat once more that the construction of P_s is universal and does not depend on the specific structure of X, Y, f or the spaces B, F. The coefficients a_s have the form (2.1) even when f is not differentiable, for example, when $f(x) = \chi_A(x)$ is the indicator function of a set $A \subset B$. The only difference in the case of a non-smooth function f is that formula (2.1) has to be interpreted in an appropriate way (see §2.3).

The Edgeworth–Cramér polynomials $E_k = E_k(m_2, \ldots, m_{k+2})$ of formal commuting variables $m_2, \ldots, m_{k+2}$ (the so-called "moment" variables) are determined as the coefficients in the formal power series expansion

$$\exp\left(t^{-2}\left[\ln\left(1 + \sum_{k=2}^{\infty} m_k t^k/k!\right) - m_2 t^2/2\right]\right) = \sum_{k=0}^{\infty} E_k(m_2, \ldots, m_{k+2}) t^k.$$

For instance,

$$\begin{aligned} E_0 &= 1, \; E_1 = m_3/6, \\ E_2 &= -m_2 m_2/8 + m_3 m_3/72 + m_4/24, \\ E_3 &= -m_2 m_3/12 - m_2 m_2 m_3/48 - m_3 m_3 m_3/1296 + m_3 m_4/144 + m_5/120. \end{aligned}$$

Write

$$E_k = \sum a_k(i_1, \ldots, i_s) m_{i_1} \cdots m_{i_s},$$

where the sum is taken over all integers $i_1, \ldots, i_s$ such that $2 \le i_1 \le \cdots \le i_s$. Clearly, only a finite number of coefficients $a_k(\ldots)$ are non-zero. Define the polynomials

$$P_k = P_k(z_1, \ldots, z_k) = \sum a_k(i_1, \ldots, i_s) z_1^{i_1} \cdots z_s^{i_s} \tag{2.2}$$

of commuting variables $z_1, \ldots, z_k$. For instance,

$$P_2 = -\frac{1}{8} z_1^2 z_2^2 + \frac{1}{72} z_1^3 z_2^3 + \frac{1}{24} z_1^4.$$

If $h \in B$, then one may introduce the differential operator $D(h)$ as follows:

$$D(h) f(x) = f'(x) h = \lim_{t \to 0} t^{-1} (f(x + th) - f(x))$$

(the so-called directional derivative). The differential operators $D(h)$, $h \in B$, and $1 \in \mathbb{R}$, generate a natural commutative algebra over the field of real numbers. Therefore one can define the random differential operators

$$P_k = P_k(D(X_1), \dots, D(X_k))$$

and coefficients $a_s = EP_s f(Y)$. For instance,

$$P_2 = -\frac{1}{8} D^2(X_1) D^2(X_2) + \frac{1}{72} D^3(X_1) D^3(X_2) + \frac{1}{24} D^4(X_1).$$

Lemma 2.1. *Let $E||X||^{k+2} < \infty$ and let function $f \in C^{3k}(B; F)$. Then there exists an $\alpha = \alpha(\mathcal{L}(Y)) > 0$ such that the coefficients*

$$a_s = EP_s f(Y), \; s = 0, 1, \dots, k,$$

are well defined when

$$\max_{0 \le i \le 3k} \sup_{x \in B} \exp(-\alpha ||x||^2) ||f^{(i)}(x)|| < \infty. \tag{2.3}$$

Since we want to avoid complications connected with measurability we assume separability of B and F. The condition $E||X||^{k+2} < \infty$ allows one to interpret expectations in the sense of Bochner instead of more complicated definitions. Condition (2.3) is connected with the well-known integrability properties of the norm of a Gaussian r.e. (the Skorohod – Fernique – Landau – Shepp Theorem). The condition $f \in C^{3k}$ is unnecessarily strong and will be weakened later on.

Theorem 2.2. *Suppose that a r.e. $X \in B$ satisfies the CLT. Let $E||X||^{k+3} < \infty$, function $f \in C^{3k+3}(B; F)$ and*

$$\sup_{x \in B} (1 + ||x||)^{-k-3} ||f(x)|| < \infty. \tag{2.4}$$

Then there exists $\alpha = \alpha(\mathcal{L}(Y)) > 0$ such that the asymptotic expansion

$$Ef(S_n) = Ef(Y) + \sum_{s=1}^{k} n^{-s/2} EP_s f(Y) + R \tag{2.5}$$

is well defined and $||R|| = O(n^{-(k+1)/2})$ provided

$$\sup_{x \in B} \exp(-\alpha ||x||_B) ||f^{(i)}(x)|| < \infty \tag{2.6}$$

for all $i = 1, \dots, 3k + 3$.

We now make some comments about Theorem 2.2 and provide some references. We are not concerned with the finite-dimensional case (see, e.g., Petrov (1975), Bhattacharya and Rao (1976), Götze and Hipp (1979), etc.). We mention here only that the formal power series expansion with Edgeworth–Cramér polynomials is contained, e.g., in Bikelis (1973). The definition of the terms $EP_k f(Y)$ of the asymptotic expansions via random differential operators P_k may be found in Götze (1981). The expansion (2.5) in the case $k = 0$ (i.e., the estimate of the convergence rate) under the condition $f \in C_b^3(B; \mathbb{R})$ was obtained by Paulauskas (1976b) and Zolotarev (1976b). Götze (1981) proved (2.5) under the assumption $f \in C_b^{3k+3}(B; \mathbb{R})$. Bentkus (1984b) proved that the short asymptotic expansion (2.5) is valid under the weaker moment and differentiability conditions $E||X||^3 < \infty$, $f \in C_b^3(H; \mathbb{R})$ (H is a Hilbert space), while the estimate of the remainder term becomes worse: $R = o(n^{-1/2})$ instead of $R = O(n^{-1})$. Theorem 2.2 in the case of a Hilbert space is contained in Bentkus (1984d). In the general case Theorem 2.2 follows from more general and more precise results due to Bentkus (1986d). It should be remarked also that all papers, just mentioned, contain more or less explicit estimates of the remainder term.

The construction of asymptotic expansions is usually based on the formula obtained by iterating (1.6), that is,

$$\mu_1 \cdots \mu_n = \nu_1 \cdots \nu_n + A_1 + \cdots + A_k + R, \tag{2.7}$$

where $\mu_1, \ldots, \mu_n, \nu_1, \ldots, \nu_n$ are arbitrary measures and

$$A_s = \sum_{card\, \alpha = s} \prod_{i \in \alpha} (\mu_i - \nu_i) \prod_{j \notin \alpha} \nu_j. \tag{2.8}$$

The summation is over all possible sets $\alpha = \{i_1, \ldots, i_s\}$ of integers $1 \le i_s < i_{s-1} < \cdots < i_1 \le n$ and the second product is over all integers $j \notin \alpha$ such that $1 \le j \le n$;

$$R = \sum_{card\, \alpha = k+1} \prod_{i} \mu_i \prod_{j \in \alpha} (\mu_j - \nu_j) \prod_{l \notin \alpha} \nu_l,$$

where the summation is similar to that in (2.8), the first product is taken over all integers $i \ge 1$, $i < i_{k+1}$ and the third product is over all integers $l \notin \alpha$ such that $i_{k+1} < l \le n$. In the identically distributed case $\mu = \mu_1 = \cdots = \mu_n$, $\nu = \nu_1 = \cdots = \nu_n$, the formulae for A_s and R reduce to

$$A_s = \binom{n}{s} \nu^{n-s} (\mu - \nu)^s, \quad \binom{n}{s} = n!/[s!(n-s)!],$$

and

$$R = \sum_{i=1}^{n-k} \binom{n-i}{k} \mu^i (\mu - \nu)^{k+1} \nu^{n-i-k-1}.$$

In the i.i.d. case, integrating (2.7) we obtain

$$Ef(S_n) = Ef(Y) + \sum_{s=1}^{k} \binom{n}{s} \varphi_s(\varepsilon) + R, \tag{2.9}$$

where

$$\varphi_s(\varepsilon) = E \int_B \cdots \int_B f(\varepsilon W_s + \varepsilon x_1 + \cdots + \varepsilon x_s) H(dx_1) \cdots H(dx_s),$$

$$R = \sum_{i=1}^{n-k} \binom{n}{s} R_i,$$

$$R_i = E \int_B \cdots \int_B f(\varepsilon T + \varepsilon x_1 + \cdots + \varepsilon x_{k+1}) H(dx_1) \cdots H(dx_{k+1}),$$

and the r.e.'s W_s and T have the following distributions:

$$\mathcal{L}(W_s) = \mathcal{L}(\sqrt{n-s}\, Y) = \mathcal{L}(Y_1 + \cdots + Y_{n-s}),$$

$$\mathcal{L}(T) = \mathcal{L}(X_1 + \cdots + X_i + Y_1 + \cdots + Y_{n-i-k-1}).$$

It seems that expansions of type (2.7), (2.8) were explicitly published for the first time by Bergström (1951). In any case, they are traditionally called "Bergström's expansions".

Let us discuss the method and results presented in the paper of Götze (1981). The main idea of his paper is that to construct the asymptotic expansion for $Ef(S_n)$, one does not need the differentiability of f. It is sufficient to have the differentiability of the functions φ_s and R_i in (2.9). The gain is based on the interchanging of the order of integration and differentiation and on the property of integration as a "smoothing operation". Put

$$U(\eta_1, \ldots, \eta_q) = Ef(Y + \eta_1 X_1 + \cdots \eta_q X_q),$$

$$U_i(\varepsilon_1, \ldots, \varepsilon_q | \eta_1, \ldots, \eta_q) = Ef(\varepsilon W_i + \varepsilon_1 Y_1 + \cdots + \varepsilon_q Y_q + \eta_1 X_1 + \ldots \eta_q X_q),$$

where the r.e. W_i satisfies

$$\mathcal{L}(W_i) = \mathcal{L}(X_1 + \cdots + X_i + Y_{i+1} + \cdots + Y_n), \ 1 \le i \le n.$$

Theorem 2.3. (Götze (1981)). *Suppose that the functions U_i are differentiable. Let*

$$C_n := \sup \left| \left(\frac{\partial}{\partial \varepsilon_1}\right)^i \left(\frac{\partial}{\partial \eta_1}\right)^{j_1} \cdots \left(\frac{\partial}{\partial \eta_q}\right)^{j_q} U_i(\varepsilon_1, 0, \ldots, 0 | 0, \eta_2, \ldots, \eta_q) \right|,$$

where sup *is taken over all $1 \le i \le n$, all $0 \le \varepsilon_1, \eta_2, \ldots, \eta_q \le n^{-1/2}$ and all $i, j_1, \ldots, j_q \le k+3$ such that $k+3 \le i + j_1 + \ldots + j_q \le 3(k+1)$. Furthermore, suppose that*

$$\left(\frac{\partial}{\partial\varepsilon_1}\right)^{i_1}\cdots\left(\frac{\partial}{\partial\varepsilon_q}\right)^{i_q}\left(\frac{\partial}{\partial\eta_1}\right)^{j_1}\cdots\left(\frac{\partial}{\partial\eta_q}\right)^{j_q}U_i(0,\dots,0|0,\dots,0)$$
$$=\left(\frac{\partial}{\partial\eta_1}\right)^{i_1+j_1}\cdots\left(\frac{\partial}{\partial\eta_q}\right)^{i_q+j_q}U_i(0,\dots,0|0,\dots,0)$$

when $i_1 j_1 = \cdots = i_q j_q = 0$ *for all* $1 \le i \le n$, *all* $i_1,\dots,i_q \le 2$ *such that* $i_1 + j_1 \le k+3,\dots,i_q+j_q \le k+3$ *and* $i_1+\cdots+i_q+j_1+\cdots j_q \le 3(k+1)$. *Then*

$$\left|Ef(S_n) - \sum_{s=0}^{k} n^{-s/2}P_s U(0,\dots,0)\right| \le c(k)C_n n^{-(k+1)/2},$$

where the differentiable (non-random) *operator*

$$P_s = P_s\left(\frac{\partial}{\partial\eta_1},\dots,\frac{\partial}{\partial\eta_s}\right)$$

is defined via the polynomials P_s *of (2.2).*

It is possible to generalize Theorem 2.3 to apply in situations which are not directly connected with sums of independent r.e.'s. It occurs that the proof of this result is not probability-theoretic at all. This was noted by Götze (1985). Suppose that a sequence h_n, $n = 1,2,\dots$, of numbers is given, for example, $h_n = Ef(S_n)$, or the h_n are probabilities related to n observations, etc. Furthermore, suppose that it is possible to introduce "weights" $\varepsilon_1,\dots,\varepsilon_n$ for the observations and to determine a sequence $h_n(\varepsilon_1,\dots,\varepsilon_n)$, $0 \le \varepsilon_1,\dots,\varepsilon_n \le n^{-1/2}$, of functions such that

$$h_n = h_n(n^{-1/2},\dots,n^{-1/2}) + O(n^{-(k+1)/2}).$$

For example, if $h_n = Ef(S_n)$, it is natural to set

$$h_n(\varepsilon_1,\dots,\varepsilon_n) = Ef(\varepsilon_1 X_1 + \cdots + \varepsilon_n X_n).$$

Götze (1985) has shown under certain natural conditions (in particular, differentiability conditions similar to that in Theorem 2.3 in case $k=0$) that the "central limit theorem" holds. That is for fixed $\eta_1,\dots,\eta_r$ there exist limits

$$h_\infty = \lim_{n\to\infty} h_n(n^{-1/2},\dots,n^{-1/2}),$$
$$h_\infty(\eta_1,\dots,\eta_r) = \lim_{m\to\infty} h_{m+r}(\eta_1,\dots,\eta_r,n^{-1/2},\dots,n^{-1/2}).$$

Using the functions $h_n(\varepsilon_1,\dots,\varepsilon_n)$ and $h_\infty(\eta_1,\dots,\eta_r)$, one can define functions similar to the functions U and U_i in Theorem 2.3. Under differentiability conditions similar to those in Theorem 2.3, this allows one to obtain an asymptotic expansion of $h_n(n^{-1/2},\dots,n^{-1/2})$ analogous to that in Theorem 2.3. For further details in this direction and some applications to mathematical statistics, see Götze (1985).

Theorem 2.3 is valid for non-differentiable f, too. It has been proved without an explicit condition similar to the classical Cramér condition which is replaced here by differentiability conditions. In applications the verification of these conditions is the difficult task.

Asymptotic expansions with the Edgeworth–Cramér polynomials have the classical structure of asymptotic power series. A disadvantage of these expansions is that their existence is guaranteed only when X has all moments up to the order $k+2$. Furthermore, the structure of these expansions is not simple and convenient in all cases. For instance, the obvious fact that $EP_k f(Y) = 0$ when $\mathcal{L}(X) = \mathcal{L}(Y)$ needs a special proof when using the definitions of the P_k only. Most general and simple are the Bergström expansions (2.7), (2.9). Unfortunately, these expansions are not too informative. Bentkus (1984d) introduced certain expansions of intermediate type

$$Ef(S_n) = Ef(Y) + a_1(\varepsilon) + \cdots + a_k(\varepsilon) + R, \tag{2.10}$$

where the functions $a_s(\varepsilon)$ satisfy

$$|a_s(\varepsilon)| \le C_s \varepsilon^s$$

with some constants $C_s < \infty$. These expansions have the following properties. They exist more frequently than Edgeworth expansions. They arise in a natural way from Bergström expansions. The closeness of $\mathcal{L}(X)$ and $\mathcal{L}(Y)$ is taken into account and therefore it is possible to pass to the case of stable Y. They can be used to derive Edgeworth expansions and they lead to more general and exact results. But a drawback of these expansions is that the terms $a_s(\varepsilon)$ are not uniquely determined in contrast to the power series expansion. In order to describe the construction of $a_s(\varepsilon)$, consider the random differential operator

$$Q_i = Q_i(l^{(i)}) = (D^{l_1}(X_1) - D^{l_1}(Y_1)) \cdots (D^{l_i}(X_i) - D^{l_i}(Y_i)),$$

where $l^{(i)} = (l_1, \dots, l_i)$ is a non-negative multi-index. If a mapping f has sufficiently many bounded derivatives, then one can define

$$a_s(\varepsilon) = \sum_{i=1}^{s} \sum_{|l^{(i)}|=2i+s} \left(n^{-i} \binom{n}{i} / l^{(i)}! \right) n^{-s/2} EQ_i f(\tau Y),$$

where $|l^{(i)}| = l_1 + \cdots l_i$, $\tau = (1 - i/n)^{1/2}$ and the second sum is over the $l^{(i)}$ satisfying $l_1 \ge 3, \dots, l_i \ge 3$. Here $l^{(i)}! = l_1! \cdots l_i!$ and $\binom{n}{i} = n!/[i!(n-i)!]$.

Put $X^t = X\chi\{||X|| \le \sqrt{n}\}$, $X_t = X - X^t$,

$$L_p = n^{-(p-2)/2} E||X^t||^p, \ \Lambda_p = n^{-(p-2)/2} E||X_t||^p.$$

Note that

$$\Lambda_{k+2} + L_{k+3} \le \varepsilon^{k+1} E||X||^{k+3}.$$

Lemma 2.4. *Under the conditions of Lemma 2.1,*

$$||a_s(\varepsilon)|| \leq C\varepsilon^s E||X||^{s+2},$$

$$\left\| \sum_{s=1}^{k} [a_s(\varepsilon) - \varepsilon^s E P_s f(Y)] \right\| \leq C(\Lambda_{k+2} + L_{k+3}),$$

where the constant $C = C(k, \mathcal{L}(Y))$.

Sometimes it is very useful to apply a truncation procedure, i.e., to replace $Ef(S_n)$ by $Ef(S_n^t)$, where $S_n^t = n^{-1/2}(X_1^t + \cdots X_n^t)$. Usually the difference $Ef(S_n) - Ef(S_n^t)$ can be easily estimated. As an example we state the following almost obvious Lemma 2.5 for bounded f (see Sazonov and Zalesskii (1985), and Bentkus (1986d) for the results concerning the case of unbounded f).

Lemma 2.5. *The following estimate holds:*

$$||Ef(S_n) - Ef(S_n^t)|| \leq 2n||f||_\infty P\{||X|| \geq \sqrt{n}\}.$$

Lemma 2.5 (and its generalizations) allows one to replace the analysis of asymptotic expansions for $Ef(S_n)$ by the analysis of the asymptotic expansion

$$Ef(S_n^t) = Ef(Y) + a_1^t(\varepsilon) + \cdots + a_k^t(\varepsilon) + R, \tag{2.11}$$

where the $a_s^t(\varepsilon)$ are determined by replacing X_1, X_2, ... in the definition of $a_s(\varepsilon)$ by X_1^t, X_2^t, ...

Lemma 2.6 (Compare with Lemma 2.1). *There exists* $\alpha = \alpha(\mathcal{L}(Y)) > 0$ *such that the condition*

$$\sup_{x \in B} \exp(-\alpha||x||^2)||f^{(i)}(x)|| < \infty, \; 0 \leq i \leq 3k,$$

ensures the existence of the asymptotic expansion (2.11). Moreover,

$$||Ef(Y)|| \leq C,$$
$$||a_s^t(\varepsilon)|| \leq C(\Lambda_2 + L_{s+2}), \; 1 \leq s \leq k,$$

where $C = C(k, \mathcal{L}(Y))$.

Theorem 2.7 (Compare with Theorem 2.2). *Suppose that a r.e.* X *satisfies the CLT and that* $E||X||^2 < \infty$. *Then there exists* $\alpha = \alpha(\mathcal{L}(Y)) > 0$ *such that the condition*

$$\sup_{x \in B} \exp(-\alpha||X||)||f^{(i)}(x)|| < \infty, \; 0 \leq i \leq 3k+3,$$

ensures the estimate

$$||R|| \leq C(\mathcal{L}(X), k)(\Lambda_2 + L_{k+3})$$

of the remainder term in (2.11). If in addition $E||X||^{k+2} < \infty$, *then*

$$||a_s^t(\varepsilon) - a_s(\varepsilon)|| \leq C\Lambda_{s+2}, \; 1 \leq s \leq k.$$

Further details and results concerning expansions (2.10), (2.11) may be found in Bentkus (1984d, 1986d).

We say that a Banach space B is of the class C_b^s (briefly $B \in C_b^s$) if the norm function $g(x) = ||x||$ is s-times continuously Fréchet differentiable on the open set $B - \{0\}$ and $\sup\{||g^{(i)}(x)|| : ||x|| = 1\} < \infty$, $1 \le i \le s$. In the next theorem concerning expansions for moments we assume that a r.e. X (or Y) is not finite-dimensional. This means that $P\{X \in E\} = 0$ if $E \subset B$ is a finite-dimensional subspace of B. This assumption does not restrict the generality since otherwise we may apply finite-dimensional results.

Theorem 2.8. *Suppose that a Banach space $B \in C_b^{3k+3}$. Then the asymptotic expansion*

$$Ef(S_n) = Ef(Y) + \sum_{s=1}^{k} \varepsilon^s EP_s f(Y) + R$$

is well defined and the remainder term satisfies $R = O(n^{-(k+1)/2})$ in the following cases:

(i) $f(x) = ||x||^p$ *for some* $p > 0$, $p < k+3$, *and* $E||X||^{k+3} < \infty$;
(ii) $f(x) = ||x||^p$ *for some* $p \ge k+3$ *and* $Ef(X) < \infty$;
(iii) $f(x) = \exp(a||x||^\alpha)$ *for some* $a > 0$, $0 < \alpha < 1$, *and* $Ef(X) < \infty$.

The first estimate of the convergence rate for the moments in an infinite-dimensional Hilbert space was given by Zalesskii and Sazonov (1984). Their proof includes integration of a nonuniform estimate for the probabilities and is based on the use of characteristic functions. Rhee and Talagrand (1984) constructed an example showing that even in a Banach space with very smooth norm, moment conditions alone cannot guarantee any convergence rate for the probabilities. Hence it turns out that the method of Zalesskii and Sazonov (1984) cannot be generalized directly to the case of Banach spaces. Using another method Bentkus (1984g, 1986d) generalized the result of Sazonov and Zalesskii to the case of a Banach space with sufficiently smooth norm and removed certain unnecessary restrictions. The asymptotic expansion of Theorem 2.8 is a consequence of more general and more exact results due to Bentkus (1986d).

§2.3. Asymptotic Expansions for Probabilities

Let A be a subset of a Banach space B. This section is devoted to asymptotic expansions for the probability $P\{S_n \in A\}$. If we choose $f(x) = \chi_A(x)$, then in the i.i.d. case we can rewrite the short Bergström expansion (1.8) in the following form

$$\begin{aligned} P\{S_n \in A\} &= P\{Y \in A\} + n \int_B Ef(Z + \varepsilon x)H(dx) + R \\ &= \nu(A) + n \int_B \Phi(A - \varepsilon x)H(dx) + R, \end{aligned} \tag{3.1}$$

with Gaussian measure $\Phi = \mathcal{L}(Z)$, $Z = \varepsilon\sqrt{n-1}Y$, and $H = \mathcal{L}(X) - \mathcal{L}(Y)$. The construction of asymptotic expansions in the case of smooth f was based on transforming $nE\int_B f(Z+\varepsilon x)H(dx)$ into $\varepsilon ED^3(X)f(Y)/6$ (if we leave the problem of estimating R aside). This, in turn, was based on expanding the smooth function $x \to f(Z+\varepsilon x)$ in a Taylor series in powers of εx. If f is non-differentiable, then such direct expansion is impossible. But assuming that the function

$$x \to \Phi(A - \varepsilon x) \tag{3.2}$$

is sufficiently smooth, we may repeat this argument with respect to the integral on the right-hand side of (3.1). Differentiability of the function (3.2) naturally leads to the notion of differentiable measure (see Averbuch, Smolyanov and Fomin (1971) and Daletskii and Fomin (1983) for more information concerning differentiable measures).

Suppose that $\Phi : \mathcal{A} \to \mathbb{R}$ is a set function (not necessarily additive or σ-additive) defined on a class $\mathcal{A}$ of subsets $A \subset B$. Let us suppose also that $\mathcal{A}$ is invariant under translations, $A \in \mathcal{A}$, $h \in B \Rightarrow A + h \in \mathcal{A}$. Then we can define (if it exists) the (directional) derivative

$$D(h)\Phi(A) = \lim_{t\to 0}\{\Phi(A - th) - \Phi(A)\}/t. \tag{3.3}$$

We mention that the traditional definition would require $\Phi(A + th)$ in (3.3). We did choose $\Phi(A - th)$ to keep the same notation for asymptotic expansions as in the previous sections. The first derivative $D(h)\Phi : \mathcal{A} \to \mathbb{R}$ is a set function. Hence we can define the successive derivatives $D(h_1)\cdots D(h_s)\Phi$ iteratively. This allows one to well define such quantities as, e.g., $EP_s\nu(A)$, where $P_s = P_s(D(X_1), \ldots, D(X_s))$ is the random differential operator determined via the Edgeworth–Cramér polynomials. In the same way all formulas from the previous section can be correctly interpreted for the case of the non-smooth function $f(x) = \chi_A(x)$.

Concerning asymptotic expansions in the infinite-dimensional case, it is generally accepted to define the set A through a function $F : B \to \mathbb{R}$, i.e., to let $A = A_{r,F} = \{x \in B : F(x) < r\}$. The results obtained so far concern cases where F is sufficiently smooth or $F(x) = ||x||^p$, $p > 0$, in a Banach space with sufficiently smooth norm. Let us start with the case $F(x) = ||x||$, the norm of Hilbert space H. Let $V_{r,a} = \{x \in H : ||x - a|| < r\}$.

Theorem 3.1. *Suppose that r.e. $X \in H$ and is not finite-dimensional. If $E||X||^p < \infty$ for some $2 \le p < 3$, then for all $a \in H$, $r \in \mathbb{R}$*

$$P\{S_n \in V_{r,a}\} = P\{Y \in V_{r,a}\} + R \tag{3.4}$$

with $R = o(n^{-(p-2)/2})$. *If* $E||X||^p < \infty$ *for some* $3 \le p < 4$, *then*

$$P\{S_n \in V_{r,a}\} = P\{Y \in V_{r,a}\} + \frac{1}{6}\varepsilon ED^3(X)\nu(V_{r,a}) + R, \tag{3.5}$$

with $R = o(n^{-(p-2)/2})$. *Furthermore,* $ED^3(X)\nu(V_{r,a}) = 0$ *if* $a = 0$ *or if* X *is symmetric.*

We note that the theorem is valid without conditions like Cramér's condition for the c.f. The paper by Bentkus and Zalesskii (1985) contains an example with an infinite-dimensional $X \in H$ which shows that the distribution function $r \to P\{||S_n(X)|| < r\}$ has a jump bigger than C/n with some $C > 0$. Therefore the asymptotic expansions longer then those in Theorem 3.1 should contain additional discontinuous (with respect to r) terms or one should impose a condition similar to Cramér's condition. The following condition was introduced by Bentkus (1984e), and Nagaev and Chebotarev (1986).

The Cramér-type condition. There exists an operator $K \ge 0$ such that the operator K *cov* X is not finite-dimensional and

$$\limsup_{r\to\infty} \sup\{|E\exp(i(x, X))| : (Kx, x) = r\} < 1. \tag{3.6}$$

Remark 3.2. All of the results below hold if the range of the operator K *cov* X is of finite but sufficiently large dimension.

We note that one can never choose K to be the identity operator since it is easy to show that every r.e. X in an infinite-dimensional H satisfies (for each $r \ge 0$)

$$\sup\{|E\exp(i(x, X))| : ||x|| = r\} = 1.$$

Consider the condition

$$\int_{n^{3/4}\le|t|\le T} \frac{1}{|t|}|E\exp(it||S_n(X)||^2)|dt = O(1/T), \tag{3.7}$$

where T will be chosen later so that $1/T$ has the order of the desired error.

Usually one can verify (3.7) if Cramér's condition (3.6) is fulfilled (see Bentkus (1984e) and Nagaev and Chebotarev (1986)). Condition (3.7) is expressed in terms of the whole sum $S_n(X)$. Generally this is not accepted as a solution of the problem. Usually an estimate of the error is preferable when it is expressed in terms of one summand. We formulate the results while imposing (3.7) because it happens to be more convenient in applications (see, e.g., § 3.1 concerning ω^2 statistics); we are able to verify (3.7) but not the Cramér condition (3.6).

Theorem 3.3. *Let a r.e.* $X \in H$ *and let integer* $k \ge 2$. *Suppose that* $E||X||^{k+2} < \infty$. *Then the asymptotic expansion*

$$P\{S_n \in V_{r,a}\} = P\{Y \in V_{r,a}\} + \sum_{s=1}^{k} \varepsilon^s EP_s\nu(V_{r,a}) + R \tag{3.8}$$

is well defined. Here the P_s are the random differential operators determined via Edgeworth–Cramér polynomials. Furthermore, suppose that condition (3.6) is fulfilled. If $E||X||^{k+2+\alpha} < \infty$, where $0 \le \alpha \le 1$, then

$$R = o(n^{-(k+\alpha)/2})$$

and if $E||X||^{k+3} < \infty$, then

$$R = O(n^{-(k+1)/2}).$$

Moreover the result remains valid if instead of (3.6) condition (3.7) is fulfilled with any $T = o(n^{(k+\alpha)/2})$, $0 \le \alpha < 1$, or $T = O(n^{(k+1)/2})$, respectively.

The method of proving Theorem 3.1 and Theorem 3.3 was developed by Götze (1979, 1981, 1984), Yurinskii (1981, 1982), Zalesskii (1982), Nagaev (1983) and Bentkus (1984e). In these and later papers, various estimates of the remainder were obtained. Let us give a short review not concerning the first part of Theorem 3.1 (see § 1.1 for the convergence rates). The paper of Götze (1979) contains the very important symmetrization inequality (see § 1.1) which allows one to estimate the c.f. $E\exp(it||S_n + a||^2)$ for $|t| \le n^{1-\varepsilon}$, $\varepsilon > 0$. Also this paper gives bounds for the remainder term exact with respect to n but under somewhat restrictive moment conditions. Zalesskii (1982) proved Theorem 3.1 for $a = 0$. Theorem 3.1 and Theorem 3.3 follow from the results due to Bentkus (1984e). In that paper Theorem 3.1 and Theorem 3.3 are generalized to the case of sets $V_{r,a} = \{x \in H : w(x+a) < r\}$, where $w : H \to \mathbb{R}$ is a polynomial, second-degree and under appropriate conditions, the estimate

$$R = O_s\left((1 + ||a||^{3k+3})(n^{-s} + E||X_t||^2 + n^{-(k+1)/2}E||X^t||^{k+3})\right) \tag{3.9}$$

is obtained, valid for every $s > 0$. We recall that $X^t = X\chi\{||X|| \le n^{1/2}\}$ and $X_t = X - X^t$. Sazonov and Zalesskii (1985) developed truncation techniques adapted to the nonuniform estimates of the remainder. Using these techniques Bentkus and Zalesskii (1985) inserted the factor $(1+\rho)^{-m}$, $m > 0$, in the estimate (3.9) provided $E||X||^m < \infty$, where ρ denotes the distance between $0 \in H$ and the boundary of the set $V_{r,a}$. Nagaev and Chebotarev (1986, 1987, 1989a, b) considerably sharpened the estimate of the remainder in Theorems 3.1 and 3.3. These papers contain also the Bergström expansions in Hilbert space (see also Bentkus (1984e) and Bentkus and Zalesskii (1985) for expansions of an intermediate type). Papers of Sazonov, Ul'yanov and Zalesskii (1987a, b), Sazonov and Ul'yanov (1991) are devoted to a detailed investigation of uniform and nonuniform estimates of the remainders in Theorems 3.1 and 3.3. The case of non-identically distributed summands was investigated by Bentkus (1984e). Ul'yanov (1987) removed certain unnecessary restrictions from Bentkus (1984e) and obtained a more exact estimate of the remainder. Asymptotic expansions are also treated by Koroliuk and Borovskikh (1984).

We now state one result concerning asymptotic expansions in Banach spaces. Let $A_r = \{x \in B : F(x) < r\}$, where F is a functional $F : B \to \mathbb{R}$. Let

$$\sigma_{k+1}^2 = E_{k+1}(F^{(k+1)}(Y)Y_1 \cdots Y_{k+1})^2,$$

where the symbol E_{k+1} means that the expectation is taken only with respect to the r.e. Y_{k+1}.

Theorem 3.4. *Suppose that a r.e. $X \in B$ satisfies the CLT. Let a number $\gamma > 0$ and an integer $k \geq 0$ be fixed. Assume that F is $3k+3$ times Fréchet differentiable and that there exists a constant $M \geq 0$ such that*

$$\sup_{x \in B}(1 + ||x||)^{-M}||F^{(j)}(x)|| < \infty$$

for all $j = 0, \ldots, 3k+3$. If

$$P\{\sigma_{k+1}^2 < \delta\} = O(\delta^m)$$

for sufficiently large $m = m(\varepsilon, k)$ as $\delta \downarrow 0$, then $E||X||^{k+2} < \infty$ ensures the existence of the asymptotic expansion

$$P\{S_n \in A_r\} = P\{Y \in A_r\} + \sum_{s=1}^{k} \varepsilon^s EP_s\nu(A_r) + R.$$

Furthermore, if $E||X||^{k+3} < \infty$, then $R = O(\varepsilon^{k+1-\gamma})$.

The theorem is contained in Götze (1989), where more general and more exact results are obtained. Earlier Götze (1983) proved the theorem for $k = 0$. In Götze (1984) the theorem is proved for $F(x) = ||x||^{k+1}$ and $B = L_{k+1}$. Vinogradova (1985) showed that the theorem is valid for symmetric F and $k = 1$.

§2.4. Asymptotic Expansions in the Local Limit Theorem

Let us consider the following Cramér-type condition: there exist a non-negative operator $K : H \to H$ and constants $\rho < \infty$, $\delta > 0$ such that for $r > 0$

$$\sup\{|E\exp(i(x, X))| : (Kx, x) \geq r^2\} \leq \rho r^{-\delta}. \tag{4.1}$$

Clearly, this is stronger than the Cramér condition (3.6) of the previous section. Put $A_{a,r} = \{x \in H : ||x + a||^2 < r\}$.

Theorem 4.1. *Suppose that* $E||X||^{k+2} < \infty$ *for some* $k = 0, 1, \ldots$ *and that for some* $\varepsilon > 0$ *the integral*

$$J_\varepsilon = \int_{|t| \geq n^{1-\varepsilon}} |E \exp(it||S_n + a||^2)| dt$$

exists. Then the asymptotic expansion

$$\frac{d}{dr} P\{S_n \in A_{a,r}\} = \frac{d}{dr} P\{Y \in A_{a,r}\} + \sum_{s=1}^{k} \varepsilon^s \frac{d}{dr} EP_s(A_{a,r}) + R$$

is well defined. If

$$J_\varepsilon = O((1 + ||a||^{k+3}) n^{-(k+1)/2}) \tag{4.2}$$

and $E||X||^{k+3} < \infty$, *then*

$$R = O((1 + ||a||^{k+3}) n^{-(k+1)/2}). \tag{4.3}$$

If Cramér's condition (4.1) *holds with an operator* $K \geq 0$ *such that the range of the operator* $K \operatorname{cov} X$ *is of sufficiently large or infinite dimension, then the integral* J_ε *exists for* n *sufficiently large,* (4.2) *holds and the remainder* R *therefore admits the estimate* (4.3).

From this result one can also derive an asymptotic expansion for $P\{||S_n + a|| < r\}$. The theorem is a consequence of more general and more exact results due to Bentkus (1985a). Earlier Chebotarev (1982) proved under certain conditions that

$$\left| \frac{d}{dr} \{P\{||S_n||^2 < r\} - P\{||Y||^2 < r\}\} \right| = O(n^{-1/2})$$

if $E||X||^3 < \infty$ and the coordinates of X are independent. Recently Bentkus and Zitikis (to appear elsewhere) have generalized Theorem 4.1 to the case of a Banach space B and sets $A_{a,r} = \{x \in B : w(a + x) < r\}$, where $w : B \to \mathbb{R}$ is a polynomial of arbitrary degree.

Chapter 3
Applications

§3.1. Cramér–von Mises Statistics

This section considers the statistics

$$\omega_n^p(q) = n^{p/2} \int_0^1 |F_n(t) - t|^p q(t) dt. \tag{1.1}$$

Here $p > 0$ is an integer, $q : [0,1] \to [0,\infty)$ denotes a Lebesgue measurable weight function, and $F_n(t)$ denotes the empirical distribution function based on an independent random sample $x_1, \ldots, x_n$ from the uniform distribution on $[0,1]$,

$$F_n(t) = n^{-1} \sum_{i=1}^{n} \mathcal{X}\{x_i < t\}.$$

We shall use the name "the Cramér–von Mises" statistics for $\omega_n^p(q)$. Denote the corresponding distribution functions by

$$U_n^p(x; q) = P\{\omega_n^p(q) < x\},$$

and let

$$U^p(x; q) = \lim_{n \to \infty} U_n^p(x; q).$$

Note that the statistics of the type

$$n^{p/2} \int_{-\infty}^{\infty} |F_n(t) - F(t)|^p q(F(t)) dF(t),$$

based on an arbitrary continuous distribution function F, can be reduced to (1.1) by changing variables.

We start with the discussion of results concerning the well-known ω_n^2-test,

$$\omega_n^2 = n \int_0^1 (F_n(t) - t)^2 dt \tag{1.2}$$

since in this case we are able to derive an almost complete set of convergence rate results from the general results obtained in Banach spaces. It is well known (see (Smirnov (1937)) that ω_n^2 converge weakly to

$$\omega^2 = \int_0^1 (W(t) - tW(1))^2 dt,$$

where $W(t)$, $0 \le t \le 1$, is the standard Wiener process. Prokhorov and Sazonov (1969) had noted that the following representation holds:

$$\omega_n^2 = ||S_n||^2, \; S_n = n^{-1/2}(X_1 + \cdots + X_n). \tag{1.3}$$

Here X, X_1, ... $\in L_2(0,1)$ is a sequence of i.i.d. r.v.'s taking values in the Hilbert space $L_2(0,1)$ and

$$X(t) = \mathcal{X}\{x < t\} - t, \tag{1.4}$$

where x is a r.v. uniformly distributed in $[0,1]$. Clearly the X in (1.4) as a r.e. in $L_2(0,1)$ has mean zero and is bounded: $||X|| \le 1/3$. Therefore we may apply the general results obtained in Hilbert space. Indeed, the following theorem is a consequence of these general results (it is sufficient to apply the result due to Götze (1979)).

Theorem 1.1. *There exists an absolute constant C such that*

$$\Delta_n := \sup_{x\in\mathbb{R}} |U_n^2(x) - U^2(x)| \le Cn^{-1}, \tag{1.5}$$

where $U_n^2(x) = P\{\omega_n^2 < x\}$ and $U^2(x) = P\{\omega^2 < x\}$. Moreover,

$$U_n^2(x) = U^2(x) + \sum_{s=1}^{k} a_s(x)n^{-s} + R_k, \tag{1.6}$$

where the $a_s(x)$ are certain known functions and $|R_k| \le C(k)n^{-k-1}$.

The proof of Theorem 1.1 depends strongly on the following estimate. For sufficiently large positive A,

$$|E\exp(it\omega_n^2)| \le C(k,A)(1+|t|)^{-A} \tag{1.7}$$

for $|t| < n$ and $k = 0$ in case (1.5), and for $|t| \le n^{k+1}$ in case (1.6). Now it is known (see Bentkus and Zitikis (1988)) that for every positive A there exist constants $C_1(A)$, $C_2(A)$ such that

$$|E\exp(it\omega_n^2)| \le C_1(A)(1+|t|)^{-A} \tag{1.8}$$

for all $t \in \mathbb{R}$ and $n \ge C_2(A)$.

The remainder term R_k in (1.6) is nonuniform in $|x|$, i.e. for all $A > 0$

$$|R_k| = |R_k(x)| \le C(k,A)n^{-k-1}(1+|x|)^{-A}.$$

This follows from known estimates similar to (1.7) for the derivatives of the characteristic function $E\exp(it\omega_n^2)$ (see, e.g., Bentkus and Zitikis (1988)) and general results in Hilbert space concerning nonuniform estimates (see Sazonov and Zalesskii (1985), Bentkus and Zalesskii (1985)).

It should be noted that the proof of Theorem 1.1 has a long history. Papers by Sazonov (1968, 1969), Rosenkratz (1969), Kiefer (1972), Nikitin (1972), Orlov (1974) and Csörgő (1976) contain results of the type $\Delta_n = O(n^{-\beta})$ with various $\beta < 1/2$. The statistics ω_n^2 were also investigated by Koroliuk and Borovskikh (1984).

Theorem 1.2. *There exist absolute positive constants C_1, C_2 and a function $I(n,t)$* (an analog of the classical "Cramér series" in large deviations results) *such that*

$$1 - U_n^2(x) = (1 - U^2(x))I(n,x)(1 + C_1\theta(1+\sqrt{x})/\sqrt{n}) \qquad (1.9)$$

when $0 \le x \le C_s n$. Furthermore, if $x \le C_2 n^{1/3}$ then

$$1 - U_n^2(x) = (1 - U^2(x))(1 + C_1\theta(1 + x^{3/2})/\sqrt{n}). \qquad (1.10)$$

The quantity θ satisfies $|\theta| \le 1$.

The pioneer papers of Osipov (1977, 1978a, 1978b) had a strong influence on the investigations of large deviations for ω_n^2 and in Hilbert space, as well. Theorem 1.2 is a special case of a general result in Hilbert space due to Yurinskii (1988). That paper contains a construction of the Cramér series $I(n,t)$. Inequality (1.10) also follows from Zalesskii (1989), Račkauskas (1988) and with a less precise estimate $O((1+\sqrt{x})n^{-1/6})$ of the remainder term from Bentkus (1986c).

Recall that C_b^k denotes the class of functions having k continuous and bounded derivatives.

Theorem 1.3 (Bentkus and Zitikis (1988)). *Let $k = n/2 - 1$ if n is even and $k = (n-1)/2$ if odd. The distribution function $U_n^2 \in C_b^k$, but $U_n^2 \notin C_b^{k+1}$. Moreover, for all $m > 0$, $p = 0, 1, \ldots, n \ge 2(p+1)$ and $k = 0, 1, \ldots$,*

$$\sup_{x \ge 0}(1 + x^m)\left|\left(\frac{d}{dx}\right)^p \left\{U_n^2(x) - U^2(x) - \sum_{s=1}^{k} a_s(x) n^{-s}\right\}\right| \le C(m,p,k)n^{-k-1}.$$

The proof of Theorem 1.3 (i.e., the local limit theorem for ω_n^2 with an asymptotic expansion) is based on a detailed analysis of the characteristic functions and strongly depends on estimates such as (1.8). The result that $U_n^2 \in C_b^k, U_n^2 \notin C_b^{k+1}$, slightly improves the corresponding result due to Csörgő and Stacho (1980).

The condition $E||X||^2 < \infty$ is necessary and sufficient for the CLT in Hilbert space (see, e.g., Araujo and Giné (1980)). In the case of $\omega_n^2(q)$ the condition

$$\int_0^1 t(1-t)q(t)dt < \infty \qquad (1.11)$$

is clearly equivalent to $E||X||^2 < \infty$ for the r.e. X in (1.4) considered as an element of the space

$$L_2(q) = L_2((0,1), q(t)dt), \; ||x||^2 = \int_0^1 x^2(t)q(t)dt.$$

Thus (1.11) guarantees the existence of

$$\lim_{n\to\infty} U_n^2(x,q) = U^2(x,q).$$

This result is wellknown and was proved by Chibisov (1964) using other methods. If (1.11) is fulfilled, then the condition $E||X||^{2m} < \infty$, $m \geq 0$, is equivalent to (see, e.g., Zitikis (1989))

$$\int_0^1 \left\{ \int_0^x tq(t)dt \right\}^m dx < \infty,$$
$$\int_0^1 \left\{ \int_x^1 (1-t)q(t)dt \right\}^m dx < \infty. \tag{1.12}$$

On the other hand, one can easily show that the existence of the integrals

$$\int_0^{1/2} \left(\int_x^{1/2} q(t)dt \right)^m dx < \infty,$$
$$\int_{1/2}^1 \left(\int_{1/2}^x q(t)dt \right)^m dx < \infty \tag{1.13}$$

is equivalent to (1.11) and (1.12) if $m \geq 1$. For example, if the function q is symmetric about the point $t = 1/2$ and $q(t) = t^{-\delta}$ with some δ for $0 < t \leq 1/2$ then (1.13) is fulfilled if and only if $\delta < 1 + 1/m$.

Theorem 1.4. *Suppose that* (1.13) *holds for some* $1 \leq m < 2$. *Then*

$$\Delta_n(q) := \sup_{x \geq 0} |U_n^2(x;q) - U^2(x;q)| \leq C(q,m)n^{1-m}.$$

Moreover, $\Delta_n(q) = o(n^{1-m})$ *as* $n \to \infty$.

Theorem 1.4 follows from the known estimates of the convergence rate in Hilbert space (see Götze (1979), Zalesskii (1982)). One has to verify that the corresponding r.e. $X \in L_2(q)$, or equivalently, the limiting Gaussian process $Y(t) = W(t) - tW(1)$, where $W(t)$, $t \in [0,1]$, is the standard Wiener process, is not concentrated in a finite-dimensional subspace of $L_2(q)$. It suffices to show that W is not finite-dimensional since $tW(1)$ is concentrated in a one-dimensional subspace. Clearly, to this end it is enough to construct linear measurable functionals $l_1, \dots, l_m : L_2(q) \to \mathbb{R}$ for every integer $m \geq 1$ such that the Gaussian r.v.'s $l_1(W), \dots, l_m(W)$ are independent and nondegenerate. Without loss of generality we may assume that $meas\{t : q(t) > 0\} > 0$. Therefore there exist pairwise disjoint intervals $(a_i, b_i) \subset (0,1)$, $1 \leq i \leq m$, such that $\int_{a_i}^{b_i} q(t)dt > 0$. Thus we can put

$$l_i(W) = \int_{a_i}^{b_i} (W(t) - W(a_i))q(t)dt.$$

Consider the condition

$$\inf\{q(t) : t \in (0,1)\} > 0. \tag{1.14}$$

We shall assume that q satisfies the following condition: there exist a finite number of points $t_0 = 0 < t_1 < \cdots < t_N < t_{N+1} = 1$ such that the function q is monotone and either convex or concave on every interval (t_{i-1}, t_i). Moreover, we suppose that there exist numbers $\alpha \geq 0$ and $c(\alpha) < \infty$ such that

$$\max\{q(t_{i-1}+\varepsilon);\ q(t_i-\varepsilon)\} \leq c(\alpha)\varepsilon^{-\alpha} \tag{1.15}$$

for all $\varepsilon > 0$ such that $2\varepsilon < t_i - t_{i-1}$.

Theorem 1.5 (Zitikis (1989)). *Suppose that conditions* (1.13) *with* $m = 2$ *and* (1.14), (1.15) *hold. Then*

$$\Delta_n(q) \leq c(q)n^{-1}.$$

Moreover, if (1.13) *holds with* $m = \mu(1+\varepsilon) \geq 2$ *for some* $\varepsilon > 0$ *and* $\mu = 1, 2, \ldots$, *then*

$$\sup_{x>0} x^{\mu}|U_n^2(x;q) - U^2(x;q)| \leq c(q,\varepsilon,\mu)n^{-1}.$$

The paper of Zitikis (1989) also contains asymptotic expansions for $U_n^2(\cdot;q)$ and its derivatives. Estimates of the remainder term obtained are uniform and nonuniform. These expansions are found under the additional condition that q' satisfies (1.15).

Concerning the large deviations results, one can replace U_n^2 by $U_n^2(\cdot;q)$ in Theorem 1.2 if the following condition is satisfied: there exists a positive C such that

$$\int_{1/2}^{1} \exp\left(C\left[\int_{1/2}^{\tau} q(t)dt\right]^{1/2}\right) d\tau < \infty,$$

$$\int_{0}^{1/2} \exp\left(C\left[\int_{\tau}^{1/2} q(t)dt\right]^{1/2}\right) d\tau < \infty. \tag{1.16}$$

This replacement is possible since the general result of Yurinskii (1988) is valid when $E\exp(c||X||) < \infty$ for some $c > 0$, which in our case is equivalent to (1.16).

Let us now discuss the situation when $p \neq 2$. Similarly to the representation (1.3), one can write

$$\omega_n^p(q) = ||S_n||^p \tag{1.17}$$

with the same r.e. X of (1.4) but considered as an element of the space $L_p = L_p(q)$, $p \geq 1$, of functions $f : [0,1] \to \mathbb{R}$ with finite integral

$$||f||^p = \int_0^1 |f(t)|^p q(t)dt.$$

The weak convergence of the distributions $U_n^p(x;q)$ to $U^p(x;q)$ was investigated by Csörgő and Horvath (1988) and Norvaiša (1990). Corresponding deep and general results on the convergence rate in the CLT in Banach spaces (see, e.g., Götze (1989), (1984) and Zalesskii (1988)) were obtained under certain "variance" conditions (see § 1.3). Unfortunately up till now this condition still has not been expressed in terms of the weight function q, with the exception of certain special cases, for example, $q(t) \equiv 1$. In this particular case we have the following result.

Theorem 1.6 (Götze (1984)). *Suppose that $p \geq 4$ is even and that $q(t) \equiv 1$. Then*

$$\sup_{x \geq 0} |U_n^p(x;1) - U^p(x;1)| \leq C(p)n^{-1}.$$

Moreover, one can expand $U_n^p(x;1)$ in an asymptotic series with a remainder term of the order $O(n^{\varepsilon - p/2})$, $\varepsilon > 0$.

For further details and results, see Götze (1984). Thus, for $U_n^p(x,q)$, $p \neq 2$, we have the Berry–Esseen estimates and large deviations results with a remainder term of order $O(n^{-1/6})$ only.

Theorem 1.7. *If $p \geq 3$ and*

$$\int_0^1 t^p(1-t)^p q(t)dt < \infty,$$

$$\int_{1/2}^1 \left(\int_{1/2}^x q(t)dt\right)^{3/p} dx < \infty,$$

$$\int_0^{1/2} \left(\int_x^{1/2} q(t)dt\right)^{3/p} dx < \infty, \tag{1.18}$$

then

$$\sup_{x \geq 0}(1 + x^{3/p})|U_n^p(x;q) - U^p(x;q)| \leq C(q)n^{-1/6}. \tag{1.19}$$

When $1 < p < 3$, estimate (1.19) is valid if there exists $\alpha > 0$, $\alpha < p/3$, such that

$$\int_0^1 t^\alpha(1-t)^\alpha q(t)dt < \infty.$$

The theorem is found in Paulauskas and Račkauskas (1991) for $p \in (1,3)$. For $p \geq 3$ the result follows from Paulauskas and Račkauskas (1991) provided that the distribution function $x \to (U^p(x;q))^{1/p}$ has a bounded density and that the corresponding random variable $X \in L_p$ has a third moment. The first condition follows from general results of Davydov and Lifshits (1984) and Rhee and Talagrand (1986). The second one is a consequence of the identity

$$E\phi(||X||) = \int_0^1 \phi\left(\left\{\int_0^x t^p q(t)dt + \int_x^1 (1-t)^p q(t)dt\right\}^{1/p}\right) dx, \tag{1.20}$$

valid for every $\phi : [0,\infty) \to [0,\infty)$. When the function $\phi(t) = t^3$, (1.20) is equivalent to (1.18).

The following theorem is a consequence of more general results due to Bentkus and Račkauskas (1990) (if $1 < p < 3$, one must use additionally the result of Paulauskas and Račkauskas (1991) about differentiable functions in L_p).

Theorem 1.8. *If*

$$\int_0^1 t(1-t)q(t)dt < \infty$$

when $p \geq 3$ *and*

$$\int_0^1 t^\alpha (1-t)^\alpha q(t)dt < \infty$$

for some $\alpha > 0$, $\alpha < p/3$, *when* $1 < p < 3$, *then the limiting distribution function* $U^p(x;q)$ *exists and the following three statements are equivalent:*

(a) *there exists* $h > 0$ *such that*

$$\int_{1/2}^1 \exp\left(h \left[\int_{1/2}^x q(t)dt \right]^{1/2p} \right) dx < \infty,$$

$$\int_0^{1/2} \exp\left(h \left[\int_x^{1/2} q(t)dt \right]^{1/2p} \right) dx < \infty;$$

(b) *there exist constants* $M_i = M_i(p,q) > 0$, $i = 1,2$, *such that*

$$\left| \frac{1 - U_n^p(x;q)}{1 - U^p(x;q)} - 1 \right| \leq M_1 (1 + x^{1/p}) n^{-1/6},$$

when $0 \leq x \leq M_2 n^{p/6}$;

(c) *for each function* $f : \mathbb{R} \to \mathbb{R}$ *such that* $f(n) \to 0$, *as* $n \to \infty$,

$$\frac{1 - U_n^p(x;q)}{1 - U^p(x;q)} \to 1 \quad \textit{as } n \to \infty,$$

uniformly for $0 \leq x \leq f(n)n^{p/6}$.

§3.2. L-Statistics

This section discusses only recent work on large deviations for L-statistics since these results were obtained by reducing the problem to results in Banach spaces or by applying appropriate techniques. We refer to Stigler (1974), Serfling (1980), Helmers (1982), Bhattacharya and Denker (1990), Norvaiša and Zitikis (1991), and Zitikis (1991a) for other results concerning L-statistics.

Let $X_1, \ldots, X_n$ be independent copies of a r.v. $X \in \mathbb{R}$ with distribution function F. Consider the L-statistic

$$\ell_n = n^{-1} \sum_{i=1}^{n} c_{in} X_{i:n},$$

where $X_{1:n} \leq \cdots \leq X_{n:n}$ are the order statistics of $X_1, \ldots, X_n$ and $c_{1n}, \ldots, c_{nn}$ are certain real coefficients.

When the coefficients are generated by a weight function $J : [0,1] \to \mathbb{R}$ (see, e.g., Stigler (1974)), that is

$$c_{in} = c_{in}^0 = n \int_{(i-1)/n}^{i/n} J(u)du,$$

we shall denote the corresponding L-statistic by ℓ_n^0. We shall assume that there exists an $L \geq 0$ such that

$$|J(u) - J(v)| \leq L|u - v|$$

for all $u, v \in [0,1]$. It is known (see Stigler (1974)) that if $EX^2 < \infty$, then there exist finite limits

$$\mu = \lim_{n\to\infty} E\ell_n^0, \ \sigma^2 = \lim_{n\to\infty} nE(\ell_n^0 - \mu)^2,$$

where

$$\mu = \int_{-\infty}^{\infty} xJ(F(x))dF(x),$$

$$\sigma^2 = \int_{-\infty}^{\infty} \int_{-\infty}^{\infty} J(F(x))J(F(y))\{F(x \wedge y) - F(x)F(y)\}dxdy.$$

We shall assume that $\sigma > 0$. This guarantees that the limiting distribution of the r.v. $n^{-1/2}(\ell_n^0 - \mu)/\sigma$ is the standard normal distribution $\Phi(x)$.

Theorem 2.1. *If for some* $h > 0$

$$E\exp(h|X|^{1/2}) < \infty, \tag{2.1}$$

then for each sequence $b_n \to 0$

$$P\{\sqrt{n}(\ell_n^0 - \mu) > \sigma x\}/[1 - \Phi(x)] \to 1 \quad \text{as } n \to \infty, \tag{2.2}$$

uniformly for $0 \leq x \leq b_n n^{1/6}$.

Note that the theorem (and the results below) remains valid for the quotient

$$P\{\sqrt{n}(\ell_n^0 - \mu) < \sigma x\}/\Phi(x)$$

when $-b_n n^{1/6} \leq x \leq 0$. Indeed, it is sufficient to replace J by $-J$.

Assertions (2.1) and (2.2) are equivalent when $J \equiv const. \neq 0$. In this case the statistic ℓ_n^0 reduces to a sum of i.i.d. r.v.'s and the equivalence is proved, for example, in Bentkus and Račkauskas (1990). The same remark applies to Theorem 2.2 below.

Let us fix a number a, $0 < a \leq 1/2$, and put $x_n = n^{a/(4-2a)}$.

Theorem 2.2. *If there exists $h > 0$ such that*

$$E\exp(h|X|^a) < \infty, \tag{2.3}$$

then for each sequence $b_n \to 0$

$$P\{\sqrt{n}(\ell_n^0 - \mu) > \sigma x\}/[1 - \Phi(x)] \to 1 \quad \text{as } n \to \infty,$$

uniformly for $0 \le x \le b_n x_n$.

Theorem 2.1 and Theorem 2.2 are stable with respect to small perturbations of the coefficients $c_{11}^0, \dots, c_{nn}^0$ in the following sense.

Theorem 2.3. *Theorems 2.1 and 2.2 remain valid if the coefficients $c_{1n}^0, \dots, c_{nn}^0$ are replaced by $c_{1n}, \dots, c_{nn}$ such that*

$$\sum_{i=1}^{n} |c_{in} - c_{in}^0|^3 = O(n^{-1}) \tag{2.4}$$

in the case of Theorem 2.1 and

$$\sum_{i=1}^{n} |c_{in} - c_{in}^0|^{(2-a)/(1-a)} = O(n^{a/(a-1)}) \tag{2.5}$$

in the case of Theorem 2.2. In both cases the statistic ℓ_n^0 should be replaced by ℓ_n.

The application of Hölder's inequality shows that (2.4) implies (2.5). Theorems 2.1–2.3 were proved by Bentkus and Zitikis (1990). The proofs are based on the representation $\ell_n = S_n + \omega_n^2 + R$, where S_n denotes a sum of i.i.d. $\mathbb{R}$-valued r.v.'s, ω_n^2 is an ω^2-statistic and R is a remainder term. The estimation of S_n is based on the well-known one-dimensional results (see, e.g., Ibragimov and Linnik (1965), Petrov (1972), and Bentkus and Račkauskas (1990)). The estimation of ω_n^2 is based on large deviations results in Hilbert space by Bentkus and Račkauskas (1990) (see also § 3.1). Using the same techniques Zitikis (1991b) showed that the speed of convergence in Theorem 2.1 and 2.2 is of the order $(1 + x^3)n^{-1/2}\ln n$. Combining the methods and techniques developed to obtain convergence rates and large deviations theorems in Banach spaces, Zitikis (1991a) proved the next result.

Theorem 2.4. *Condition (2.3) guarantees that there exist constants $C = C(a, J)$, and $A = A(a, J) > 0$ such that*

$$|1 - P\{\sqrt{n}(\ell_n^0 - \mu) > \sigma x\}/[1 - \Phi(x)]| \le C(1 + x^3)n^{-1/2}$$

for $0 \le x \le A x_n$.

Cramér-type large deviations for L-statistics were investigated by Vandamaele and Veraverbeke (1982), Puri and Seoh (1987), and Aleškevičienė (1989).

§3.3. Kolmogorov–Smirnov Statistics

Let F_n be the empirical distribution function based on a sample $x_1, \ldots, x_n$ of observations from a uniform distribution on $[0,1]$ and let

$$D_n(t) = n^{1/2}(F_n(t) - t).$$

In the previous sections we have seen that it is useful to consider $\{D_n, n \geq 1\}$ as a sequence of r.e.'s in L_p. Here we shall consider D_n as a r.e. in the Skorokhod space $D[0,1]$,

$$D_n(t) = n^{-1/2} \sum_{i=1}^{n} X_i(t),$$

where X, X_1, X_2, ... are i.i.d. r.e.'s in $D[0,1]$ having the same distribution as the process $X(t) = \mathcal{X}\{x < t\} - t$, $t \in [0,1]$, with r.v. x uniformly distributed on $[0,1]$. Therefore the weak convergence of D_n (that is, the classical Donsker theorem) is nothing more than the CLT in $D[0,1]$ for i.i.d. summands having the special structure. Any of Theorems 4.7–4.10 of Chapter 1 can be easily applied to D_n. For example, to apply Theorem 4.7 the following two conditions are sufficient:

$$E(X(t) - X(s))^2 \leq t - s,$$

$$E(X(u) - X(t))^2(X(t) - X(s))^2 \leq (u - s)^3, \ 0 \leq s \leq t \leq u \leq 1.$$

Of course, this application has only methodological importance. Here we shall apply the results of § 1.4 to weighted empirical processes. To this end some new definitions are required. Let Q be the class of functions $q : [0,1] \to \mathbb{R}^+$ continuous and increasing on $[0, 1/2]$ and symmetric about the point $1/2$: $q(t) = q(1-t)$, $0 \leq t \leq 1/2$. Let

$$W_0(t) = W(t) - tW(1), \ 0 \leq t \leq 1,$$

be the Brownian bridge, where W is the standard Wiener process. The weighted empirical process and the weighted Brownian bridge are given by

$$D_{n,q}(t) = D_n(t)/q(t) = n^{-1/2} \sum_{i=1}^{n} X_{i,q}(t),$$

$$W_{0,q}(t) = W_0(t)/q(t),$$

respectively, where $X_{i,q}(t) = X_i(t)/q(t)$. Chibisov (1964) (see also O'Reilly (1974)) proved that for $q \in Q$ the distributions $\mathcal{L}(D_{n,q})$ converge weakly to $\mathcal{L}(W_{0,q})$ if and only if

$$\int_0^{1/2} \exp(-\varepsilon q^2(t)/t) t^{-1} dt < \infty \tag{3.1}$$

for all $\varepsilon > 0$. We shall compare this result with the one obtained by applying Theorem 4.8 of Chapter 1. Suppose in addition that $t^{1/2}/q(t)$ is a nondecreasing function. Performing some simple estimates (for details, see Paulauskas and Stieve (1990)), we have for all $0 \le s \le t \le u \le 1$

$$E(X_q(t) - X_q(s))^2 \le C \int_s^t v^{-1/2}/q(v)dv, \tag{3.2}$$

$$E\left((X_q(t) - X_q(s))^2 \wedge 1\right)(X_q(u) - X_q(t))^2 \le C\left(\int_s^u v^{-1/2}/q(v)dv\right)^2 \tag{3.3}$$

where $X_q(t) = X(t)/q(t)$. Therefore, if

$$\int_0^{1/2} v^{-1/2}/q(v)dv < \infty, \tag{3.4}$$

then the conditions of Theorem 4.8 are satisfied with $\alpha_1 = 1$, $\alpha_2 = 2$ and $F_1(t) = F_2(t) = C\int_0^t v^{-1/2}/q(v)dv$ and this implies the weak convergence of $\mathcal{L}(D_{n,q})$. It follows that condition (3.4) is very close to the optimal O'Reilly–Chibisov condition.

We close this section with a result implied by Theorem 4.11 of Chapter 1 applied to weighted empirical processes. Let

$$\Delta_{n,q}(x) = |P\{||D_{n,q}||_\infty < x\} - P\{||W_{0,q}||_\infty < x\}|.$$

Theorem 3.1 (Paulauskas and Stieve (1990)). *Suppose that $q \in Q$ and for some $0 \le \delta \le 2/3$ and $0 < \gamma < 1$ the function $t^{1/(3+\delta)}/q(t)$ is increasing and*

$$\sup_{0 \le t \le 1/2} t^{(1-\gamma)/(3+\delta)}/q(t) < C.$$

Then there exists a constant $C_0 = C_0(q)$ such that

$$\Delta_{n,q}(x) \le C_0 n^{-(1+\delta)/10}(1+x)^{-3}(\ln n + \ln(1+x))^2. \tag{3.5}$$

To prove this one needs to verify the conditions of Theorem 4.12 of Chapter 1, among them condition (4.3) for $W_{0,q}$. See Paulauskas (1990) and Paulauskas and Stieve (1990) for details.

Here it is necessary to note that there exists a vast literature on weighted empirical and quantile processes (see, for example, Csörgő et al. (1986) and the references therein). Most of these papers employ the so-called Hungarian (or KMT) construction. Using these results, one can easily deduce the following: if $q(t) = t^\alpha$, $1/4 < \alpha < 1/2$, and condition (4.3) is fulfilled for $W_{0,q}$, then

$$\sup_{x \in \mathbb{R}} \Delta_{n,q}(x) \le Cn^{-1/2+\alpha}. \tag{3.6}$$

Hence the rate $n^{-1/6}$ in (3.6) corresponds to the rather natural weight function $t^{1/3}$ while in (3.5) for the same rate we have to choose the function $t^{3/11}$.

§3.4. Empirical Processes

In this section we shall state two estimates of the convergence rate in the CLT for general empirical processes. There exist a large number of papers devoted to the CLT itself. We refer the reader to survey papers by Dudley (1978), Gaensler and Stute (1979), and Giné and Zinn (1984).

Let $X, X_1, X_2, \ldots$ be i.i.d. r.e.'s taking values in a measurable space $(\mathcal{X}, \mathcal{A})$. Let $\mu = \mathcal{L}(X)$ and $\mu f = \mu(f) = \int_{\mathcal{X}} f(x)\mu(dx)$. Let μ_n and E_n denote the empirical measure and empirical process, respectively, associated with μ, i.e.,

$$\mu_n = n^{-1}\sum_{i=1}^{n} \delta_{X_i},\ E_n = \sqrt{n}(\mu_n - \mu),$$

where δ_x denotes the point mass at $x \in \mathcal{X}$. We shall consider the empirical process $E_n(f)$, $f \in \mathcal{F}$, indexed by some class $\mathcal{F} \subset L_2(\mathcal{X}, \mathcal{A}, \mu)$ of measurable functions $f : \mathcal{X} \to \mathbb{R}$. Let us define the pseudo-metrics ρ_μ and $e_{\mu,p}$, $1 \le p \le \infty$, on $\mathcal{F}$ as follows:

$$\rho_\mu(f,g) = (\mu(f-g)^2 - \mu^2(f-g))^{1/2},$$

$$e_{\mu,p}(f,g) = (\mu|f-g|^p)^{1/p}, \text{ if } 1 \le p < \infty,$$

$$e_{\mu,\infty}(f,g) = ess\ sup|f-g|,$$

where the *ess sup* is taken with respect to the measure μ. We suppose that the class $\mathcal{F}$ is μ-pre-Gaussian. This means that there exists a zero-mean Gaussian process $B_\mu(f)$, $f \in \mathcal{F}$, with covariance

$$EB_\mu(f)B_\mu(g) = \mu(fg) - \mu(f)\mu(g),\ f,g \in \mathcal{F},$$

which has a version with bounded and uniformly continuous sample functions with respect to ρ_μ. In order to avoid measurability problems we suppose that the class $\mathcal{F}$ is countably determined (for μ). That is, we suppose that there exists a countable subclass $\mathcal{F}_0 \subset \mathcal{F}$ such that $||E_n||_{\mathcal{F}} := \sup_{f\in\mathcal{F}} |E_n(f)| = ||E_n||_{\mathcal{F}_0}$ a.e. for all $n \ge 1$. If the class $\mathcal{F}$ is totally bounded under a pseudo-metric d, then the covering number $N_d(u)$ and the metric entropy $H_d(u)$ of $\mathcal{F}$ are defined as

$$N_d(u) = N(\mathcal{F}, d, u) := \min\{k : \exists f_1, \ldots, f_k \in \mathcal{F} : \min_{1\le i\le k} d(f, f_i) < u \forall f \in \mathcal{F}\},$$

$$H_d(u) = \log N_d(u),\ u \ge 0.$$

The last notion that will be needed to formulate results is the Vapnik–Chervonenkis class. A class $\aleph$ of subsets of $\mathcal{X}$ is said to shatter a finite subset Γ_0 of $\mathcal{X}$ if every $\Gamma \subset \Gamma_0$ is of the form $C \cap \Gamma_0$ for some $C \in \aleph$. $\aleph$ is called a Vapnik–Chervonenkis class (VC for short) if for some $n \ge 1$ no n-element subset of $\mathcal{X}$ is shattered by $\aleph$. The least such n is called the index of $\aleph$ and is denoted by $v(\aleph)$. For a non-negative function $f : \mathcal{X} \to \mathbb{R}$, the set defined by $sub(f) := \{(x,t) \in \mathcal{X} \times \mathbb{R} : 0 \le 1 \le f(x)\}$ is called a subgraph of f. Let

$$D(\mathcal{F}) = \{sub(f) : f \in \mathcal{F}\}.$$

Following Dudley (1987), we call $\mathcal{F}$ a VC-subgraph class (for D) if $D(\mathcal{F})$ is a VC class.

As usual in order to estimate the convergence rate in the CLT, we need the ε-strip condition on the limiting Gaussian process: there exists a constant C such that for all $\varepsilon > 0$ and $r \ge 0$,

$$P\{r \le ||B_\mu||_{\mathcal{F}} \le r + \varepsilon\} \le C\varepsilon(1+r)^{-3}. \tag{4.1}$$

Theorem 4.1 (Norvaiša and Paulauskas (1991)). *Let $\mathcal{F}$ be a countably determined VC subgraph class of functions f such that $0 \le f \le 1$. Assume that (4.1) holds. Then*

$$\begin{aligned}\Delta_n(\mathcal{F}, r) &:= |P\{||E_n||_{\mathcal{F}} < r\} - P\{||B_\mu|| < r\}| \\ &= O((1+r)^{-3} n^{-1/6} \ln^2 n). \end{aligned} \tag{4.2}$$

Theorem 4.2 (Norvaiša and Paulauskas (1991)). *Let $\mathcal{F}$ be a countably determined class of functions f such that $0 \le f \le 1$. Assume that (4.1) is satisfied and that for some C and some $s \in (0, 1/2)$*

$$H(\mathcal{F}, e_{\mu,\infty}, u) \le C u^{-s}, \ u > 0.$$

Then

$$\Delta_n(\mathcal{F}, r) = O((1+r)^{-3} n^{-(1-2s)/6} \ln^s n). \tag{4.3}$$

Let us consider as an example the multivariate empirical process. Namely let

$$\mathcal{X} = \mathbb{R}^d, \ (-\infty, t] = \{s = (s_1, \dots, s_d) \in \mathbb{R}^d : s_1 \le t_1, \dots, s_d \le t_d\},$$
$$\mathcal{F} = \{\chi_{(-\infty,t]}, \ t \in \mathbb{R}^d\}, \ E_n(t) = \sqrt{n}(F_n(t) - F(t)), \ t \in \mathbb{R}^d,$$

where $F(t)$ denotes a distribution function with law μ and F_n denotes the empirical distribution function related to F. Norvaiša and Paulauskas (1991) proved the validity of (4.1) under very mild conditions on F. Combining this with the known fact that the class of lower-left orthants of $\mathbb{R}^d$ is a VC class and is always countably determined (see Wenocur and Dudley (1981)), by Theorem 4.1 we obtain the following result.

Theorem 4.3. *Suppose there is an increasing sequence $t_k \in \mathbb{R}^d$, $k \ge 1$ such that the distribution function F satisfies $0 < F(t_1) < F(t_2) < \cdots < 1$. Then*

$$|P\{\sup_{t \in \mathbb{R}^d} |E_n(t)| < r\} - P\{\sup_{t \in \mathbb{R}^d} |B_F(t)| < r\}| = O((1+r)^{-3} n^{-1/6} \ln^2 n). \tag{4.4}$$

One can compare Theorem 4.3 with results which follow from the weak (or strong) invariance principle (see, e.g., Massart (1986, 1989) and references therein). Here we note only that in (4.4) the order of the remainder term does not depend on d, whereas usually the convergence rates obtained by the Hungarian construction are of order $n^{-1/2d}$. Hence (4.4) is better for $d > 3$. For more details on the comparison of these results and for a more complete list of references on this topic we refer the reader to the above-mentioned papers of Massart (1986, 1989) and Norvaiša and Paulauskas (1991).

References*

Acosta, A de, and Giné, E. (1979): Convergence of moments and related functionals in the general central limit theorem in Banach spaces. Z. Wahrscheinlichkeitstheorie Verw. Geb. **48**, 213–231. Zbl. 395.60008

Aleškevičienė, A. (1989): On large deviations for linear combinations of order statistics. Liet. Mat. Rink. (Litov. Mat. Sb.) **29**, 212–222 (in Russian). Zbl. 682.60021

Aliev, F.A. (1987): A lower bound for the convergence rate in the central limit theorem in Hilbert space. Teor. Veroyatn. Primen. **31**, No. 4, 825–828. English transl.: Theory Probab. Appl. **31**, No. 4, 730–733. Zbl. 623.60013

Aliev, F.A. (1989): On the convergence rate in the central limit theorem in Hilbert space. Teor. Veroyatn. Primen. **34**, No. 2, 407–409. English transl.: Theory Probab. Appl. **34**, No. 2, 361–363. Zbl. 672.60006

Araujo, A., and Giné, E. (1980): The Central Limit Theorem for Real and Banach-Valued Random Variables. John Wiley, New York. Zbl. 457.60001

Asriev, A.V., and Rotar', V.J. (1985): On the convergence rate in the infinite-dimensional central limit theorem for probabilities of hitting parallelepipeds. Teor. Veroyatn. Primen. **30**, No. 4, 652–661. English transl.: Theory Probab. Appl. **30**, No. 4, 691–701. Zbl. 586.60005

Averbukh, V.I., Smolyanov, O.G., and Fomin, S.V. (1971): Generalized functions and differential equations in linear spaces. Tr. Mosk. Mat. O.-va **24**, 133–174. English transl.: Trans. Mosc. Math. Soc. **24**, 140–184. Zbl. 234.28005

Badrikian, A., and Chevet, S. (1974): Measures Cylindriques. Espaces de Wiener et Functions Aléatoires Gaussiennes. Lect. Notes Math 379, Springer-Verlag, Berlin. Zbl. 288.60009

Barsov, S.S. (1985): On the accuracy of the normal approximation to the distribution of a random sum of random vectors. Teor. Veroyatn. Primen. **30**, No. 2, 351–354. English transl.: Theory Probab. Appl. **30**, No. 2, 376–379. Zbl.569.60021

Barsov, S.S. (1987): Rates of convergence to the normal distribution and decrease of the tail of the summand distribution. Teor. Veroyatn. Primen. **32**, No. 2, 356–358. English transl.: Theory Probab. Appl. **32**, No. 2, 329–331. Zbl. 623.60028

* For the convenience of the reader, references to reviews in Zentralblatt für Mathematik (Zbl.), compiled by means of the MATH database, and Jahrbuch über die Fortschritte der Mathematik (Jbuch) have, as far as possible, been included in this bibliography.

Bass, R., and Pyke, R. (1985): The space $D(A)$ and weak convergence for set-indexed processes. Ann. Probab. **13**, 860–864. Zbl. 585.60007

Bass, R., and Pyke, R. (1987): A central limit theorem for $D(A)$-valued processes. Stoch. Processes Appl. **24**, 109–131. Zbl. 617.60020

Bentkus, V. (1982): Estimates of the rate of convergence in the central limit theorem in the space $C(S)$. Dokl. Akad. Nauk SSSR **266**, 526–529. English transl.: Soviet Math. Dokl. **26**, 349–352. Zbl. 521.60005

Bentkus, V. (1984a): Lower bounds for the sharpness of a normal approximation in Banach spaces. Liet. Mat. Rink. **24**, No. 1, 12–18. English transl.: Lith. Math. J. **24**, No. 1, 6–10. Zbl. 584.60034

Bentkus, V. (1984b): Asymptotic analysis of the remainder term in the central limit theorem in Hilbert space. Liet. Mat. Rink **24**, No. 1, 5–11. English transl.: Lith. Math. J. **24**, No. 1, 2–6. Zbl. 586.60022

Bentkus, V. (1983): Differentiable functions defined in the spaces c_0 and $\mathbb{R}^k$. Liet. Mat. Rink. **24**, No. 2, 26–36. English transl.: Lith. Math. J. **24**, No. 2, 146–154. Zbl. 527.46036

Bentkus, V. (1984d): Asymptotic expansions in the central limit theorem in Hilbert space. Liet. Mat. Rink. **24**, No. 3, 29–50. English transl.: Lith. Math. J. **24**, No. 3, 210–225. Zbl. 568.60019

Bentkus, V., (1984e): Asymptotic expansions for the distributions of sums of independent random elements of a Hilbert space. Liet. Mat. Rink. **24**, No. 4, 29–48. English transl.: Lith. Math. J. **24**, No. 4, 305–319. Zbl. 573.60007

Bentkus, V. (1984f): The Asymptotic Analysis of Sums of Independent Banach-Space-Valued Random Elements. Doctoral Dissertation, Vilnius (in Russian).

Bentkus, V. (1984g): Asymptotic analysis of moments in the central limit theorem in Banach spaces. Liet. Mat. Rink. **24**, No. 2, 49–64. English transl.: Lith. Math. J. **24**, No. 2, 113–125. Zbl. 558.60007

Bentkus, V. (1985a): Asymptotic expansions in the local limit theorem in a Hilbert space. Liet. Mat. Rink. **25**, No. 1, 9–22. English transl.: Lith. Math. J. **25**, No. 1, 1–10. Zbl. 567.60003

Bentkus, V. (1985b): Concentration functions of sums of independent random elements of a Banach space. Liet. Mat. Rink. **25**, No. 2, 32–39. English transl.: Lith. Math. J. **25**, No. 2, 32–39. Zbl. 577.60005

Bentkus, V. (1986a): Dependence of the Berry-Esseen estimate on the dimension. Liet. Mat. Rink. **26**, No. 2, 205–210. English transl.: Lith. Math. J. **26**, No. 2, 110–114. Zbl. 612.60022

Bentkus, V. (1986b): Lower bounds for the rate of convergence in the central limit theorem in Banach spaces. Liet Mat. Rink. **25**, No. 4, 10–25. English transl.: Lith. Math. J. **26**, 312–320. Zbl. 588.60010

Bentkus, V. (1986c): Large deviations in Banach space. Teor. Veroyatn. Primen. **31**, No. 4, 710–716. English transl.: Theor. Probab. Appl. **31**, No. 4, 627–632. Zbl. 623.60012

Bentkus, V. (1986d): Asymptotic expansions for moments in the central limit theorem in Banach spaces. Liet. Mat. Rink. **26**, No. 1, 10–26. English transl.: Lith. Math. J. **26**, No. 1, 6–18. Zbl. 649.60024

Bentkus, V. (1987): Lower estimates of the convergence rate in the central limit theorem in Banach spaces. Probab. Theory Math. Stat. (Vilnius 1985). English transl.: VNU Science Press, Utrecht, The Netherlands **1**, 171–187. Zbl. 657.60013

Bentkus, V. (1990): Smooth approximations of the norm and differentiable functions with bounded support in Banach space. Liet. Mat. Rink. **30**, No. 3, 489–499. English transl.: Lith. Math. J. **30**, No. 3, 223–230. Zbl. 725.46009

Bentkus, V., and Liubinskas, K. (1987): Rate of convergence in the invariance principle in Banach spaces. Liet. Mat. Rink. **27**, No. 3, 423–434. English transl.: Lith. Math. J. **27**, No. 3, 205–213. Zbl. 635.60007

Bentkus, V., and Račkauskas, A. (1981): Convergence rate in the central limit theorem in infinite-dimensional spaces. Liet. Mat. Rink. **21**, No. 4, 9–18. English transl.: Lith. Math. J. **21**, No. 4, 271–276. Zbl. 493.60009

Bentkus, V., and Račkauskas, A. (1982): Estimates of the convergence rate of sums of independent random variables in a Banach space, I, II. Liet. Mat. Rink. **22**, No. 3, 12–28; No. 4, 8–20. English transl: Lith. Math. J. **22**, No. 3, 223–234; No. 4, 344–353. I. Zbl. 522.60007, II. Zbl. 522.60007

Bentkus, V., and Račkauskas, A. (1984): Estimates of the distance between sums of independent random elements in Banach spaces. Teor. Veroyatn. Primen. **29**, No. 1, 49–64. English transl.: Theory Probab. Appl. **29**, No. 1, 50–65. Zbl. 534.60008

Bentkus, V., and Račkauskas, A. (1990): On probabilities of large deviations. Probab. Theory Relat. Fields **86**, 131–154. Zbl. 678.60005

Bentkus, V., and Zalesskii, B. (1985): Asymptotic expansions with nonuniform remainders in the central limit theorem in Hilbert space. Liet. Mat. Rink. **25**, No. 3, 3–16. English transl.: Lith. Math. J. **25**, No. 3, 199–208. Zbl. 585.60011

Bentkus, V., and Zitikis, R. (1988): A remark on the Cramér–von Mises–Smirnov test. Liet. Mat. Rink. **28**, No. 1, 14–22. English transl.: Lith. Math. J. **28**, No. 1, 8–13. Zbl. 647.62032

Bentkus, V., and Zitikis, R. (1990): Probabilities of large deviations for L-statistics. Liet. Mat. Rink. **30**, No. 3, 479–488. English transl.: Lith. Math. J. **30**, No. 3, 215–222. Zbl. 706.62015

Bentkus, V., Götze, F., and Zitikis, R. (1993): Asymptotic expansions in the integral and local limit theorems in Banach spaces with applications to ω-statistics. J. Theoretical Probab. **6**, 4, 727–780. Zbl. 807.60007

Bergström, H. (1944): On the central limit theorem. Skand. Aktuarietidskrift **27**, 139–153. Zbl. 706.62015

Bergström, H. (1951): On asymptotic expansions of probability functions. Skand. Aktuarietidskrift **1**. No. 2, 1–34. Zbl. 045.07301

Bernotas, V. (1979): Uniform and nonuniform proximity bounds for the distribution of two normalized sums of independent random variables with values in Banach spaces. Liet. Mat. Rink. **19**, No. 4, 55–68. English transl.: Lith. Math. J. **19**, No. 4, (1980), 482–489. Zbl. 424.60007

Berry, A.C. (1941): The accuracy of the Gaussian approximation to the sum of independent variates. Trans. Am. Math. Soc. **49**, 122–136. Zbl. 025.34603

Bézandry, P.H., and Fernique, X. (1992): Sur la proprieté de la limite centrale dans $D[0,1]$. Ann. Inst. Henri Poincaré, Probab. Stat. **28**, No. 1, 31–46. Zbl. 749.60003

Bhattacharya, R.N. (1977): Refinements of the multidimensional central limit theorem and its application. Ann. Probab. **5**, 1–27. Zbl. 361.60001

Bhattacharya, R.N., and Denker, M. (1990): Asymptotic Statistics, Birkhäuser, Basel-Boston-Berlin. Zbl. 706.62049

Bhattacharya, R.N., and Rango Rao, R. (1976): Normal Approximation and Asymptotic Expansions. John Wiley, New York. Zbl. 331.41023

Bickel, P.J., Götze, F., and van Zwet, W.R. (1986): The Edgeworth expansion for U-statistics of degree two. Ann. Stat. **14**, 1463–1484. Zbl. 614.62015

Bikelis, A. (1967): Remainder terms in asymptotic expansions for characteristic functions and their derivatives. Liet. Mat. Rink. **7**, No. 4, 571–582. English transl.: Selected Transl. Math. Stat. Probab. **11**, 149–162 (1973). Zbl. 167.17401

Billingsley, P. (1969): Convergence of Probability Measures. John Wiley, New York. Zbl. 172.21201

Bloznelis, M. (1989): A lower bound for the convergence rate in the central limit theorem in Hilbert space. Liet. Mat. Rink. **29**, No. 4, 674–681. English transl.: Lith. Math. J. **29**, No. 4, 333–338. Zbl. 705.60010

Bloznelis, M., and Paulauskas, V. (1994): A note on the central limit theorem for stochastically continuous processes. Stoch. Proc. Appl. **53**, 351–361. Zbl. 838.60018

Bogachev, V.I. (1988): Subspaces of differentiability of smooth measures on infinite-dimensional spaces. Dokl. Akad. Nauk SSSR **299**, No. 1, 18–22. English transl.: Sov. Math. Dokl. **37**, 304–308. Zbl. 721.46030

Borisov, I.S. (1983): Problem of accuracy of approximation in the central limit theorem for empirical measures. Sib. Mat. Zh. **24**, No. 6, 14–25. English transl.: Sib. Math. J. **24**, 833–843. Zbl. 541.60020

Borisov, I.S. (1985): A remark on the speed of convergence in the central limit theorem in Banach spaces. Sib. Mat. Zh. **26**, No. 2, 29–35. English transl.: Sib. Math. J. **26**, 180–185. Zbl. 569.60004

Borisov, I.S. (1989): Approximation for distributions of smooth functionals of sums of independent random variables in Banach spaces. In: Asymptotic Analysis of Distributions of Stochastic Processes, Novosibirsk, Tr. Inst. Mat. **13**, 7–40. Engl. transl.: Sib. Adv. Math. **1**, No. 1, 1–38 (1991). Zbl. 718.60005, 705.60007

Borovkov, A.A. (1984): On the convergence rate in the invariance principle for a Hilbert space. Teor. Veroyatn. Primen. **29**, No. 3, 532–535. English transl.: Theory Probab. Appl. **29**, No. 3, 550–553. Zbl. 568.60008

Borovkov, A.A., and Sakhanenko, A.I. (1980): On estimates of the rate of convergence in the invariance principle for Banach spaces. Teor. Veroyatn. Primen. **25**, No. 4, 734–744. English transl.: Theory Probab. Appl. **25**, No. 4, 721–731. Zbl. 454.60033

Borovskikh, Yu.V., and Račkauskas, A. (1979): Asymptotic analysis of distributions in Banach spaces. Liet. Mat. Rink. **19**, No. 4, 39–54. English transl.: Lith. Math. J. **19**, 472–481. Zbl. 441.60009

Butzer, P.L., Hahn, L., and Roeckerath, M.T. (1979): General theorems on "little-o" rates of convergence of two weighted sums of independent Hilbert-space-valued random variables with applications. J. Multivariate Anal. **9**, 487–510. Zbl. 428.60028

Cartan, H. (1967): Calcul des formes differentielles. Hermann, Paris. Zbl. 184.12701

Chebotarev, V.I. (1982): Estimates of the convergence rate in the local limit theorem for the square of the norm in ℓ_2. In: Limit Theorems of Probability and Related Questions. Tr. Inst. Mat **1**, 122–126. Engl. transl.: Transl. Ser. Math. Eng. (1984), 219–225. New York, Optimization Software. Zbl. 508.60013

Chibisov, D.M. (1964): Some theorems on the limiting behavior of empirical distribution functions. Tr. Mat. Inst. Steklova **71**, 104–112 (1964). Engl. transl.: Selected Transl. Math. Stat. Probab. **6**, 147–156 (1966). Zbl. 163.40602

Csörgő, M., Csörgő, S., Horvath, L., and Mason, D.M. (1986): Weighted empirical and quantile processes. Ann. Probab. **14**, 31–85. Zbl. 589.60029

Csörgő, M., and Horvath, L. (1988): On the distributions of L_p norms of weighted uniform empirical and quantile processes. Ann. Probab. **16**. 142–161. Zbl. 646.62015

Csörgő, S. (1976): On an asymptotic expansion for the von Mises ω^2-statistics. Acta Sci. Math. **38**, 45–67. Zbl. 333.62020

Csörgő, S., and Stachó, L. (1980): A step toward asymptotic expansions for the Cramér–von Mises statistic. In: Analytic Function Methods in Probability Theory. Colloq. Math. Soc. James Bolyai **21**, 53–65. Zbl. 416.62015

Daletskii, Yu.L., and Fomin, S.V. (1983): Measures and Differential Equations in Infinite-Dimensional Spaces. Nauka, Moscow. Engl. transl.: Kluwer, Dordrecht (1991). Zbl. 536.46031

Daniels, H.E. (1945): The statistical theory of the strength of bundles of threads. Proc. R. Soc. Ser. A **183**, 405–435. Zbl. 063.01035

Davydov, Yu., and Lifshits, M.A. (1984): The fiber method in some probability problems. Itogi Nauki Tekh. Ser. Teor. Veroyatn. Mat. Stat. Kibern. **22**, 61–158. English transl.: J. Sov. Math. **31** (1985), 2796–2858. Zbl. 566.60040

Dehling, H. (1983): Limit theorems for sums of weakly dependent Banach-space-valued random variables. Z. Wahrscheinlichkeitstheorie Verw. Geb. **63**, 393–432. Zbl. 509.60012

Dudley, R.M. (1978): Central limit theorem for empirical measures. Ann. Probab. **6**, 899–923. Zbl. 404.60016

Dudley, R.M. (1987): Universal Donsker classes and metric entropy. Ann. Probab. **15**, 1306–1326. Zbl. 631.60004

Esseen, C.G. (1942): On the Liapounoff limit of error in the theory of probability. Ark. Mat., Astro. Fysik. **28A**, 1–19. Zbl. 027.33902

Esseen, C.G. (1945): Fourier analysis of distribution functions. Acta Math. **77**, 1–125. Zbl. 060.28705

Esseen, C.G. (1968): On the concentration function of a sum of independent random variables. Z. Wahrscheinlichkeitstheorie Verw. Geb. **9**, 290–308. Zbl. 195.19303

Feller, W. (1971): An Introduction to Probability Theory and Its Applications, Vol. II, 2nd. edn., John Wiley, New York. Zbl. 219.60003

Fernique, X. (1971): Regularité de processus Gaussien. Invent. Math. **12**, 304–320. Zbl. 217.21104

Fernique, X. (1994): Les functions aléatoires cadlag, la compacité de leurs lois. Liet. Mat. Rink. **34**, 288–306. Zbl. 829.60031

Fisz, M. (1959): A central limit theorem for stochastic processes with independent increments. Stud. Math. **18**, 223–227. Zbl. 152.16404

Gaenssler, P., and Stute, W. (1979): Empirical processes: a survey of results for independent and identically distributed random variables. Ann. Probab. **7**, 193–242. Zbl. 402.60031

Giné, E. (1976): Bounds for the speed of convergence in the central limit theorem in $C(S)$. Z. Wahrscheinlichkeitstheorie Verw. Geb. **36**, 317–331. Zbl. 351.60029

Giné, E., and Zinn, J. (1984): Some limit theorems for empirical processes. Ann. Probab. **12**, 929–989. Zbl. 553.60037

Götze, F. (1979): Asymptotic expansions for bivariate von Mises functionals. Z. Wahrscheinlichkeitstheorie Verw. Geb. **50**, 333–355. Zbl. 415.60008

Götze, F. (1981): On Edgeworth expansions in Banach spaces. Ann. Probab. **9**, 852–859. Zbl. 473.60009

Götze, F. (1983): On the rate of convergence in the CLT under moment conditions. Preprints in Statistics, Univ. Cologne.

Götze, F. (1984): Expansions for von Mises functionals. Z. Wahrscheinlichkeitstheorie Verw. Geb. **65**, 599–625. Zbl. 531.60037

Götze, F. (1985): Asymptotic expansion in a functional limit theorem. J. Multivariate Anal. **16**, 1–20. Zbl. 557.60027

Götze, F. (1986): On the rate of convergence in the central limit theorem in Banach spaces. Ann. Probab. **14**, 922–942. Zbl. 599.60009

Götze, F. (1987): Approximations for multivariate U-statistics. J. Multivariate Anal. **22**, 212–229. Zbl. 624.62028

Götze, F. (1989): Edgeworth expansions in functional limit theorems. Ann. Probab. **17**, 1602–1634. Zbl. 689.60038

Götze, F., and Hipp, C. (1982): Asymptotic expansions in the CLT under moment conditions. Z. Wahrscheinlichkeitstheorie Verw. Geb. **42**, 67–87. Zbl. 369.60027

Hahn, M.G. (1978): Central limit theorem in $D[0,1]$. Z. Wahrscheinlichkeitstheorie Verw. Geb. **44**, 89–101. Zbl. 364.60002

Hardy, G.H., and Littlewood, J.E. (1920): A new solution of Waring's problem. Quarterly J. Math. **48**, 272–293. Zbl. 47.0114.01

Helmers, R. (1982): Edgeworth expansions for linear combinations of order statistics. Amsterdam Math. Center Tracts **105**. Zbl. 485.62017

Ibragimov, I.A., and Linnik, Yu. (1971): Independent and Stationary Sequences of Random Variables. Wolters-Noordhoff Publishing, Groningen. Zbl. 154.42201

Juknevičienė, D. (1985): Central limit theorem in the space $D[0,1]$. Liet. Mat. Rink. **25**, No. 3, 198–205. English transl.: Lith. Math. J. **25**, No. 3, 293–298. Zbl. 593.60032

Kandelaki, N.P. (1965): On a limit theorem in Hilbert space. Tr. Vychisl. Tsentra Akad. Nauk Gruz. SSR **5**, 46–55 (in Russian). Zbl. 253.60014

Kandelaki, N.P., and Vakhaniya, N.N. (1969): On a bound for the convergence rate in the central limit theorem in Hilbert space. Tr. Vychisl. Tsentra Akad. Nauk Gruz. SSR **9**, 150–160 (in Russian). Zbl. 198.22901

Kiefer, J. (1972): Skorokhod embedding of multivariate r.v.'s and the sample d.f. Z. Wahrscheinlichkeitstheorie Verw. Geb. **24**, 1–35. Zbl. 267.60034

Koroliuk, V.S., and Borovskikh, Yu.V. (1984): Asymptotic Analysis of Distributions of Statistics. Naukova Dumka, Kiev (in Russian). Zbl. 565.62003

Kuelbs, J., and Kurtz, T. (1974): Berry-Esseen estimates in Hilbert space and an application to the law of the iterated logarithm. Ann. Probab. **2**, 387–407. Zbl. 298.60017

Kukuš, A.G. (1981): The weak convergence of measures and the convergence of cumulants. Teor. Veroyatn. Mat. Stat. **23**, 74–80 (1980). English transl.: Theory Probab. Math. Stat. **25**, 79–86. Zbl. 482.28010

Kukuš, A.G. (1982): The central limit theorem in Hilbert space in terms of the Lévy–Prokhorov metric. Teor. Veroyatn. Mat. Stat. **25**, 55–63. English transl.: Theory Probab. Math. Stat. **25**, 61–68. Zbl. 457.60005

Lapinskas, R. (1978): Approximation of partial sums in certain Banach spaces. Liet. Mat. Rink. **18**, No. 4, 65–71. English transl.: Lith. Math. J. **18**, No. 4, 494–498. Zbl. 403.60016

Liapunov, A.M. (1900): Sur une proposition de la théorie des probabilités. Bull. Acad. Sci., St. Petersbourg (5) **13**, 359–386. Jbuch 31.0228.02

Liapunov, A.M. (1901): Nouvelle forme du théoreme sur la limite de théorie des probabilités. Mem. Acad. Sci. St. Petersbourg (8) **12**, 1–24. Jbuch 33.0248.07

Lifshits, M.A. (1983): Distribution (density) of the maximum of a Gaussian process. Teor. Veroyatn. Primen. **31**, No. 1, 131–142. English transl.: Theory Probab. Appl. **31**, No. 1, 125–132. Zbl. 602.60040

Lindeberg, J.W. (1920): Über das Exponentialgesetz in der Wahrscheinlichkeitsrechnung. Ann. Acad. Sci. Fenn., Ser. A. **16**, 1–23. Jbuch 47.0485.01

Lindeberg, J.W. (1922): Eine neue Herleitung des Exponentialgesetzes in der Wahrscheinlichkeitsrechnung. Math. Z. **15**, 211–225. Jbuch 48.0602.04

Liubinskas, K. (1987): On the closeness of moments in the central limit theorem in Banach spaces. Liet. Mat. Rink. **27**, No. 2, 285–302. Zbl. 635.60006

Marcus, M.B., and Pisier, G. (1981): Random Fourier Series with Applications to Harmonic Analysis. Ann. Math. Stud. **101**. Zbl. 474.43004

Marcus, M.B., and Shepp, L.A. (1972): Sample behavior of Gaussian processes. Proc. Sixth Berkeley Symp. Math. Stat. Probab. **3**, 423–442. Zbl. 379.60040

Martynov, G.V. (1978): Omega-Square Tests, Nauka, Moscow (In Russian).

Massart, P. (1986): Rates of convergence in the central limit theorem for empirical processes. Ann. Inst. Henri Poincaré, Probab. Stat. **22**, 381–423. Zbl. 615.60032

Massart. P. (1989): Strong approximation for multivariate empirical and related processes, via KMT construction. Ann. Probab. **17**, 266–291. Zbl. 675.60026

Nagaev, S.V. (1976): An estimate of the remainder term in the multi-dimensional central limit theorem. Proc. Third Japan-USSR Symp. Probab. Theory; Lect. Notes Math. **550**, 419–438. Zbl. 363.60024

Nagaev, S.V. (1983): On accuracy of normal approximation for the distribution of a sum of independent Hilbert-space-valued random variables. Lect. Notes Math. **1021**, 461–473. Zbl. 526.60008

Nagaev, S.V. (1985): On the rate of convergence to the normal law in Hilbert space. Teor. Veroyatn. Primen. **30**, No. 1, 19–32. English transl.: Theory Probab. Appl. **30**, No. 1, 19–37 (1986). Zbl. 657.60009

Nagaev, S.V. (1989a): A Berry-Esseen-type estimate for sums of random variables with values in a Hilbert space. Sib. Mat. Zh. **30**, No. 3, 84–96. English transl.: Sib. Math. J. **30**, No. 3, 413–423. Zbl. 675.60010

Nagaev, S.V. (1989b): A new approach to the study of the distribution of the norm of a random element in Hilbert space. Abst., Fifth Vilnius Conf. Probab.Th. Math. Stat. 77–78. English transl.: Vol. II, 214–226 (1990). Zbl. 733.60013

Nagaev, S.V., and Chebotarev, V.I. (1978): Bounds for the convergence rate in the central limit theorem in the Hilbert space ℓ_2. In: Mathematical Analysis and Related Topics, Novosibirsk, 153–182 (In Russian).

Nagaev, S.V., and Chebotarev, V.I. (1986): A refinement of the error estimate of the normal approximation in a Hilbert space. Sib. Mat. Zh. **27**, No. 3, 154–173. English transl.: Sib. Math. J. **27**, No. 3, 434–450. Zbl. 604.60009

Nagaev, S.V., and Chebotarev, V.I. (1987): Asymptotic expansions for the distribution of the sum of i.i.d. Hilbert-space-valued r.v.'s. Probab. Theory Math. Stat. (1985), Vilnius. English transl.: VNU Science Press, Utrecht, The Netherlands **2**, 357–363. Zbl. 673.60007

Nagaev, S.V., and Chebotarev, V.I. (1989a): On asymptotic expansions of Bergström type in Hilbert space. In: Asymptotic Analysis of Distributions of Stochastic Processes. Tr. Inst. Mat. **13**, 66–77 (1989). English transl.: Sib. Adv. Math. **1**, No. 2, 130–145 (1991). Zbl. 705.60008

Nagaev, S.V., and Chebotarev, V.I. (1989b): On Edgeworth expansions in Hilbert space. Tr. Inst. Mat. **20**, 170–203 (1993). English transl.: Sib. Adv. Math. **3**, No. 3, 89–122 (1993). Zbl. 866.60012

Nikitin, Ya.Yu. (1972): On a boundary problem for an empirical process. Dokl. Akad. Nauk SSSR **205**, No. 5, 1043–1045. English transl.: Sov. Math. Dokl. **13**, No. 4, 1081–1084. Zbl. 306.60014

Norvaiša, R. (1993): The central limit theorem for empirical and quantile processes in some Banach spaces (Preprint). English transl.: Stochastic Processes Appl. **46**, No. 1, 1–27 (1993). Zbl. 780.60011

Norvaiša, R., and Paulauskas, V. (1991): Rate of convergence in the central limit theorem for empirical processes. J. Theor. Probab. **4**, No. 3, 511–534. Zbl. 734.60020

Norvaiša, R., and Zitikis, R. (1991): Asymptotic behavior of linear combinations of functions of order statistics. J. Stat. Plann. Inference, 305–317. Zbl. 62046

O'Reilly, N. (1974): On the convergence of empirical processes in supernorm metrics. Ann. Probab. **2**, 642–651. Zbl. 60007

Orlov, A.I. (1974): The rate of convergence of the distribution of the Smirnov-von Mises statistic. Teor. Veroyatn. Primen. **19**, No. 4, 766–786. English transl.: Theory Probab. Appl. **19**, No. 4, 737–757. Zbl. 301.60007

Osipov, L.V. (1977): On large deviations of sums of independent random vectors. Abst., 2nd Vilnius Conf. Probab. Th. Math. Stat. **2**, 95–96 (in Russian).

Osipov, L.V. (1978a): On large deviation probabilities for sums of independent random vectors. Teor. Veroyatn. Primen. **23**, No. 3, 510–526. English transl.: Theory Probab. Appl. **23**, No. 3, 490–506 (1979). Zbl. 437.60005

Osipov, L.V. (1978b): Large Deviation Probabilities for Sums of Independent Random Vectors. Doct. Dissertation, Leningrad (in Russian).

Osipov, L.V., and Rotar', V.I. (1984): On an infinite-dimensional central limit theorem. Teor. Veroyatn. Primen. **29**, No. 2, 366–373. English transl.: Theory Probab. Appl. **29**, No. 2, 375–382. Zbl. 544.60015

Paulauskas, V. (1973): On the concentration function of finite-dimensional and infinite-dimensional random vectors. Liet. Mat. Rink. **13**, No. 1, 137–157. English transl.: Lith. Math. J. **13**, No. 1, 97–111. Zbl. 264.60033

Paulauskas, V. (1976a): Estimate of the convergence rate in the central limit theorem in $C(S)$. Liet. Mat. Rink. **16**, No. 4, 167–201. English transl.: Lith. Math. J. **16**, No. 4, 587–611. Zbl. 392.60022

Paulauskas, V. (1976b): On the rate of convergence in the central limit theorem in certain Banach spaces. Teor. Veroyatn. Primen. **21**, No. 4, 775–791. English transl.: Theory Probab. Appl. **21**, No. 4, 754–769. Zbl. 403.60007

Paulauskas, V. (1981): Estimation of the convergence rate in the central limit theorem in ℓ_p. Liet. Mat. Rink. **21**, No. 1, 109–119. English transl.: Lith. Math. J. **21**, No. 1, 55–62. Zbl. 455.60031

Paulauskas, V. (1984): On the central limit theorem in Banach space c_0. Probab. Math. Stat. **3**, 127–141.

Paulauskas, V. (1990): On the rate of convergence for the weighted empirical process. Probability in Banach Spaces **7**. Proc. 7th Int. Conf. Oberwolfach, FRG 1988, Prog. Probab. **21**, 147–158 (1990). Zbl. 704.60009

Paulauskas, V., and Juknevičienė, D. (1988): On the rate of convergence in the central limit theorem in $D[0, 1]$. Liet. Mat. Rink. **28**, No. 3, 507–519. English transl.: Lith. Math. J. **28**, No. 3, 229–239. Zbl. 711.60008, Zbl. 657.60011

Paulauskas, V., and Račkauskas, A. (1987): Approximation Theory in the Central Limit Theorem. Exact Results in Banach Spaces, Mokslas, Vilnius. English transl.: Kluwer, Dordrecht (1989). Zbl. 708.60005

Paulauskas, V., and Račkauskas, A. (1991): Nonuniform estimates in the central limit theorem in Banach spaces. Liet. Mat. Rink. **31**, No. 3, 483–496. Zbl. 786.60009

Paulauskas, V., and Stieve, Ch. (1990): On the central limit in D[0, 1] and D([0, 1], H). Liet. Mat. Rink. **30**, No. 3, 567–579. English transl.: Lith. Math. J. **30**, No. 3, 267–276. Zbl. 722.60023

Petrov, V.V. (1975): Sums of Independent Random Variables, Nauka, Moscow (1972). English transl.: Springer-Verlag, Berlin Heidelberg New York

Phoenix, S.L., and Taylor, H. (1973): The asymptotic strength distribution of a general fiber bundle. Adv. Appl. Probab. **5**, 200–216. Zbl. 272.60006

Prokhorov, Yu.V., and Sazonov, V.V. (1969): On estimates of the convergence rate in the central limit theorem in the infinite-dimensional case. In: Soviet-Japanese Khabarovsk Symposium on Probability Theory **1**, 223–230.

Puri, M.L., and Seoh, M. (1987): On the rate of convergence in normal approximation and large deviation probabilities for a class of statistics. Teor. Veroyatn. Primen. **33**, No. 4, 736–750. Already in English: Theory Probab. Appl. **33**, No. 4, 682–697. Zbl. 665.62017

Rachev, S.T., and Rüschendorf, L. (1992): Rate of convergence for sums and maxima and doubly ideal metrics. Teor. Veroyatn. Primen. **37**, No. 2. Already in English: Theory Probab. Appl. **37**, No. 2, 222–235. Zbl. 787.60006

Rachev, S.T, and Yukich, J.E. (1989): Rates for the CLT via new ideal metrics. Ann. Probab **17**, 775–788. Zbl. 675.60018

Račkauskas, A. (1981): Approximation in the uniform metric of sums of independent random variables with values in Hilbert space. Liet. Mat. Rink. **21**, No. 3, 83–90. English transl.: Lith. Math. J. **21**, No. 3, 258–263. Zbl. 497.60008

Račkauskas, A. (1988): Probabilities of large deviations in Linnik zones in Hilbert space. Liet. Mat. Rink. **28**, No. 3, 520–533. English transl.: Lith. Math J. **28**, No. 3, 239–248. Zbl. 657.60008

Račkauskas, A. (1991): On the convergence rate in martingale CLT in Hilbert space. Liet. Mat. Rink. **31**, No. 3, 497–512. English transl.: Lith. Math. J. **31**, No. 3, 345–355. Zbl. 777.60027

Rhee, W.S., and Talagrand, M. (1984): Bad rates of convergence for the CLT in Hilbert space. Ann. Probab. **12**, 843–850. Zbl. 545.60014

Rhee, W.S., and Talagrand, M. (1986): Uniform convexity and the distribution of the norm of a Gaussian measure. Probab. Theory Relat. Fields **71**, 59–68. Zbl. 554.60007

Rosenkrantz, W.A. (1969): A rate of convergence for the von Mises statistic. Trans. Am. Math. Soc. **139**, 329–337. Zbl. 182.52301

Sakalauskas, V. (1983): Approximation by a stable law in nonuniform metrics of Lévy–Prokhorov and χ type. Liet. Mat. Rink. **23**, No. 4, 40–49. English transl.: Lith. Math. J. **23**, No. 4, 384–391. Zbl. 561.60013

Sakhanenko, A.I. (1988): Simple method of obtaining estimates in the invariance principle. Probab. Theory Math. Stat. (Kyoto 1986). Lect. Notes Math. **1299**, 430–443. Zbl. 637.60010

Sazonov, V.V. (1968): On ω^2-test. Sankhya, Ser. A. **30**, 205–209. Zbl. 177.47201

Sazonov, V.V. (1969): An improvement of a convergence-rate estimate. Teor. Veroyatn. Primen. **14**, No. 4, 667–678. English transl.: Theory Probab. Appl. **14**, No. 4, 640–651. Zbl. 185.46402

Sazonov, V.V. (1981): Normal Approximation – Some Recent Advances. Lect. Notes Math. **879**. Springer-Verlag. Zbl. 462.60006

Sazonov, V.V and Ul'yanov, V.V. (1990): Speed of convergence in the central limit theorm in Hilbert space under weakened moment conditions. In: Probab. Theory Math. Stat.; Proc. Fifth Vilnius Conf. **2**, 394–410. Vilnius, Mokslas. Zbl. 734.60008

Sazonov, V.V., and Ul'yanov, V.V. (1991): An improved estimate of the accuracy of the Gaussian approximation in Hilbert space. In: New Trends in Probab., and Stat., Mokslas, VSP, 123–136. Zbl. 803.60006

Sazonov, V.V., Ul'yanov, V.V., and Zalesskii, B.A. (1987a): On normal approximation in Hilbert space. In: Probab. Theory Math. Stat. (Vilnius 1985). Proc. 4th Vilnius Conf., VNU Science Press, Utrecht, The Netherlands, 561–580. Zbl. 652.60011

Sazonov, V.V., Ul'yanov, V.V., and Zalesskii, B.A. (1987b): Asymptotic expansions refining the central limit theorem in Hilbert space. In: Probab. Theory Appl. (Yu.A. Prokhorov and V.V. Sazonov, Eds.), VNU Science Press, Utrecht, The Netherlands **1**, 679–688. Zbl. 697.60007

Sazonov, V.V., Ul'yanov, V.V., and Zalesskii, B.A. (1988): Normal approximation in Hilbert space, I, II. Teor. Veroyatn. Primen. **33**, No. 2, 225–245; No. 4, 733–754. English transl.: Theory Probab. Appl. **33**, No. 2, 207–227; No. 4, 473–487, No. 2. Zbl. 649.60005, No. 4. Zbl. 662.60009

Sazonov, V.V., Ul'yanov, V.V., and Zalesskii, B.A. (1989a): Asymptotically precise estimate of the accuracy of Gaussian approximation in Hilbert space. J. Multivariate Anal. **28**, 304–330. Zbl. 675.60011

Sazonov, V.V., Ul'yanov, V.V., and Zalesskii, B.A. (1989b): A precise estimate of the convergence rate in the central limit theorem in Hilbert space. Mat. Sb. **180**, No. 12, 1587–1613. English transl.: Math USSR Sb. **68** (1991), No. 2, 453–482. Zbl. 709.60006

Sazonov, V.V., and Zalesskii, B.A. (1985): On the CLT in Hilbert space. J. Multivariate Anal. **24**, 495–526. Zbl. 603.60018

Schmidt, W. (1984): Bounds for exponential sums. Acta Arith. **44**, 281–297. Zbl. 544.10036

Senatov, V.V. (1981): Some lower convergence rate bounds in the central limit theorem. Dokl. Akad. Nauk SSSR **256**, No. 6, 1318–1321. English transl.: Sov. Math. Dokl. **23**, 188–192. Zbl. 603.60018

Senatov, V.V. (1983): On estimating the rate of convergence in the central limit theorem over a system of balls in R^k. Teor. Veroyatn. Primen. **28**, No. 2, 440–445. English transl.: Theory Probab. Appl. **28**, No. 2, 463–467. Zbl. 569.60022

Senatov, V.V. (1985a): On the dependency of estimates of the convergence rate in the central limit theorem on the covariance operator of the summands. Teor. Veroyatn. Primen. **30**, No. 2, 354–357. English transl.: Theory Probab. Appl. **30**, No. 2, 380–383. Zbl. 569.60023

Senatov, V.V. (1985b): Four examples of lower estiimates in the multi-dimensional central limit theorem. Teor. Veroyatn. Primen. **30**, No. 4, 750–758. English transl.: Theory Probab. Appl. **30**, No. 4, 797–805. Zbl. 579.60020

Senatov, V.V. (1986): On the dependence of estimates of the convergence rate in the central limit theorem for balls with center zero on the covariance operator of the summands. Teor. Veroyatn. Primen. **31**, No. 1, 128–132. English transl.: Theory Probab. Appl. **31**, No. 1, 119–122. Zbl. 589.60004

Senatov, V.V. (1989a): On bounds for the convergence rate in the central limit theorem in Hilbert space. Abst., Fifth Vilnius Conf. Theory Probab. Math. Stat. **4**. (In Russian)

Senatov, V.V. (1989b): On estimating the rate of convergence in the central limit theorem in Hilbert space. Lect. Notes Math. **1412**, 309–327. Zbl. 692.60005

Serfling, R.J. (1980): Approximation Theorems of Mathematical Statistics, John Wiley, New York.

Siegel, G. (1981): Upper estimates for the concentration function in Hilbert space. Teor. Veroyatn. Primen. **26**, No. 2, 335–349. English transl. Theory Probab. Appl. **26**, No. 2, 328–343. Zbl. 487.60014

Smirnov, N.V. (1937): On the distribution of the ω^2-test of von Mises. Rec. Math. (NS) **2**, 973–993 (In Russian). Zbl. 018.41202

Stigler, S.M. (1974): Linear functions of order statistics with smooth weight functions. Ann. Stat **2**, 676–693. Zbl. 286.62028

Sweeting, T.J. (1977): Speed of convergence for the multi-dimensional central limit theorem. Ann. Probab. **5**, 28–41. Zbl. 362.60041

Sweeting, T.J. (1980): Speeds of convergence and asymptotic expansions in the CLT – a treatment by operators. Ann. Probab. **8**, 279–281. Zbl. 444.60017

Thomasian, A. (1969): The Structure of Probability Theory with Applications, McGraw-Hill, New York. Zbl. 204.50101

Trotter, H.F. (1959): Elementary proof of the central limit theorem. Arch. Math. **10**, 226–234. Zbl. 086.34002

Ul'yanov, V.V. (1981): An estimate for the rate of convergence in the central limit theorem in a separable Hilbert space. Mat. Zametki **29**, No. 1, 145–153. English transl.: Math. Notes **29**, No. 1, 78–82. Zbl. 458.60008

Ul'yanov, V.V. (1986): Asymptotic expansions for distributions of sums of independent random variables in H. Teor. Veroyatn. Primen. **31**, No. 1, 31–46. English transl.: Theory Probab. Appl. **31**, No. 1, 25–39. Zbl. 625.60045

Vakhaniya, N.N., Tarieladze, V.I., and Chobanyan, S.A. (1987): Probability Distributions in Banach Spaces, Nauka, Moscow (1985). English transl.: D. Riedel Publishing Company, Dordrecht. Zbl. 572.60003

Vandemaele, M., and Veraverbeke, N. (1982): Cramér-type large deviations for linear combinations of order statistics. Ann. Probab. **10**, 423–434. Zbl. 482.60026

Vinogradov, I.M. (1934): A new evaluation of $G(n)$ in Waring's problem. Dokl. Akad. Nauk SSSR **5**, 249–251. English transl.: Version 251–253. Zbl. 011.00803

Vinogradova, T.R. (1985): On the accuracy of normal approximation on sets defined by a smooth function, I, II. Teor. Veroyatn. Primen. **30**, No. 2, 219–229; **30**, No. 3, 554–557. English transl.: Theory Probab. Appl. **30**, No. 2, 235–246; **30**, No. 3, 590–593. No. 2 Zbl. 573.60010, No. 3 Zbl. 658.60048

Wenocur, R.S., and Dudley, R.M. (1981): Some special Vapnik-Chervonenkis classes. Discrete Math. **33**, 313–318. Zbl. 459.60008

Weyl, H. (1916): Über die Gleichverteilung der Zahlen mod-Eins, Ann. **77**, 313–352. Jbuch 46.0278.06

Yurinskii, V.V. (1977): On the error in Gaussian approximation of convolutions. Teor. Veroyatn. Primen. **22**, No. 2, 242–253. English transl.: Theory Probab. Appl. **22**, No. 2, 236–247. Zbl. 378.60008

Yurinskii, V.V. (1981): An error estimate for the normal approximation of the probability of landing in a ball. Dokl. Akad. Nauk SSSR **258**, No. 3, 557–558. English transl.: Sov. Math. Dokl. **23**, 576–578. Zbl. 508.60026

Yurinskii, V.V. (1982): On the accuracy of normal approximation of the probability of hitting a ball. Teor. Veroyatn. Primen. **27**, No. 2, 270–278. English transl.: Theory Probab. Appl. **27**, No. 2, 280–289. Zbl. 565.60005

Yurinskii, V.V. (1983): Error of normal approximation. Sib. Mat. Zh. **24**, No. 6, 188–199. English transl.: Sib. Math. J. **24**, No. 6, 977–987. Zbl. 541.60019

Yurinskii, V.V. (1991): On asymptotic analysis of large deviations in Hilbert space, I, II. Teor. Veroyatn. Primen. **36**, No. 1, 78–92; No. 3, 535–541. English transl.: Theory Probab. Appl. **36**, No. 1, 99–114; No. 3, 548–554. No. 1 Zbl. 727.60026, No. 3 Zbl. 813.60007

Zaĭtsev, A.Yu. (1987): On the Gaussian approximation of convolutions under multidimensional analogues of S.N. Bernstein's inequality conditions. Probab. Theory Relat. Fields **74**, 535–566. Zbl. 612.60031

Zalesskii, B.A. (1982): Estimation of the accuracy of normal approximation in Hilbert space. Teor. Veroyatn. Primen. **27**, No. 2, 279–285. English transl.: Theory Probab. Appl. **27**, No. 2, 290–298. Zbl. 565.60004

Zalesskii, B.A. (1985): On the convergence rate in the central limit theorem on a class of sets in Hilbert space. Teor. Veroyatn. Primen. **30**, No. 4, 662–670. English transl.: Theory Probab. Appl. **30**, No. 4, 702–711. Zbl. 586.60010

Zalesskii, B.A. (1988): On the accuracy of normal approximation in Banach spaces. Teor. Veroyatn. Primen. **33**, No. 2, 257–265. English transl.: Theory Probab. Appl. **33**, No. 2, 239–247. Zbl. 666.60009

Zalesskii, B.A. (1989): Probabilities of large deviations in Hilbert space. Teor. Veroyatn. Primen. **34**, No. 4, 650–655. English transl.: Theory Probab. Appl. **34**, No. 4, 591–596. Zbl. 695.60029

Zalesskii, B.A., and Sazonov, V.V. (1984): Closeness of moments for normal approximation in a Hilbert space. Teor. Veroyatn. Primen. **28**, No. 2, 251–263. English transl.: Theory Probab. Appl. **28**, No. 2, 263–277. Zbl. 515.60013

Zalesskii, B.A., Sazonov, V.V., and Ul'yanov, V.V. (1988): An asymptotically regular estimate of the accuracy of normal approximation in Hilbert space. Teor. Veroyatn. Primen. **33**, No. 4, 753–754. English transl.: Theory Probab. Appl. **33**, No. 4, 700–701. Zbl. 662.60009

Zitikis, R. (1988): Asymptotic expansions in the local limit theorem for ω_n^2 statistics. Liet. Mat. Rink. **28**, No. 3, 461–474. Zbl. 662.62052

Zitikis, R. (1989): Asymptotic expansions for the derivatives of the distribution function of the Anderson-Darling statistic. Liet. Mat. Rink. **29**, No. 1, 35–53. Zbl. 777.62026

Zitikis, R. (1990a): Smoothness of the distribution function of the FL-statistic I, II. Liet. Math. Rink. **30**, No. 2, 233–246; **30**, No. 3, 500–512. English transl.: Lith. Math. J. **30**, No. 2, 97–106; **30**, No. 3, 231–239. No. 2 Zbl. 716.62028, No. 3. Zbl. 716.62029

Zitikis, R. (1990b): A uniform limit theorem for the densities of L-statistics. Liet. Mat. Rink **30**, No. 4, 728–740. English transl.: Lith. Math. J. **30**, No. 4, 331–341. Zbl. 761.62058

Zitikis, R. (1991a): On large deviations for L-statiistics. NewTrends in Probability and Statistics **1**, (V.V. Sazonov and T. Shervashidze, eds.), VSP Mokslas, Vilnius, 137–164. Zbl. 768.60021

Zitikis, R. (1991b): Cramér-type large deviations for a class of statistics. Liet. Mat. Rink. **31**, No. 2, 302–310. English transl.: Lith. Math. J. **31**, No. 2, 204–210. Zbl. 738.62028

Zolotarev, V.M. (1976a): Metric distances in spaces of random variables and their distributions. Mat. Sb. **101**, No. 3, 416–450. English transl.: Math. USSR, Sb. **101**, No. 3, 373–401. Zbl. 376.60003

Zolotarev, V.M. (1976b): Approximations of distributions of sums of independent random variables with values in infinite-dimensional spaces. Teor. Veroyatn. Primen. **21**, No. 4, 741–758. English transl.: Theory Probab. Appl. **21**, No. 4, 721–737. Zbl. 378.60003

Zolotarev, V.M. (1977): Ideal metrics in the problem of approximating distributions of sums of independent random variables. Teor. Veroyatn. Primen. **22**, No. 3, 449–465. English transl.: Theory Probab. Appl. **22**, No. 3, 433–449. Zbl. 385.60025

III. Approximation of Distributions of Sums of Weakly Dependent Random Variables by the Normal Distribution

J. Sunklodas

Contents

This article is a review of known results on approximating distributions of sums of weakly dependent random variables (r.v.'s) and random fields defined on the integer lattice Z^d by the normal distribution. It also contains some new results of the author. Attention is given mainly to methods of proof due to Stein, Tikhomirov and Heinrich.

§1. Weak Dependence Conditions and Covariance Inequalities

Let $X_1, X_2, \ldots$ be a sequence of real r.v.'s, $\mathcal{F}_a^b$ the σ-algebra generated by X_i, $a \le i \le b$, R the real line, and $N = \{1, 2, \ldots\}$.

Later on we shall assume that the sequence $X_1, X_2, \ldots$ satisfies one of the following weak dependence conditions:

1. *m-dependence*: $\mathcal{F}_1^r$ and $\mathcal{F}_{r'}^\infty$ are independent for all integers r and r' such that $1 \le r < r' < \infty$, $r' - r > m$;
2. *ψ-mixing*:
$$\psi(\tau) = \sup_{t \in N} \sup_{\substack{A \in \mathcal{F}_1^t,\ B \in \mathcal{F}_{t+\tau}^\infty \\ \mathbf{P}\{A\}>0,\ \mathbf{P}\{B\}>0}} |\mathbf{P}\{AB\} - \mathbf{P}\{A\}\mathbf{P}\{B\}| / \mathbf{P}\{A\}\mathbf{P}\{B\} \downarrow 0,\ \tau \to \infty;$$
3. *uniformly strong mixing* (u.s.m.):
$$\varphi(\tau) = \sup_{t \in N} \sup_{\substack{A \in \mathcal{F}_1^t,\ B \in \mathcal{F}_{t+\tau}^\infty \\ \mathbf{P}(A)>0}} |\mathbf{P}\{AB\} - \mathbf{P}\{A\}\mathbf{P}\{B\}| / \mathbf{P}\{A\} \downarrow 0,\ \tau \to \infty;$$
4. *absolute regularity* (a.r.):
$$\beta(\tau) = \sup_{t \in N} \mathbf{E}\left(\sup_{B \in \mathcal{F}_{t+\tau}^\infty} |\mathbf{P}\{B/\mathcal{F}_1^t\} - \mathbf{P}\{B\}| \right) \downarrow 0,\ \tau \to \infty;$$
5. *strong mixing* (s.m.):
$$\alpha(\tau) = \sup_{t \in N} \sup_{A \in \mathcal{F}_1^t,\ B \in \mathcal{F}_{t+\tau}^\infty} |\mathbf{P}\{AB\} - \mathbf{P}\{A\}\mathbf{P}\{B\}| \downarrow 0,\ \tau \to \infty.$$

Strong mixing was introduced by Rosenblatt (1956), absolute regularity by Kolmogorov (Volkonskii and Rozanov (1959)), uniformly strong mixing by Ibragimov (1962), ψ-mixing by Blum, Hanson and Koopmans (1963) and m-dependence by Hoeffding and Robbins (1948) (and goes back to Bernstein (1926)).

All of these conditions are requirements of weak dependence between the start and end of a sequence of r.v.'s.

Information about these and other measures of weak dependency (and the corresponding mixing coefficients) as well as appropriate references may be found, for example, in Ibragimov and Linnik (1965), Ibragimov and Rozanov (1970), Statulevičius (1974), Philipp and Stout (1975), Bradley, Bryc and Janson (1985), Bradley (1986) and Bulinskii (1987), (1989). The mixing coefficients defined above are interrelated by the following inequalities (Ibragimov and Linnik (1965), Hipp (1979a)):

$$2\alpha(\tau) \le \varphi^{1/2}(\tau),$$
$$\alpha(\tau) \le \beta(\tau) \le \varphi(\tau) \le \psi(\tau),$$
$$\alpha(\tau) \le 1/4,\ \varphi(\tau) \le 1.$$

Important refinements of these inequalities, a comparison of the mixing coefficients and other measures of dependency are found in Kolmogorov and Rozanov (1960), Ibragimov and Linnik (1965), Ibragimov and Rozanov (1970), Peligrad (1982), Bradley (1983), (1986), Bradley and Bryc (1985), Bradley, Bryc and Janson (1985), Bulinskii (1985b), (1989) and Liptser and Shiryaev (1986). We denote by $\operatorname{cov}(\xi,\eta) = \mathbf{E}\xi\eta - \mathbf{E}\xi\mathbf{E}\eta$ the covariance of real r.v.'s ξ and η. We shall subsequently need bounds for the covariance in terms of the mixing coefficients α, β, φ and ψ.

Lemma 1. *Suppose that ξ is $\mathcal{F}_1^t$-measurable and η is $\mathcal{F}_{t+\tau}^{\infty}$-measurable; $t, \tau \in N$.*

1. *If* $\mathbf{E}|\xi| < \infty$ *and* $\mathbf{E}|\eta| < \infty$, *then* (see Philipp (1969a)):
$$|\operatorname{cov}(\xi,\eta)| \le \mathbf{E}|\xi|\mathbf{E}|\eta|\psi(\tau); \tag{1}$$
2. *If* $\mathbf{E}|\xi| < \infty$ *and* $\mathbf{P}\{|\eta| > C\} = 0$, *then* (see Billingsley (1977))
$$|\operatorname{cov}(\xi,\eta)| \le 2C\mathbf{E}|\xi|\varphi(\tau); \tag{2}$$
3. *If* $\mathbf{E}|\xi|^q < \infty$ *and* $\mathbf{E}|\eta|^r < \infty$, q, $r > 1$, $q^{-1} + r^{-1} = 1$, *then* (see Ibragimov (1962))
$$|\operatorname{cov}(\xi,\eta)| \le 2\mathbf{E}^{1/q}|\xi|^q\mathbf{E}^{1/r}|\eta|^r\varphi^{1/q}(\tau); \tag{3}$$
4. *If* $\mathbf{P}\{|\xi| > C_1\} = \mathbf{P}\{|\eta| > C_2\} = 0$, *then* (see Volkonskii and Rozanov (1959), (1961), Ibragimov (1962))
$$|\operatorname{cov}(\xi,\eta)| \le 4C_1C_2\alpha(\tau); \tag{4}$$
5. *If* $\mathbf{E}|\xi|^r < \infty$, $r > 1$, *and* $\mathbf{P}\{|\eta| > C\} = 0$, *then* (see Davydov (1968), Hipp (1979a))
$$|\operatorname{cov}(\xi,\eta)| \le 4C\mathbf{E}^{1/r}|\xi|^r(\alpha(\tau))^{1-r^{-1}}; \tag{5}$$
6. *If* $\mathbf{E}|\xi|^q < \infty$ *and* $\mathbf{E}|\eta|^r < \infty$, q, $r > 1$, $q^{-1} + r^{-1} < 1$, *then* (see Davydov (1968), Hipp (1979a))
$$|\operatorname{cov}(\xi,\eta)| \le 6\mathbf{E}^{1/q}|\xi|^q\mathbf{E}^{1/r}|\eta|^r(\alpha(\tau))^{1-q^{-1}-r^{-1}}. \tag{6}$$

Inequalities (1–6) may be extended easily by induction to finitely many r.v.'s (see, for example, Volkonskii and Rozanov (1959), (1961), Roussas and Ionnides (1987)). We shall need a result that follows from (4) and (5).

Lemma 2. *Suppose that ξ_j is $\mathcal{F}_{s_j}^{t_j}$-measurable, $j = 1, 2, \ldots, k$, where $1 \le s_1 \le t_1 < s_2 \le t_2 < \ldots < s_k \le t_k < \infty$, and that $\tau = \min_{1\le j<k}(s_{j+1} - t_j)$.*

1. *If* $P\{|\xi_j| > C_j\} = 0$, $j = 1, 2, \ldots, k$, *then for* $k \ge 2$
$$\left|\mathbf{E}\left(\prod_{j=1}^{k}\xi_j\right) - \prod_{j=1}^{k}\mathbf{E}\xi_j\right| \le 4(k-1)\prod_{j=1}^{k}C_j\alpha(\tau); \tag{7}$$

2. *If* $\mathbf{E}|\xi_1|^r < \infty$, $r > 1$, *and* $\mathbf{P}\{|\xi_j| > C_j\} = 0$, $j = 2, 3, \ldots, k$, *then for* $k \geq 2$

$$\left| \mathbf{E}\left(\prod_{j=1}^{k} \xi_j\right) - \prod_{j=1}^{k} \mathbf{E}\xi_j \right| \leq 4(k-1)\mathbf{E}^{1/r}|\xi_1|^r \prod_{j=2}^{k} C_j(\alpha(\tau))^{1-r^{-1}}. \quad (8)$$

Sometimes it is advantageous to have an estimate of the covariance of r.v.'s in terms of the mixing coefficients when the σ-algebras $\mathcal{F}_1^t$ (the "past") and $\mathcal{F}_{t+\tau}^{\infty}$ (the "future") are replaced by σ-algebras of a more complex structure (see, for example, Lemma 3).

Let $\sigma(\mathcal{G} \cup \mathcal{H})$ be the σ-algebra generated by the σ-algebras $\mathcal{G}$ and $\mathcal{H}$.

Lemma 3 (Takahata (1981)). *Let* ξ *be* $\mathcal{F}_{t+\tau}^{t+\tau+n-1}$*-measurable and* η *be* $\sigma\{\mathcal{F}_1^t \cup \mathcal{F}_{t+2\tau+n-1}^{\infty}\}$*-measurable,* t, τ, $n \in N$.

1. *If* $\mathbf{E}|\xi|^q < \infty$ *and* $\mathbf{E}|\eta|^r < \infty$, $q, r > 0$, $q^{-1} + r^{-1} = 1$, *then*

$$|\operatorname{cov}(\xi, \eta)| \leq 6\mathbf{E}^{1/q}|\xi|^q \mathbf{E}^{1/r}|\eta|^r \psi^{1/q}(\tau); \quad (9)$$

2. *If* $\mathbf{E}|\xi|^q < \infty$ *and* $\mathbf{E}|\eta|^r < \infty$, $q, r > 0$, $q^{-1} + r^{-1} < 1$, *then*

$$|\operatorname{cov}(\xi, \eta)| \leq 18\mathbf{E}^{1/q}|\xi|^q \mathbf{E}^{1/r}|\eta|^r (\beta(\tau))^{1-q^{-1}-r^{-1}}. \quad (10)$$

If ξ and η in inequalities (1)–(10) are complex r.v.'s, then the right-hand sides of these inequalities must be multiplied by 4 (see, for example, Ibragimov and Linnik (1965), Roussas and Ionnides (1987)).

§2. Estimation of the Rate of Convergence in the Central Limit Theorem for Weakly Dependent Random Variables

2.1. Introduction and Notation.

We shall concentrate on estimating the rate of convergence in the one-dimensional central limit theorem (CLT).

Conditions for the applicability of the CLT to weakly dependent r.v.'s have been studied by many authors.

Devoted specifically to m-dependent r.v.'s are the papers by Hoeffding and Robbins (1948), Diananda (1955), Kallianpur (1955), Orey (1958), Zaremba (1958), and Berk (1973); to functionals defined on Markov chains, the papers by Bernstein (1926), Sirazhdinov (1955), Dobrushin (1956a,b), Nagaev (1957), (1962), Statulevičius (1961), (1969a,b), (1970a), Gudynas (1977), and Lifshits (1984). Ibragimov (1962) is basic as far as strictly stationary sequences of weakly dependent r.v.'s are concerned. The results in that paper were subsequently refined and generalized by Rosén (1967), Serfling (1968), Philipp (1969a), Berk (1973), Ibragimov (1975), Bradley (1981), (1988), Peligrad (1982), (1985), (1986a,b), (1986), (1990), Herrndorff (1983a,b), (1985), Utev

(1984), (1990a,b), Samur (1984), Grin' (1990a,b), etc. Bergström (1970), (1971), (1972) studied this problem by means of his comparison method. The latest results as well as a bibliography on this topic may be found, for example, in the above papers by Peligrad, Bradley, Utev and Grin'. The advances by Utev in (1990a,b) were made by taking advantage of Peligrad's bounds for large deviation probabilities (1985) which were also used in a modified form by Grin' (1990a) (see also (1990b)).

As of today, there exists much research devoted to studying the rate of convergence in the CLT for weakly dependent r.v.'s.

This question was first investigated by Petrov for m-dependent r.v.'s in (1960). He obtained a bound in the uniform metric of order $O(n^{-(s-2)/(6s-4)})$ under the assumption that the s-th absolute moments of the variables, $2 < s \leq 3$, are uniformly bounded and the variance of their sum increases linearly; here and elsewhere n is the number of terms. The problem was later investigated for m-dependent r.v.'s by Ibragimov (1967), Egorov (1970), Stein (1972), Erickson (1973)–(1975), Tikhomirov (1980), Shergin (1976), (1979), (1983), Maejima (1978), Yudin (1981), (1989a), Zuparov (1981), Heinrich (1982), (1984), (1985a,b,d), Sunklodas (1982), (1984), (1989), Rhee (1985), (1986a), Zuev (1986) and others. The first optimal estimate, that is, of order $O(n^{-(s-2)/2})$, in the uniform metric for m-dependent r.v.'s in the CLT was obtained by Ibragimov (1967). He considered a sequence of r.v.'s of the form $f(\varepsilon_i, \varepsilon_{i+1}, \ldots, \varepsilon_{i+m-1})$, $i = 1, 2, \ldots$, where $\varepsilon_1, \varepsilon_2, \ldots$ are independent and identically distributed r.v.'s with $\mathbf{E}|\varepsilon_1|^s < \infty$, $2 < s \leq 3$.

The rate of convergence in the one-dimensional CLT has been estimated under other weak dependence conditions by Statulevičius (1962), (1974), (1977a,b), Iosifescu (1968), Philipp (1969b), Stein (1972), Dubrovin (1971), Bulinskii (1977), Yoshihara (1978a), Negishi (1977), Tikhomirov (1976), (1980), Dubrovin and Moskvin (1979), Schneider (1981), Zuparov (1981), Yudin (1984), (1987), (1989b), (1990), (1991), Sunklodas (1977a), (1982), (1984), (1989), Lappo (1986) and others.

These papers made use of diverse methods of proof.

The method of cumulants (semi-invariants) involving the logarithmic derivatives of the characteristic function (c.f.) of the original sum was developed by Statulevičius in (1961), (1969a,b), (1970a) in which he found exact bounds for nonhomogeneous Markov chains. This method was investigated under other weak dependence conditions in (1962), (1974), (1977a,b). Many of the papers mentioned above (see, for example, Petrov (1960), Iosifescu (1968), Philipp (1969b), Egorov (1970) and Bulinskii (1977)) employed methods of proof based on Bernstein's fruitful idea of splitting the sum Z_n of r.v.'s into two sums $Z_n = U_n + V_n$. Then by virtue of the weak dependency, U_n behaves as if its terms are independent and V_n has a relatively small variance. For example, when Z_n is centered and normalized, this was accomplished as a rule via the following inequality (Petrov (1960)): For any positive ε and $x \in R$,

$$|\mathbf{P}\{U_n + V_n < x\} - \Phi(x)| \leq \sup_x |\mathbf{P}\{U_n < x\} - \Phi(x)| + \frac{\varepsilon}{\sqrt{2\pi}} + \mathbf{P}\{|V_n| \geq \varepsilon\}, \tag{11}$$

where $\Phi(x)$ is the standard normal distribution function. However the estimates obtained in this way for the one-dimensional case are not optimal and the order is no better than $O(n^{-1/4})$ in the uniform metric.

Nevertheless, Bernstein's classical method has received further development in recent years in the estimation of the rate of convergence in the CLT for weakly dependent n-dimensional and infinite-dimensional r.v.'s. It is precisely in the infinite-dimensional case that this method has manifested its generality. But since we are confining ourselves to only the one-dimensional (that is, the R-valued) case, the reader interested in Bernstein's method in the n-dimensional and infinite-dimensional cases is referred to the papers by Hipp (1979b) and Zuparov (1983), (1984) (see also Dubrovin (1974), Lapinskas (1976), Gabbasov (1977), Sunklodas (1978)). Multi-dimensional and infinite-dimensional analogues of (11) may be found in Lapinskas (1976), Sunklodas (1978), Hipp (1979b) and Zuparov (1984). The passage to independent infinite-dimensional r.v.'s under absolute regularity can be accomplished by means of appropriate approximating inequalities in Gudynas (1989) and Eberlein (1984) (see also Yoshihara (1976), Hipp (1979b)), Zuparov (1983), (1984)).

In (1972), Stein gave a new way of estimating the rate of convergence in the CLT for weakly dependent r.v.'s. Stein's method involves using a linear differential equation in terms of the difference between the distribution function (d.f.) of the sum of weakly dependent r.v.'s and the normal distribution. By means of this method, Stein was able to derive a uniform bound for the rate of convergence in the CLT of order $O(n^{-1/2}\ln^2 n)$ under complete regularity (this is weaker than u.s.m.), and the optimal order $O(n^{-1/2})$ for m-dependency. However, he required the eight-order moments of the terms to be finite and that the original sequence of r.v.'s be strictly stationary.

Developing Stein's idea further, Tikhomirov (1980) constructed a similar differential equation for the c.f. of a sum of weakly dependent one-dimensional r.v.'s. Using this, he succeeded in obtaining the wanted bound for how close this c.f. is to the c.f. of the normal distribution. Finally, under minimal restrictions on the moments (the finiteness of the s-th order absolute moments, $2 < s \leq 3$), in particular Tikhomirov derived the following uniform estimates for the rate of convergence in the CLT:

(a) when the s.m. coefficient decreases exponentially, the order is $O(n^{-(s-2)/2}\,(\ln n)^{s-1})$;
(b) for m-dependent r.v.'s, the optimal estimate is of order $O(n^{-(s-2)/2})$.

Tikhomirov (1980) also obtained a nonuniform estimate for strongly mixing r.v.'s (Theorem 11).

However the class of r.v.'s considered in (1980) is narrowed down by the requirement of strict stationarity. Schneider (1981) considered a uniformly strong mixing sequence and he replaced strict stationarity and finiteness of the s-th order absolute moments of the variables, $2 < s \leq 3$, by uniform boundedness of the third-order absolute moments of the variables and linear growth of the variance of their sum. Then assuming the exponential decay of the u.s.m. coefficient, he obtained a uniform estimate for the rate of convergence in the CLT of order $O(n^{-1/2} \ln n)$.

Specifically devoted to the m-dependency case are the papers by Shergin (1976), (1979), (1983) and Heinrich (1982), (1984), (1985a,b,d). They employ the c.f. to work out differing ways of estimating the rate of convergence in the one-dimensional CLT to optimal order.

Yudin solved a more general problem in (1981)–(1991). He gives a general method of estimating the rate of convergence of the distributions of sums of weakly dependent r.v.'s to infinitely divisible distributions. His method is also applicable to the normal case. In addition, he studies how to approximate the distributions of sums of weakly dependent r.v.'s by distributions in the class L (1989a,b), (1991). We point out also that Yudin uses Bernstein's method of partitioning a sum and the basic idea behind Tikhomirov's method (1987), (1989)–(1991) when approximating the distributions of sums of weakly dependent r.v.'s by the distributions in class L. These studies are represented in detail in Yudin's book (1990).

Because they yield optimal estimates for the rate of convergence in the one-dimensional CLT for weakly dependent r.v.'s under minimal restrictions on the moments of the terms and because they are simple to use and are capable of solving other problems, the methods of Stein, Tikhomirov and Heinrich are currently some of the most extensively used techniques. Therefore we shall give a more detailed presentation below of precisely these three methods. Before doing this, we introduce some notation.

$\mathcal{N}$ will denote a real r.v. with standard normal d.f. $\Phi(x)$ and density $\varphi(x) = \Phi'(x)$.

Later on, we shall estimate the difference $\mathbf{E}h(Z_n) - \mathbf{E}h(\mathcal{N})$, where Z_n is a normalized sum of weakly dependent centered r.v.'s and h is either the indicator of an interval or satisfies a smoothness condition.

For any function $g : R \to R$, let

$$\mathcal{L}(g; p, \alpha) = \sup_{x \neq y} \frac{|g(x) - g(y)|}{|x - y|^{\alpha}(1 + |x|^p + |y|^p)}, \quad ||g||_{\infty} = \sup_x |g(x)|,$$

and g' be the derivative of g.

In §2, 2.3. and §3, we shall assume that $h : R \to R$ satisfies $||h||_{\infty} < \infty$ (except in Theorems 3 and 16) and one of the conditions $H_1^{(p)} = \mathcal{L}(h; p, \alpha) < \infty$ or $H_2^{(p)} = \mathcal{L}(h'; p+1, \alpha) < \infty$ with $p \geq 0$ and $0 < \alpha \leq 1$.

We now single out the space $BL(R)$ of bounded functions $h : R \to R$ satisfying a Lipschitz condition, that is, such that

$$||h||_{\infty} < \infty \quad \text{and} \quad ||h||_L = \sup_{x \neq y} \frac{|h(x) - h(y)|}{|x - y|} < \infty.$$

Write

$$||h||_{BH_i^{(p)}} = ||h||_\infty + H_i^{(p)}, \quad ||h||_{BL} = ||h||_\infty + ||h||_L.$$

Let

$$X_1, X_2, \ldots \tag{12}$$

be a sequence of real r.v.'s with $\mathbf{E}X_j = 0$ and $\mathbf{E}X_j^2 < \infty$, $j = 1, 2, \ldots, n$. Let

$$S_n = \sum_{j=1}^n X_j, \quad B_n^2 = \mathbf{E}S_n^2, \quad Z_n = S_n/B_n, \quad F_n(x) = \mathbf{P}\{Z_n < x\},$$

$$A_j = X_j/B_n, \quad \overline{A}_j = A_j 1_{(|A_j| \le t)}, \quad \overline{\overline{A}}_j = A_j 1_{(|A_j| > t)},$$

$$L_{r,n} = \sum_{j=1}^n \mathbf{E}|A_j|^r, \quad \overline{L}_{r,n} = \sum_{j=1}^n \mathbf{E}|\overline{A}_j|^r, \quad \overline{\overline{L}}_{r,n} = \sum_{j=1}^n \mathbf{E}|\overline{\overline{A}}_j|^r,$$

$$L_{s,n}^* = nd_{s,n}, \quad d_{s,n} = \max_{1 \le j \le n} \mathbf{E}|A_j|^s,$$

$$\Delta_n(x) = \mathbf{P}\{Z_n < x\} - \Phi(x), \quad \Delta_n = \sup_x |\Delta_n(x)|, \quad ||\cdot||_1 = \int_{-\infty}^{\infty} |\cdot|\,dx,$$

$$d_i^{(p)}(F_n, \Phi) = \sup_{h \in \mathcal{H}_i^{(p)}} |\mathbf{E}h(Z_n) - \mathbf{E}h(\mathcal{N})|/||h||_{BH_i^{(p)}},$$

$$d_{BL}(F_n, \Phi) = \sup_{h \in BL(R)} |\mathbf{E}h(Z_n) - \mathbf{E}h(\mathcal{N})|/||h||_{BL},$$

where $\mathcal{H}_i^{(p)} = \{h : ||h||_\infty < \infty,\ H_i^{(p)} < \infty\}$, $i = 1, 2$, and 1_A is the indicator of event A, $t > 0$ and $B_n > 0$.

The quantities $d_i^{(p)}(F_n, \Phi)$ and $d_{BL}(F_n, \Phi)$ may be expressed as follows:

$$d_i^{(p)}(F_n, \Phi) = \sup_{\substack{h \in \mathcal{H}_i^{(p)} \\ ||h||_{BH_i^{(p)}} \le 1}} |\mathbf{E}h(Z_n) - \mathbf{E}h(\mathcal{N})|,$$

$$d_{BL}(F_n, \Phi) = \sup_{\substack{h \in BL(R) \\ ||h||_{BL} \le 1}} |\mathbf{E}h(Z_n) - \mathbf{E}h(\mathcal{N})|.$$

The quantity $d_{BL}(F_n, \Phi)$ is known as the bounded Lipschitz distance between the d.f.'s F_n and Φ.

Later on, we shall omit the subscript n in the notation $L_{r,n}$, $\overline{L}_{r,n}$, $\overline{\overline{L}}_{r,n}$, $L_{s,n}^*$ and $d_{s,n}$ and instead of $d_i^{(p)}(F_n, \Phi)$ and $d_{BL}(F_n, \Phi)$ we shall write $d_i^{(p)}$ and d_{BL}.

The letter $C(\cdot)$ with or without a subscript will denote a finite positive constant (not always the same one) that depends on the quantities indicated in the parentheses. C is an absolute positive constant; Θ is a complex function not exceeding one in modulus; $0 < K < \infty$ and $\lambda > 0$ are constants.

We shall show later how the methods of Stein, Tikhomirov and Heinrich yield upper bounds for $||\Delta_n(x)||_1$, $d_i^{(p)}$, d_{BL} and Δ_n if the sequence (12) satisfies one of the above weak dependence conditions.

The bounds for $||\Delta_n(x)||_1$ (Theorem 1 and Corollary 1), $d_i^{(p)}$ (Theorems 3–8 and Corollaries 2, 3) and Δ_n (Theorem 10) for the mixing coefficients described above are new and are due to the author. More general results are contained in Propositions 2–4.

To describe the methods of Stein and Tikhomirov, we shall make use of the r.v.'s

$$Z_j^{(i)} = \sum_{|p-j|\le im} A_p \quad \text{and} \quad z_j^{(i)} = Z_n - Z_j^{(i)}$$

with $2mi+1 < n$ and $m = 1, 2, \ldots$; in the case of m-dependence, $m = 0, 1, \ldots$. The bounds derived by Stein's method make use of a subsidiary r.v. J which is uniformly distributed on the set $\{1, 2, \ldots, n\}$ and independent of the r.v.'s $X_1, X_2, \ldots, X_n$.

Let

$$c_* = \begin{cases} m+1 & \text{under } m\text{-dependency,} \\ 1+2\psi_n & \text{under } \psi\text{-mixing,} \\ 1+4\Phi_n & \text{under u.s.m.,} \\ 1+12B_{n,r} & \text{under a.r.,} \\ 1+12A_{n,r} & \text{under s.m.} \end{cases}$$

where $\psi_n = \sum_{\tau=1}^{n-1} \psi(\tau)$, $\Phi_n = \sum_{\tau=1}^{n-1} \varphi^{1/2}(\tau)$, $B_{n,r} = \sum_{\tau=1}^{n-1} (\beta(\tau))^{(r-2)/r}$, $A_{n,r} = \sum_{\tau=1}^{n-1} (\alpha(\tau))^{(r-2)/r}$ and $2 < r < \infty$.

Applying the respective inequalities (1), (3) and (6), we find when the sequence (12) is either m-dependent, ψ-mixing or u.s.m. that

$$\mathbf{E}Z_n^2 \le c_* L_2. \tag{13}$$

For a.r. or s.m.,

$$\mathbf{E}Z_n^2 \le c_* \sum_{j=1}^{n} \mathbf{E}^{2/r}|A_j|^r, \quad 2 < r < \infty. \tag{14}$$

However, for $2 < r < \infty$,

$$\sum_{j=1}^{n} \mathbf{E}^{2/r}|A_j|^r \le n^{(r-2)/r} L_r^{2/r}. \tag{15}$$

Therefore for the weakly dependent r.v.'s defined above in (12) with ψ_n, Φ_n, $B_{n,r}$ and $A_{n,r}$ respectively finite, the estimates (13)–(15) imply that

$$n^{-(r-2)/2} \le c_*^{r/2} L_r \tag{16}$$

for $2 < r < \infty$. Consequently if $C_1(\cdot)(m+1) \ge n$, then

$$1 \le (C_1(\cdot))^{(r-2)/2} c_*^{r/2} (m+1)^{(r-2)/2} L_r. \tag{17}$$

By virtue of (17), it suffices to consider the case where $C_1(\cdot)(m+1) < n$ when estimating $||\Delta_n(x)||_1$, $d_i^{(p)}$, d_{BL} and Δ_n.

2.2. Estimation of $||\Delta_n(x)||_1$.

Theorem 1. *Suppose that the sequence* (12) *is m-dependent. Then for $m \ge 0$ and $t > 0$,*

$$||\Delta_n(x)||_1 \le C\{\overline{\overline{L}}_1 + (m+1)\overline{\overline{L}}_2 + (m+1)^2\overline{L}_3 + (m+1)^3\overline{L}_4\}.$$

If the truncation level in this estimate is taken to be $t = 1$ and $\overline{\overline{L}}_1$ is replaced by $(m+1)\overline{\overline{L}}_1$, then one obtains Erickson's result (1974) (see also Sunklodas (1982)).

Corollary 1. *Let the hypotheses of Theorem 1 hold and let $\mathbf{E}|X_j|^s < \infty$, $2 < s \le 3$ and $j = 1, 2, \ldots, n$. Then for $m \ge 0$,*

$$||\Delta_n(x)||_1 \le C(m+1)^{s-1} L_s.$$

By means of (24) below, similar bounds may be obtained for $||\Delta_n(x)||_1$ in the case of the mixing coefficients ψ, φ and β. Bounds were found by Sunklodas (1982), (1986) and Takahata (1983) for other mixing coefficients. For the sake of simplifying the presentation, we confine ourselves here to estimating $||\Delta_n(x)||_1$ just for m-dependent r.v.'s.

Heinrich's results for m-dependent r.v.'s (1985d) imply this particular one.

Theorem 2 (Heinrich (1985d)). *Suppose that the sequence* (12) *is m-dependent and that $\mathbf{E}|X_j|^s < \infty$ with $2 < s < 3$ and $j = 1, 2, \ldots, n$. Then for $1 \le p < \infty$ and $n \ge 1$,*

$$\left(\int_{-\infty}^{\infty} (1+|x|)^{sp-1} |\Delta_n(x)|^p \, dx \right)^{1/p} \le C(m+1)^{s-1} L_s^*.$$

Shergin (1983) considered m-dependent r.v.'s (12) with finite k-th absolute moments, $k \ge 2$, such that $\sum_{j=1}^n \mathbf{E}X_j^2 \le M_0 B_n^2$ for all $n \ge n_0$, where M_0 and n_0 are positive constants. He used Stein's technique to derive a bound for $\int_{-\infty}^{\infty} |x|^\ell |\Delta_n(x)| dx$ in terms of Lyapunov's quotients for $0 \le \ell \le k-1$.

Proof of Theorem 1.

Stein pointed out the following characterization of the standard normal law: If the r.v. $W = \mathcal{N}$, then $\mathbf{E}f'(W) - \mathbf{E}[Wf(W)] = 0$ for a fairly broad class of functions f, and conversely (see Stein (1972), (1981)). Consequently, the quantity $\mathbf{E}f'(W) - \mathbf{E}[Wf(W)]$ is a good measure of the proximity of the r.v. W to $\mathcal{N}$.

Therefore to estimate the difference $\mathbf{E}h(W)-\mathbf{E}h(\mathcal{N})$, it suffices to estimate the right-hand side of the relation

$$\mathbf{E}h(W)-\mathbf{E}h(\mathcal{N})=\mathbf{E}f'(W)-\mathbf{E}[Wf(W)).$$

To this end, consider the linear differential equation

$$f'(y)-yf(y)=h(y)-\mathbf{E}h(\mathcal{N}). \tag{18}$$

Its solution is

$$f(y)=e^{y^2/2}\int_{-\infty}^{y}h_0(u)e^{-u^2/2}\,du=-e^{y^2/2}\int_{y}^{\infty}h_0(u)e^{-u^2/2}\,du, \tag{19}$$

where $h_0(y)=h(y)-\mathbf{E}h(\mathcal{N})$.

For instance, to estimate $\mathbf{P}\{W<x\}-\Phi(x)$, it is necessary to take h in (18) to be the indicator function of the interval, i.e., $h(y)=h_x(y)=1_{(-\infty,x)}(y)$. Then (18) and (19) become respectively

$$f_x'(y)-yf_x(y)=1_{(-\infty,x)}(y)-\Phi(x) \tag{20}$$

and

$$f_x(y)=\begin{cases}\Phi(-x)\Phi(y)/\varphi(y), & \text{if } y<x,\\ \Phi(x)\Phi(-y)/\varphi(y), & \text{if } y\ge x.\end{cases} \tag{21}$$

To estimate $||\Delta_n(x)||_1$, we make use of the following result.

Lemma 4 (Erickson (1974), Ho and Chen (1978)). *Let f_x be given by (21). Then for all real y,*

$$\int_{-\infty}^{\infty}|f_x(y)|dx=1 \quad \textit{and} \quad \int_{-\infty}^{\infty}|f_x'(y)|dx\le 1.$$

In order that all of the r.v.'s occurring in (22) and (24) below should be well defined, it is assumed further that $6m+1<n$.

For brevity, put $Q_j=A_jZ_j^{(1)}-\mathbf{E}(A_jZ_j^{(1)})$.

It follows from (20) that

$$\begin{aligned}\Delta_n(x)&=\mathbf{E}\left\{f_x'(Z_n)-\sum_{j=1}^{n}A_jf_x(Z_n)\right\}\\ &=\mathbf{E}\left\{f_x'(Z_n)+\sum_{j=1}^{n}A_j\left[f_x(z_j^{(1)})-f_x(Z_n)\right]-\sum_{j=1}^{n}A_jf_x(z_j^{(1)})\right\}.\end{aligned}$$

Applying the Newton–Leibniz formula to the differences $f_x(z_j^{(1)})-f_x(Z_n)$ and $f_x(u)-f_x(Z_n)$ and noting (20), we find that

$$\Delta_n(x) = \mathbf{E}f'_x(Z_n) + \sum_{j=1}^{n} \mathbf{E}\left\{A_j \int_{Z_n}^{z_j^{(1)}} u \int_{Z_n}^{u} f'_x(v)\,dv\,du\right\}$$
$$+ \frac{1}{2}\sum_{j=1}^{n} \mathbf{E}\{A_j(Z_j^{(1)})^2 f_x(Z_n)\} - \sum_{j=1}^{n} \mathbf{E}\{A_j f_x(z_j^{(1)})\}$$
$$+ \sum_{j=1}^{n} \mathbf{E}\left\{A_j \int_{Z_n}^{z_j^{(1)}} [1_{(-\infty,x)}(u) - 1_{(-\infty,x)}(Z_n)]du\right\}$$
$$- \sum_{j=1}^{n} \mathbf{E}\{A_j Z_j^{(1)} f'_x(Z_n)\}.$$

Similar reasoning leads to

$$\sum_{j=1}^{n} \mathbf{E}\{Q_j[f'_x(Z_n) - f'_x(z_j^{(2)})]\} = \sum_{j=1}^{n} \mathbf{E}\{Q_j Z_j^{(2)} f_x(Z_n)\}$$
$$+ \sum_{j=1}^{n} \mathbf{E}\left\{Q_j z_j^{(2)} \int_{z_j^{(2)}}^{Z_n} f'_x(u)du\right\}$$
$$+ \sum_{j=1}^{n} \mathbf{E}\{Q_j[1_{(-\infty,x)}(Z_n) - 1_{(-\infty,x)}(z_j^{(2)})]\}.$$

Since $\mathbf{E}Z_n^2 = 1$, we have

$$1 - \sum_{j=1}^{n} A_j Z_j^{(1)} = -\sum_{j=1}^{n} Q_j + \sum_{j=1}^{n} \mathbf{E}\{A_j z_j^{(1)}\}.$$

From the last three equalities, we conclude that (Sunklodas (1982))

$$\Delta_n(x) = \sum_{k=1}^{9} E_k(x), \tag{22}$$

where

$$E_1(x) = \sum_{j=1}^{n} \mathbf{E}\left\{A_j \int_{Z_n}^{z_j^{(1)}} u \int_{Z_n}^{u} f'_x(v)\,dv\,du\right\},$$
$$E_2(x) = \sum_{j=1}^{n} \mathbf{E}\left\{Q_j z_j^{(2)} \int_{Z_n}^{z_j^{(2)}} f'_x(u)\,du\right\},$$
$$E_3(x) = \frac{1}{2}\sum_{j=1}^{n} \mathbf{E}\{A_j(Z_j^{(1)})^2 f_x(Z_n)\},$$
$$E_4(x) = -\sum_{j=1}^{n} \mathbf{E}\{Q_j Z_j^{(2)} f_x(Z_n)\},$$

$$E_5(x) = \sum_{j=1}^{n} \mathbf{E}\left\{A_j \int_{Z_n}^{z_j^{(1)}} [1_{(-\infty,x)}(u) - 1_{(-\infty,x)}(Z_n)]\,du\right\},$$

$$E_6(x) = -\sum_{j=1}^{n} \mathbf{E}\left\{Q_j[1_{(-\infty,x)}(Z_n) - 1_{(-\infty,x)}(z_j^{(2)})]\right\},$$

$$E_7(x) = -\sum_{j=1}^{n} \mathbf{E}\{A_j f_x(z_j^{(1)})\},$$

$$E_8(x) = -\sum_{j=1}^{n} \mathbf{E}\{Q_j f_x'(z_j^{(2)})\},$$

$$E_9(x) = \sum_{j=1}^{n} \mathbf{E}\{A_j z_j^{(1)}\}\mathbf{E}f'(Z_n).$$

It is precisely (22) that plays a basic role in estimating $||\Delta_n(x)||_1$ for weakly dependent r.v.'s. A similar relation was used by Erickson (1974) for m-dependent r.v.'s.

Applying Lemma 4 and the fact that

$$\int_{-\infty}^{\infty} |1_{(-\infty,x)}(u) - 1_{(-\infty,x)}(v)|dx = |u - v|, \tag{23}$$

for any $u, v \in R$, we can deduce the next result from (22).

Basic Inequality (Sunklodas (1982)):

$$||\Delta_n(x)||_1 \leq I_1(m) + I_2(m) \tag{24}$$

with

$$I_1(m) = \frac{5}{6}n\mathbf{E}|A_J(Z_J^{(1)})^3| + n\mathbf{E}|A_J(Z_J^{(1)})^2| + 2n\mathbf{E}|Q_J Z_J^{(2)}| + \\ + \frac{n}{2}\mathbf{E}|A_J(Z_J^{(1)})^2(Z_J^{(2)} - Z_J^{(1)})| + n\mathbf{E}|Q_J Z_J^{(2)}(Z_J^{(3)} - Z_J^{(2)})|$$

and

$$I_2(m) = \frac{n}{2}\mathbf{E}|A_J(Z_J^{(1)})^2 z_J^{(2)}| + n\mathbf{E}|Q_J Z_J^{(2)} z_J^{(3)}| + \\ + \sum_{j=1}^{n}[||\mathbf{E}\{A_j f_x(z_j^{(1)})\}||_1 + ||\mathbf{E}\{Q_j f_x'(z_j^{(2)})\}||_1] + n|\mathbf{E}\{A_J z_J^{(1)}\}|.$$

Since $I_1(m)$ involves the moments of "close" r.v.'s, its estimation does not require the use of the weak dependence of $X_1, X_2, \ldots$

It is easy to see that for any real $r \geq 1$

$$\mathbf{E}|Z_J^{(i)}|^r \leq (2mi+1)^r n^{-1} L_r, \quad i = 0, 1, \ldots, \tag{25}$$

and that

$$\mathbf{E}|Z_J^{(i)} - Z_J^{(i-1)}|^r \le (2m)^r n^{-1} L_r, \quad i = 2, 3, \ldots \tag{26}$$

Therefore the application of Hölder's inequality and the estimates (25) and (26) yield

$$|I_1(m)| \le (2m+1)(18m+5)L_3 + \frac{1}{6}(2m+1)(128m^2+50m+5)L_4. \tag{27}$$

$I_2(m)$ has to be estimated separately for each weak dependence condition. Let the sequence (12) be m-dependent. Then

$$I_2(m) = \frac{n}{2}\mathbf{E}|A_J(Z_J^{(1)})^2 z_J^{(2)}| + n\mathbf{E}|Q_J Z_J^{(2)} z_J^{(3)}|.$$

Put

$$t_j^i = \mathbf{E}|Z_j^{(2)}|^i, \quad \delta_j^i = \mathbf{E}|Z_j^{(3)}|^i.$$

Then

$$\mathbf{E}|A_J(Z_J^{(1)})^2 z_J^{(2)}| \le \mathbf{E}|A_J(Z_J^{(1)})^2| + \mathbf{E}|A_J(Z_J^{(1)})^2 t_J^1|,$$
$$\mathbf{E}|Q_J Z_J^{(2)} z_J^{(3)}| \le \mathbf{E}|Q_J Z_J^{(2)}| + \mathbf{E}|Q_J Z_J^{(2)} \delta_J^1|.$$

If $\xi_j = \sum_{p\in B_j} A_p$ and $\tau_j^i = \mathbf{E}|\xi_j|^i$, where $B_j \subset \{1, 2, \ldots, n\}$ for any $j = 1, 2, \ldots, n$, it is easy to see that

$$\mathbf{E}|\tau_J^i|^{j/i} \le \mathbf{E}|\xi_J|^j, \quad 0 < i \le j. \tag{28}$$

Then (25), (26), (28) and Hölder's inequality imply that

$$\mathbf{E}|A_J(Z_J^{(1)})^2| \le (2m+1)^2 n^{-1} L_3,$$
$$\mathbf{E}|A_J(Z_J^{(1)})^2 t_J^1| \le (2m+1)^2(4m+1)n^{-1}L_4,$$
$$\mathbf{E}|Q_J Z_J^{(2)}| \le 2(2m+1)(4m+1)n^{-1}L_3,$$
$$\mathbf{E}|Q_J Z_J^{(2)}\delta_J^1| \le 2(2m+1)(4m+1)(6m+1)n^{-1}L_4.$$

Consequently,

$$I_2(m) \le (2m+1)[(9m+2.5)L_3 + (52m^2+23m+2.5)L_4). \tag{29}$$

Therefore substituting (27) and (29) in (24) and dropping the condition $6m+1 < n$ by means of (17), we obtain the following result.

Proposition 1. *Suppose that the sequence* (12) *is m-dependent and* $\mathbf{E}X_j^4 < \infty$, $j = 1, 2, \ldots, n$. *Then for* $m \ge 0$,

$$||\Delta_n(x)||_1 \le (2m+1)[(27m+7.5)L_3 + (10/3)(22m^2+9.4m+1)L_4).$$

Truncation. To complete the proof of Theorem 1, we truncate the r.v.'s A_j, $j = 1, 2, \dots, n$, at the level $t > 0$ (§2, 2.1); in addition we put

$$\overline{A}_j^{(0)} = \overline{A}_j - \mathbf{E}\overline{A}_j, \quad \overline{\overline{A}}_j^{(0)} = \overline{\overline{A}}_j - \mathbf{E}\overline{\overline{A}}_j, \quad \overline{Z}_n^{(0)} = \sum_{j=1}^{n} \overline{A}_j^{(0)}.$$

Since any pair of r.v.'s ξ and η satisfies the inequality (see Erickson (1974), p. 527)

$$||\mathbf{P}\{\xi < x\} - \mathbf{P}\{\eta < x\}||_1 \le \mathbf{E}|\xi - \eta|, \tag{30}$$

it follows that

$$||\Phi(ax) - \Phi(x)||_1 \le \sqrt{2/\pi}\,|1 - a^{-1}| \tag{31}$$

for any positive a.

Therefore when $\mathbf{E}(\overline{Z}_n^{(0)})^2 > 0$, the sequence (12) satisfies

$$\begin{aligned} ||\Delta_n(x)||_1 \le\ & 2\overline{\overline{L}}_1 + \sqrt{2/\pi}\,|1 - \mathbf{E}(\overline{Z}_n^{(0)})^2| + \\ & + \mathbf{E}^{1/2}(\overline{Z}_n^{(0)})^2 ||\mathbf{P}\{\overline{Z}_n^{(0)} < x\mathbf{E}^{1/2}(\overline{Z}_n^{(0)})^2\} - \Phi(x)||_1. \end{aligned} \tag{32}$$

We note that the truncation inequality (32) holds for any dependency of the sequence (12).

Since

$$|1 - \mathbf{E}(\overline{Z}_n^{(0)})^2| \le 9\overline{\overline{L}}_2 + \sum_{1 \le i \ne j \le n} |\mathbf{E}(\overline{\overline{A}}_i^{(0)}\overline{\overline{A}}_j^{(0)}) + 2\mathbf{E}(\overline{A}_i^{(0)}\overline{\overline{A}}_j^{(0)})|, \tag{33}$$

any m-dependent sequence (12) satisfies

$$|1 - \mathbf{E}(\overline{Z}_n^{(0)})^2| \le 9(2m+1)\overline{\overline{L}}_2 = \varepsilon_1. \tag{34}$$

To prove Theorem 1, it is assumed that $\varepsilon_1 \le 1/2$ (otherwise the estimates are trivial). From (34), we have that $\frac{1}{2} \le \mathbf{E}(\overline{Z}_n^{(0)})^2 \le \frac{3}{2}$ and consequently,

$$\sum_{j=1}^{n} \mathbf{E}|\overline{A}_j^{(0)}/\mathbf{E}^{1/2}(\overline{Z}_n^{(0)})^2|^r \le 2^{3r/2}\overline{L}_r \tag{35}$$

for positive r.

By virtue of the truncation inequality (32) and the bounds (34) and (35), to complete the proof of Theorem 1, it is sufficient to estimate $||\Delta_n(x)||_1$ for untruncated, r.v.'s. In other words, we make use of Proposition 1.

Corollary 1 follows from Theorem 1 with $t = (m+1)^{-1}$.

2.3. Estimation of $d_i^{(p)}$ and d_{BL}.

Theorem 3. *Suppose that the sequence* (12) *is m-dependent with* $\mathbf{E}|X_j|^{2+p+\alpha} < \infty$, $j = 1, 2, \ldots, n$. *If* $6m + 1 < n$, *then*

$$|\mathbf{E}h(Z_n) - \mathbf{E}h(\mathcal{N})| \le C(p, \alpha) H_i^{(p)} \{(m+1)^{1+\alpha} L_{2+\alpha}(1 + \mathbf{E}|Z_n|^p) + (m+1)^{1+p+\alpha} L_{2+p+\alpha}\}.$$

Thus by Theorem 3, the estimation of $|\mathbf{E}h(Z_n) - \mathbf{E}h(\mathcal{N})|$, where h satisfies either the condition $H_1^{(p)} < \infty$ or $H_2^{(p)} < \infty$, reduces to estimating the absolute moment $\mathbf{E}|Z_n|^p$, $p \ge 0$.

Bounds for the absolute moments of a sum of weakly dependent r.v.'s were obtained by Utev (1984). For ψ-mixing or absolutely regular r.v.'s, estimating $d_i^{(p)}$, $i = 1, 2$, can also be reduced to estimating the absolute moments of Z_n (although of a higher order than p). However, for the sake of simplicity, we shall only consider here the estimation of $d_i^{(0)}$, $i = 1, 2$.

Theorem 4. *Let the sequence* (12) *be* ψ*-mixing with* $\psi(\tau) \le K\tau^{-\mu}$, *where* $\mu \ge (r-1)r$ *and let* $\mathbf{E}|X_j|^r < \infty$, $4 \le r < \infty$ *and* $j = 1, 2, \ldots, n$. *Then*

$$d_i^{(0)} \le C(K, \mu, r, \alpha)\{n^{(1+\alpha)(r-1)/\mu} L_{2+\alpha} + L_r\}.$$

Theorem 5. *Let the sequence* (12) *be absolutely regular and let* $\mathbf{E}|X_j|^r < \infty$, $4 \le r < \infty$ *and* $j = 1, 2, \ldots, n$.

1. *If* $\beta(\tau) \le Ke^{-\lambda\tau}$, *then*

$$d_i^{(0)} \le C(K, \lambda, r, \alpha)\{L_{2+\alpha} \ln^{1+\alpha}(n+1) + L_r\}.$$

2. *If* $\beta(\tau) \le K\tau^{-\mu}$ *with* $\mu \ge 2(r-1)r$, *then*

$$d_i^{(0)} \le C(K, \mu, r, \alpha)\{n^{2(1+\alpha)(r-1)/\mu} L_{2+\alpha} + L_r\}.$$

Truncation of the r.v.'s leads, for example, to the following results.

Theorem 6. *Let the sequence* (12) *be m-dependent. Then for* $m \ge 0$ *and* $t > 0$

$$d_1^{(0)} \le C(\alpha)\{\overline{\overline{L}}_1^{\alpha} + (m+1)^\alpha \overline{\overline{L}}_2^{\alpha} + (m+1)^{1+\alpha} \overline{L}_{2+\alpha}\}.$$

Corollary 2. *Let the hypotheses of Theorem 6 hold.*

1. *If* $\mathbf{E}|X_j|^{2+\alpha} < \infty$, $j = 1, 2, \ldots, n$, *then for* $m \ge 0$

$$d_1^{(0)} \le C(\alpha)(m+1)^{\alpha(1+\alpha)} L_{2+\alpha}^{\alpha}.$$

2. *If* $\mathbf{E}|X_j|^s < \infty$, $2 < s \le 3$ *and* $j = 1, 2, \ldots, n$, *then for* $m \ge 0$

$$d_{BL} \le C(m+1)^{s-1} L_s.$$

Theorem 7. *Suppose that* (12) *is ψ-mixing with $\psi(\tau) \le Ke^{-\lambda\tau}$. Then*

$$d_1^{(0)} \le C(K, \lambda, r, \alpha)\{\overline{\overline{L}}_1^{\alpha} + \overline{\overline{L}}_2^{\alpha} + \overline{L}_{2+\alpha} \ln^{1+\alpha}(n+1) + \overline{L}_r\}$$

for $t > 0$ and $4 \le r < \infty$.

Corollary 3. *Suppose that the hypotheses of Theorem 7 hold.*

1. *If* $\mathbf{E}|X_j|^{2+\alpha} < \infty$, $j = 1, 2, \ldots, n$, *then*

$$d_1^{(0)} \le C(K, \lambda, \alpha) L_{2+\alpha}^{\alpha} \ln^{\alpha(1+\alpha)}(n+1).$$

2. *If* $\mathbf{E}|X_j|^s < \infty$, $2 < s \le 3$ *and* $j = 1, 2, \ldots, n$, *then*

$$d_{BL} \le C(K, \lambda) L_s \ln^{s-1}(n+1).$$

Theorem 8. *Suppose that* (12) *is ψ-mixing with $\psi(\tau) \le K\tau^{-\mu}$, where $\mu \ge 12$.*

1. *If* $\mathbf{E}|X_j|^{2+\alpha} < \infty$, $j = 1, 2, \ldots, n$, *then*

$$d_1^{(0)} \le C(K, \mu, \alpha) n^{3(1+\alpha)/\mu} L_{2+\alpha}^{\alpha}.$$

2. *If* $\mathbf{E}|X_j|^s < \infty$, $2 < s \le 3$ *and* $j = 1, 2, \ldots, n$, *then*

$$d_{BL} \le C(K, \mu) n^{3(s-1)/\mu} L_s.$$

The reader can learn more details about estimating with the bounded Lipschitz metric d_{BL} by consulting Sunklodas (1989).

Proofs of Theorems 3–8.
The following result is true.

Lemma 5. *Let f and f' be given by* (18) *and* (19).

1. *If* $||h||_\infty < \infty$, *then* (see Erickson (1974), Barbour and Eagleson (1985))

$$|f(y)| \le c_1 ||h_0||_\infty \quad \text{and} \quad |f'(y)| \le 2||h_0||_\infty \tag{36}$$

for $\forall y \in R$, where $c_1 = \sup_{x \ge 0} \Xi(x)$ and $\Xi(x) = \Phi(-x)/\varphi(x)$.

2. *If* $H_1^{(p)} < \infty$ *or* $H_2^{(p)} < \infty$, *then* (Barbour (1986))

$$\mathcal{L}(f'; p, \alpha) \le C_i(p, \alpha) H_i^{(p)}, \tag{37}$$

where $i = 1$ if $H_1^{(p)} < \infty$ and $i = 2$ if $H_2^{(p)} < \infty$.

Furthermore, let $6m+1<n$. Then (18) implies that

$$\mathbf{E}h_0(Z_n)=\sum_{j=1}^{n}\{\mathbf{E}(A_jZ_j^{(1)})\mathbf{E}f'(Z_n)-\mathbf{E}[A_jf(Z_n)]+\mathbf{E}(A_jz_j^{(1)})\mathbf{E}f'(Z_n)\}.$$

In the summation, we first add and subtract $\mathbf{E}(A_jZ_j^{(1)})\mathbf{E}f'(z_j^{(1)})$, $\mathbf{E}[A_jf(z_j^{(1)})]$ and $\mathbf{E}[A_jZ_j^{(1)}f'(z_j^{(1)})]$ and then $\mathbf{E}[Q_jf'(z_j^{(2)})]$. Then using the identity

$$f(Z_n)-f(z_j^{(1)})-f'(z_j^{(1)})Z_j^{(1)}=\int_0^{Z_j^{(1)}}[f'(z_j^{(1)}+u)-f'(z_j^{(1)})]du,$$

we find that

$$\mathbf{E}h(Z_n)-\mathbf{E}h(\mathcal{N})=I_1+I_2, \tag{38}$$

where

$$\begin{aligned}I_1=&\sum_{j=1}^{n}\mathbf{E}(A_jZ_j^{(1)})\mathbf{E}[f'(Z_n)-f'(z_j^{(1)})]\\&-\sum_{j=1}^{n}\mathbf{E}\{Q_j[f'(z_j^{(1)})-f'(z_j^{(2)})]\}\\&-\sum_{j=1}^{n}\mathbf{E}\left\{A_j\int_0^{Z_j^{(1)}}[f'(z_j^{(1)}+u)-f'(z_j^{(1)})]du\right\}\end{aligned}$$

and

$$\begin{aligned}I_2=&-\sum_{j=1}^{n}\mathbf{E}\{Q_jf'(z_j^{(2)})\}+\sum_{j=1}^{n}\mathbf{E}(A_jz_j^{(1)})\mathbf{E}f'(Z_n)\\&-\sum_{j=1}^{n}\mathbf{E}[A_jf(z_j^{(1)})].\end{aligned}$$

Put

$$\begin{aligned}\nu_j^i&=\mathbf{E}|Z_j^{(1)}|^i,\ \gamma_j^i=\mathbf{E}|Z_j^{(2)}-Z_j^{(1)}|^i,\ t_j^i=\mathbf{E}|Z_j^{(2)}|^i,\ \delta_j^i=\mathbf{E}|Z_j^{(3)}|^i,\\ \omega_j&=\mathbf{E}|(Z_j^{(1)})^\alpha(Z_j^{(2)}-Z_j^{(1)})^p|,\ w_j=\mathbf{E}|(Z_j^{(2)}-Z_j^{(1)})^\alpha(Z_j^{(3)}-Z_j^{(2)})^p|.\end{aligned}$$

Estimating $|I_1|$ with the help of (37), we obtain

$$|I_1|\le C(p,\alpha)H_i^{(p)}(nI_1'+I_1''), \tag{39}$$

where

$$\begin{aligned} I_1' = {} & \mathbf{E}|A_J Z_J^{(1)} \nu_J^\alpha| + \mathbf{E}|A_J Z_J^{(1)} \nu_J^{p+\alpha}| + \mathbf{E}|A_J Z_J^{(1)} \omega_J| \\ & + \mathbf{E}|A_J Z_J^{(1)} \gamma_J^\alpha| + \mathbf{E}|A_J Z_J^{(1)} \gamma_J^{p+\alpha}| + \mathbf{E}|A_J Z_J^{(1)} w_J| \\ & + \mathbf{E}|A_J (Z_J^{(1)})^{1+\alpha}| + \mathbf{E}|A_J (Z_J^{(1)})^{1+p+\alpha}| \\ & + \mathbf{E}|A_J (Z_J^{(1)})^{1+\alpha} (Z_J^{(2)} - Z_J^{(1)})^p| + \mathbf{E}|A_J Z_J^{(1)} (Z_J^{(2)} - Z_J^{(1)})^\alpha| \\ & + \mathbf{E}|A_J Z_J^{(1)} (Z_J^{(2)} - Z_J^{(1)})^{p+\alpha}| \\ & + \mathbf{E}|A_J Z_J^{(1)} (Z_J^{(2)} - Z_J^{(1)})^\alpha (Z_J^{(3)} - Z_J^{(2)})^p| \end{aligned}$$

and

$$\begin{aligned} I_1'' = {} & \sum_{j=1}^n \mathbf{E}|A_j Z_j^{(1)}| [\mathbf{E}|(Z_j^{(1)})^\alpha (z_j^{(2)})^p| + \mathbf{E}|(Z_j^{(2)} - Z_j^{(1)})^\alpha (z_j^{(3)})^p|] \\ & + \sum_{j=1}^n [\mathbf{E}|A_j (Z_j^{(1)})^{1+\alpha} (z_j^{(2)})^p| + \mathbf{E}|A_j Z_j^{(1)} (Z_j^{(2)} - Z_j^{(1)})^\alpha (z_j^{(3)})^p|]. \end{aligned}$$

Applying (25), (26), (28) and Hölder's inequalities, we find that $\mathbf{E}|A_J Z_J^{(1)} \nu_J^\alpha|$, $\mathbf{E}|A_J Z_J^{(1)} \gamma_J^\alpha|$, $\mathbf{E}|A_J (Z_J^{(1)})^{1+\alpha}|$ and $\mathbf{E}|A_J Z_J^{(1)} (Z_J^{(2)} - Z_J^{(1)})^\alpha|$ are bounded above by $(2m+1)^{1+\alpha} n^{-1} L_{2+\alpha}$ and the remaining terms of I_1' are bounded above by $(2m+1)^{1+p+\alpha} n^{-1} L_{2+p+\alpha}$. Therefore

$$I_1' \le C(p, \alpha) n^{-1} [(m+1)^{1+\alpha} L_{2+\alpha} + (m+1)^{1+p+\alpha} L_{2+p+\alpha}). \tag{40}$$

We point out that it would have been possible to manage without the weak dependency of the r.v.'s (12) because I_1' involves moments of close r.v.'s.

The quantities I_1'' and $|I_2|$ will be estimated separately for each weak dependency.

Let the sequence (12) be m-dependent. Since

$$\mathbf{E}|z_j^{(i)}|^p \le (1 \vee 2^{p-1})(\mathbf{E}|Z_n|^p + \mathbf{E}|Z_j^{(i)}|^p), \quad i = 1, 2,$$

the m-dependence implies that

$$\begin{aligned} I_1'' \le {} & (1 \vee 2^{p-1}) n \big([\mathbf{E}|A_J Z_J^{(1)} \nu_J^\alpha| + \mathbf{E}|A_J Z_J^{(1)} \gamma_J^\alpha| + \mathbf{E}|A_J (Z_J^{(1)})^{1+\alpha}| \\ & + \mathbf{E}|A_J Z_J^{(1)} (Z_J^{(2)} - Z_J^{(1)})^\alpha|] \mathbf{E}|Z_n|^p + \mathbf{E}|A_J Z_J^{(1)} \nu_J^\alpha t_J^p| \\ & + \mathbf{E}|A_J Z_J^{(1)} \gamma_J^\alpha \delta_J^p| + \mathbf{E}|A_J (Z_J^{(1)})^{1+\alpha} t_J^p| \\ & + \mathbf{E}|A_J Z_J^{(1)} (Z_J^{(2)} - Z_J^{(1)})^\alpha \delta_J^p| \big). \end{aligned} \tag{41}$$

The expression multiplying $\mathbf{E}|Z_n|^p$ has already been estimated. The estimates (25), (26), (28) and Hölder's inequality can be used to show that each of the remaining terms in (41) is bounded above by at least $(6m + 1)^{1+p+\alpha} n^{-1} L_{2+p+\alpha}$. Consequently,

$$I_1'' \le C(p, \alpha) [(m+1)^{1+\alpha} L_{2+\alpha} \mathbf{E}|Z_n|^p + (m+1)^{1+p+\alpha} L_{2+p+\alpha}). \tag{42}$$

$I_2 = 0$ by virtue of the m-dependence. Therefore to complete the proof of Theorem 3, it suffices to substitute (40) and (42) in (39).

When $p = 0$,

$$I_1'' = n[\mathbf{E}|A_J Z_J^{(1)} \nu_J^\alpha| + \mathbf{E}|A_J Z_J^{(1)} \gamma_J^\alpha| + \mathbf{E}|A_J (Z_J^{(1)})^{1+\alpha}| \\ + \mathbf{E}|A_J Z_J^{(1)} (Z_J^{(2)} - Z_J^{(1)})^\alpha|]$$

occurs in the expression for nI_1'. Thus from (39) and (40), we find that

$$|I_1| \le C(\alpha) H_i^{(0)} m^{1+\alpha} L_{2+\alpha}. \tag{43}$$

If $H_1^{(0)} < \infty$ or $H_2^{(0)} < \infty$, it follows from (38) and (43) that any weakly dependent sequence (12) satisfies

$$|\mathbf{E}h(Z_n) - \mathbf{E}h(\mathcal{N})| \le C(\alpha) H_i^{(0)} m^{1+\alpha} L_{2+\alpha} + |I_2|. \tag{44}$$

Suppose that (12) is ψ-mixing. Then estimating with (9) and using (36), we obtain

$$|I_2| \le 6||h_0||_\infty n[2\mathbf{E}^{1/2}(A_J Z_J^{(1)})^2 + (2^{3/2} + c_1)\mathbf{E}^{1/2} A_J^2 \\ + 2^{3/2}\mathbf{E}^{1/2}(A_J^2 \nu_J^2)]\psi^{1/2}(m+1). \tag{45}$$

By virtue of (25), (28) and Hölder's inequality,

$$|I_2| \le C||h||_\infty n^{1/2}[L_2^{1/2} + mL_4^{1/2}]\psi^{1/2}(m+1). \tag{46}$$

The estimates (44) and (46) yield the following result.

Proposition 2. *Let sequence* (12) *be ψ-mixing and let* $\mathbf{E}X_j^4 < \infty$, $j = 1, 2, \ldots, n$. *Then*

$$d_i^{(0)} \le C(\alpha)\{m^{1+\alpha} L_{2+\alpha} + n^{1/2}[L_2^{1/2} + mL_4^{1/2}]\psi^{1/2}(m+1)\}$$

providing $6m + 1 < n$.

Under a.r., the quantity $|I_2|$ can be estimated with the help of (10) and in similar fashion we deduce the following

Proposition 3. *Let sequence* (12) *be absolutely regular and let* $\mathbf{E}X_j^4 < \infty$, $j = 1, 2, \ldots, n$. *Then*

$$d_i^{(0)} \le C(\alpha)\{m^{1+\alpha} L_{2+\alpha} + [n^{3/4} L_4^{1/4} + mn^{1/2} L_4^{1/2}]\beta^{1/4}(m+1)\}.$$

Theorems 4 and 5 are consequences of the respective Propositions 2 and 3.

Truncation. As in the estimation of $||\Delta_n(x)||_1$, the r.v.'s A_j, $j = 1, 2, \ldots, n$, are truncated at level $t > 0$ (§2, 2.1). Recall the subsidiary variables

$$\overline{A}_j^{(0)} = \overline{A}_j - \mathbf{E}\overline{A}_j \quad \text{and} \quad \overline{Z}_n^{(0)} = \sum_{j=1}^{n} \overline{A}_j^{(0)}.$$

Let $h \in \mathcal{H}_1^{(0)}$. Then any real r.v.'s ξ and η satisfy

$$|\mathbf{E}h(\xi) - \mathbf{E}h(\eta)| \le 3H_1^{(0)}\mathbf{E}|\xi - \eta|^\alpha. \tag{47}$$

It is easy to see that $\mathbf{E}|Z_n - \overline{Z}_n^{(0)}|^\alpha \le 2^\alpha \overline{\overline{L}}_1^\alpha$ and, when $\mathbf{E}(\overline{Z}_n^{(0)})^2 > 0$, that

$$\mathbf{E}|\overline{Z}_n^{(0)}(1 - \mathbf{E}^{-1/2}(\overline{Z}_n^{(0)})^2)|^\alpha \le |1 - \mathbf{E}(\overline{Z}_n^{(0)})^2|^\alpha.$$

Therefore estimating according to (47), we conclude that

$$\begin{aligned} |\mathbf{E}h(Z_n) - \mathbf{E}h(\mathcal{N})| \le\ & 3H_1^{(0)}[2^\alpha \overline{\overline{L}}_1^\alpha + |1 - \mathbf{E}(\overline{Z}_n^{(0)})^2|^\alpha] \\ & + |\mathbf{E}h(\overline{Z}_n^{(0)}/\mathbf{E}^{1/2}(\overline{Z}_n^{(0)})^2) - \mathbf{E}h(\mathcal{N})| \end{aligned} \tag{48}$$

when $h \in \mathcal{H}_1^{(0)}$ and $\mathbf{E}(\overline{Z}_n^{(0)})^2 > 0$.

The truncation inequality (48) clearly holds for any dependency of the sequence (12). Therefore, the proofs of Theorems 6–8 are completed on merely applying it to Theorems 3–5.

2.4. Estimation of Δ_n.

Theorem 9. *Let the sequence* (12) *be* m*-dependent and let* $\mathbf{E}|X_j|^s < \infty$, $2 < s \le 3$ *and* $j = 1, 2, \ldots, n$. *Then for* $m \ge 0$ *and* $n \ge 1$

$$\Delta_n \le C(m+1)^{s-1} L_s^*.$$

Theorem 10. *Let* (12) *be a strongly mixing sequence and let* $\mathbf{E}|X_j|^s < \infty$, $2 < s \le 3$ *and* $j = 1, 2, \ldots, n$.

1. *If* $\alpha(\tau) \le Ke^{-\lambda\tau}$, *then*

$$\Delta_n \le C(K, \lambda, s) L_s^* \ln^{s-1}(n+1).$$

2. *If* $\alpha(\tau) \le K\tau^{-\mu}$ *with* $\mu \ge \dfrac{2(s-1)}{s(s-2)^2}[\beta(5s-6) - (s-2)(3-s)]$ *and* $\beta > 1$, *then*

$$\Delta_n \le C(K, \beta, s) n^c d_s,$$

where $c = (4\beta + s^2 - 4)/(2(2\beta + s - 4))$.

For the exponential s.m. coefficient, the power of the logarithm may be lowered but under more stringent conditions on the moments of the terms. We now give a nonuniform bound in the case of strict stationarity.

Theorem 11 (Tikhomirov (1980)). *Suppose that* (12) *is a strictly stationary and strongly mixing sequence with* $\alpha(\tau) \leq Ke^{-\lambda\tau}$ *and that* $\mathbf{E}|X_1|^{4+\nu} < \infty$ *for some positive* ν. *Then*

$$\Delta_n \leq C(K, \lambda, \nu) n^{-1/2} \ln n$$

and

$$|\Delta_n(x)| \leq C(K, \lambda, \nu) \frac{\ln^3 n}{\sqrt{n}(1 + |x|)^4}.$$

Heinrich (1985d) obtained a nonuniform bound for $|\Delta_n(x)|$ for m-dependent r.v.'s. In particular, he proved the following result.

Theorem 12 (1985d). *Suppose that the sequence* (12) *is* m*-dependent and that* $\mathbf{E}|X_j|^s < \infty$, $2 < s \leq 3$ *and* $j = 1, 2, \ldots, n$. *Then*

$$|\Delta_n(x)| \leq C \frac{(m+1)^{s-1} L_s^*}{(1 + |x|)^s}$$

for all real x *and* $n \geq 1$.

When $s = 3$, the rate of convergence in Theorem 12 cannot be improved in relation to n, x (see Petrov (1987)) and m (see Berk (1973)).

We shall limit ourselves to proving Theorems 9 and 10.

Shergin (1979), (1990) was able to replace L_s^* by Lyapunov's quotient L_s in Theorems 9 and 12.

Tikhomirov (1986) obtained an estimate for the rate at which the d.f. of the $\max_{1 \leq k \leq n}(X_1 + \ldots + X_k)$ for a strictly stationary and strongly mixing sequence converges to the d.f.

$$\sqrt{2/\pi} \int_0^{x^+} e^{-u^2/2} \, du$$

in the uniform metric, where $x^+ = \max(0, x)$.

Proofs of Theorems 9 *and* 10.

Tikhomirov's method (presented in detail in Tikhomirov (1980) for a strictly stationary sequence) consists in deriving a linear differential equation for the c.f. $f_n(t) = \mathbf{E}e^{itZ_n}$ of a centered and normalized sum Z_n. This differential equation is "close" to the homogeneous differential equation $f'(t) = -tf(t)$ whose solution is the c.f. of the standard normal law, that is, $f(t) = e^{-t^2/2}$.

It will be recalled that Theorems 9 and 10 do not assume the stationarity of the r.v.'s (12). Similar bounds for Δ_n were found by Sunklodas (1984) for a sequence of either m-dependent or strongly mixing r.v.'s but under the additional condition $B_n^2 \geq c_0 n$, $0 < c_0 < \infty$.

In addition to the notation of §2, 2.1, write

$$\xi_j^{(l)} = e^{it(z_j^{(l-1)} - z_j^{(l)})} - 1, \quad \eta_j^{(r)} = e^{-itZ_j^{(r)}} - 1,$$

$$a_j^{(r-1)} = \mathbf{E}\left(iA_j \prod_{l=1}^{r-1} \xi_j^{(l)}\right), \quad z_j^{(0)} = Z_n.$$

Since $f_n'(t) = \sum_{j=1}^n \mathbf{E}(iA_j e^{itZ_n})$, by successively adding and subtracting the quantities $\mathbf{E}(iA_j e^{itz_j^{(1)}})$, $\mathbf{E}(iA_j \xi_j^{(1)} e^{itz_j^{(2)}})$, …, $\mathbf{E}(iA_j \prod_{l=1}^{k-1} \xi_j^{(l)} e^{itz_j^{(k)}})$ in the summation, one finds that (Tikhomirov (1980))

$$f_n'(t) = \sum_{j=1}^n \mathbf{E}(iA_j e^{itz_j^{(1)}}) + \sum_{j=1}^n \sum_{r=2}^k \mathbf{E}\left(iA_j \prod_{l=1}^{r-1} \xi_j^{(l)} e^{itz_j^{(r)}}\right)$$
$$+ \sum_{j=1}^n \mathbf{E}\left(iA_j \prod_{l=1}^{k} \xi_j^{(l)} e^{itz_j^{(k)}}\right).$$

The relation

$$\mathbf{E}e^{itz_j^{(r)}} = \mathbf{E}(\eta_j^{(r)} + 1) f_n(t) + \mathbf{E}[(\eta_j^{(r)} - \mathbf{E}\eta_j^{(r)}) e^{itZ_n}]$$

can be utilized to prove that the derivative of $f_n(t)$ with respect to t is (Sunklodas (1984))

$$f_n'(t) = (E_1 + E_2) f_n(t) + E_3 + E_4 + E_5 + E_6, \tag{49}$$

in which

$$E_1 = \sum_{j=1}^n a_j^{(1)}, \quad E_2 = \sum_{j=1}^n \left[a_j^{(1)} \mathbf{E}\eta_j^{(2)} + \sum_{r=3}^k a_j^{(r-1)} \mathbf{E}(\eta_j^{(r)} + 1)\right],$$

$$E_3 = \sum_{r=2}^k \sum_{j=1}^n a_j^{(r-1)} \mathbf{E}[(\eta_j^{(r)} - \mathbf{E}\eta_j^{(r)}) e^{itZ_n}],$$

$$E_4 = \sum_{j=1}^n \mathbf{E}\left(iA_j \prod_{l=1}^k \xi_j^{(l)} e^{itz_j^{(k)}}\right), \quad E_6 = \sum_{j=1}^n \mathbf{E}(iA_j e^{itz_j^{(1)}}),$$

$$E_5 = \sum_{r=2}^k \sum_{j=1}^n \mathbf{E}\left[iA_j \prod_{l=1}^{r-1} \xi_j^{(l)} (e^{itz_j^{(r)}} - \mathbf{E}e^{itz_j^{(r)}})\right].$$

It is assumed here that $2km + 1 < n$.

The linear differential equation (49) is the starting point for proving Theorems 9–11. The passage from (49) to the difference of the c.f.'s is accomplished by means of the following lemma.

Lemma 6 (Sunklodas (1984)). *Let the linear differential equation*

$$f'(t) = (-t + \theta a(t))f(t) + \theta b(t), \quad f(0) = 1, \tag{50}$$

be given for $|t| \le T_1$*, where*

$$\begin{aligned} a(t) &= a^{(0)} + a^{(1)}|t| + a^{(2)}t^2 + a^{(3)}|t|^{s-1}, \quad 2 < s \le 3, \\ b(t) &= b^{(0)} + b^{(2)}t^2. \end{aligned}$$

Here the coefficients $a^{(i)} \ge 0$ $(i = 0,1,2,3)$ *and* $b^{(j)} \ge 0$ $(j = 0,2)$ *are independent of* t*; and* θ *is a complex function such that* $|\theta| \le 1$.

If $a^{(1)} \le 1/6$*, then*

$$\begin{aligned} |f(t) - e^{-t^2/2}| \le{} & C[a^{(0)}|t| + a^{(1)}t^2 + a^{(2)}|t|^3 + a^{(3)}|t|^s]e^{-t^2/4} \\ & + C[b^{(0)}\min(|t|^{-1}, |t|) + b^{(2)}|t|] \end{aligned} \tag{51}$$

for $|t| \le \min(T_1, T_2)$*, where*

$$T_2 = \min\left\{\frac{1}{a^{(0)}}, \frac{1}{6a^{(2)}}, \left(\frac{1}{6a^{(3)}}\right)^{1/(s-2)}\right\}.$$

Lemma 6 and Essen's inequality (see, for example, Petrov (1987), p. 154) lead easily to the next assertion:

If the c.f. $f(t) = \mathbf{E}e^{it\xi}$ of a real r.v. ξ satisfies the linear differential equation (50) for $|t| \le T_1$, then

$$\sup_x |\mathbf{P}\{\xi < x\} - \Phi(x)| \le C(a^{(0)} + a^{(1)} + a^{(2)} + a^{(3)} + b^{(0)} + b^{(2)}T_1 + T_1^{-1}). \tag{52}$$

Therefore to estimate Δ_n, all attention will be directed to deriving linear differential equation (50) for $f_n(t)$ so that its coefficients and T_1 assure a bound for Δ_n as close as possible to being optimum.

Let us proceed to prove Theorem 9 and 10.

Assume that $\mathbf{E}|X_j|^s < \infty$, $2 < s \le 3$ and $j = 1, 2, \ldots, n$. For any dependency,

$$E_1 = -t\sum_{j=1}^{n} \mathbf{E}(A_j Z_j^{(1)}) + \theta(2^{3-s}/(s-1))(2m+1)^{s-1}L_s|t|^{s-1}. \tag{53}$$

Now let the sequence (12) be m-dependent. Then

$$E_1 = -t + \theta(2^{3-s}/(s-1))(2m+1)^{s-1}L_s|t|^{s-1}. \tag{54}$$

It is easily seen that

$$\mathbf{E}^{1/2}|\xi_j^{(l)}|^2 \le \sqrt{2}(m+1)d_s^{1/s}|t| \tag{55}$$

and that

$$\mathbf{E}^{1/2}|\eta_j^{(r)}|^2 \le \sqrt{2r}(m+1)d_s^{1/s}|t|. \tag{56}$$

By virtue of the m-dependence and (55),

$$|a_j^{(r-1)}| \le \mathbf{E}^{1/2} A_j^2 \prod_{l=1}^{r-1} \mathbf{E}^{1/2} |\xi_j^{(l)}|^2$$
$$\le d_s^{1/s} (\sqrt{2}(m+1) d_s^{1/s} |t|)^{r-1} = a^{(r-1)}. \tag{57}$$

Since

$$|E_2| \le \sum_{j=1}^{n} \left[|a_j^{(1)}| \mathbf{E}|\eta_j^{(2)}| + \sum_{r=3}^{k} |a_j^{(r-1)}| \right), \tag{58}$$

(56) and (57) can be used to show that

$$|E_2| \le C(m+1)^2 n d_s^{3/s} t^2 \tag{59}$$

for $|t| \le (\sqrt{2}e^2(m+1)d_s^{1/s})^{-1} = T_1$.

By the m-dependence and the estimates (56) and (57), we find for $|t| \le T_1$ that

$$|E_3| \le \sum_{r=2}^{k} a^{(r-1)} \left(\sum_{j=1}^{n} \sum_{|p-j| \le 3rm} \mathbf{E}^{1/2} |\eta_j^{(r)}|^2 \mathbf{E}^{1/2} |\eta_p^{(r)}|^2 \right)^{1/2}$$
$$\le C(m+1)^{5/2} n^{1/2} d_s^{3/s} t^2. \tag{60}$$

By virtue of (57),

$$|E_4| \le n a^{(k)} \le C(m+1)^2 n^{1/2} d_s^{3/s} t^2 \tag{61}$$

for $k \ge 2 + \frac{1}{4} \ln n$ and $|t| \le T_1$. Since $E_5 = E_6 = 0$ because of the m-dependence, on substituting (54), (59)–(61) in (49), we find that

$$f_n'(t) = (-t + \theta a_n(t)) f_n(t) + \theta b_n(t) \tag{62}$$

providing $2km + 1 < n$, $k \ge 2 + \frac{1}{4} \ln n$ and $|t| < T_1$, where

$$a_n(t) = a_n^{(2)} t^2 + a_n^{(3)} |t|^{s-1}, \qquad b_n(t) = b_n^{(2)} t^2,$$
$$a_n^{(2)} = C(m+1)^2 n d_s^{3/s}, \qquad a_n^{(3)} = C(m+1)^{s-1} L_s^*,$$
$$b_n^{(2)} = C(m+1)^{5/2} n^{1/2} d_s^{3/s}.$$

Relations (52) and (62) yield Theorem 9 if $m+1 < n/(C\ln(n+1))$, $C > 1$. Theorem 9 follows for all $m \ge 0$ from the third inequality in Lemma 7 of §2, 2.5 and Esseen's inequality.

Now suppose that (12) is a s.m. sequence. Write $z_j^{(i)} = \hat{z}_j^{(i)} + \hat{\hat{z}}_j^{(i)}$, where $\hat{z}_j^{(i)}$ is the sum of those A_p in $z_j^{(i)}$ for which $p < j - im$ and $\hat{\hat{z}}_j^{(i)}$ is the sum of A_p for which $p > j + im$.

Then according to (6),

$$\sum_{j=1}^{n} |\mathbf{E}(A_j z_j^{(1)})| \le \sum_{j=1}^{n} [|\mathbf{E}(A_j \hat{z}_j^{(1)})| + |\mathbf{E}(A_j \hat{\hat{z}}_j^{(1)})|]$$
$$\le 6n^2 d_s^{2/s} (\alpha(m+1))^{(s-2)/s}.$$

Therefore (53) yields

$$E_1 = -t + \theta(2^{3-s}/(s-1))(2m+1)^{s-1} L_s^* |t|^{s-1}$$
$$+ \theta 6n^2 d_s^{2/s} (\alpha(m+1))^{(s-2)/s} |t|. \tag{63}$$

Put $\hat{\xi}_j^{(l)} = e^{itx} - 1$, $\hat{\hat{\xi}}_j^{(l)} = e^{ity} - 1$, where $x = \sum_{p=j-lm}^{j-(l-1)m-1} A_p$, and $y = \sum_{p=j+(l-1)m+1}^{j+lm} A_p$. Then by Minkowski's inequality, for $1 \le \mu \le s$

$$\max\{\mathbf{E}^{1/\mu} |\hat{\xi}_j^{(l)}|^\mu,\ \mathbf{E}^{1/\mu} |\hat{\hat{\xi}}_j^{(l)}|^\mu\} \le (m+1) d_s^{1/s} |t| \tag{64}$$

and

$$\mathbf{E}^{1/\mu} |\eta_j^{(r)}|^\mu \le (2rm+1) d_s^{1/s} |t| = \eta^{(r)}. \tag{65}$$

Similarly,

$$a_j^{(1)} \le (2m+1) d_s^{2/s} |t|. \tag{66}$$

We next estimate $a_j^{(r-1)}$ for $r = 3, 4, \ldots, k$. Since $|\xi_j^{(l)}| \le |\hat{\xi}_j^{(l)}| + |\hat{\hat{\xi}}_j^{(l)}|$, we have

$$|a_j^{(r-1)}| \le \sum_{p=0}^{r-1} \sum\nolimits^* \mathbf{E} \left| A_j \prod_{\nu=1}^{p} \hat{\xi}_j^{(l_\nu)} \prod_{\mu=p+1}^{r-1} \hat{\hat{\xi}}_j^{(l_\mu)} \right|, \tag{67}$$

in which $\sum^*$ denotes summation over all collections of indices $1 \le l_1 < l_2 < \ldots < l_p \le r-1$ and $1 \le l_{p+1} < l_{p+2} < \ldots < l_{r-1} \le r-1$ such that $l_\nu \ne l_\mu$ for $\nu \ne \mu$ (see Tikhomirov (1980)).

By Hölder's inequality,

$$\mathbf{E} \left| A_j \prod_{\nu=1}^{p} \hat{\xi}_j^{(l_\nu)} \prod_{\mu=p+1}^{r-1} \hat{\hat{\xi}}_j^{(l_\mu)} \right|$$
$$\le \mathbf{E}^{1/s} \left| \prod_\nu{}' \hat{\xi}_j^{(l_\nu)} \prod_\mu{}' \hat{\hat{\xi}}_j^{(l_\mu)} \right|^s \mathbf{E}^{(s-1)/s} \left| A_j \prod_\nu{}'' \hat{\xi}_j^{(l_\nu)} \prod_\mu{}'' \hat{\hat{\xi}}_j^{(l_\mu)} \right|^{s/(s-1)},$$

where $\prod''$ and $\prod'$ are respectively products over all even and all odd l from 1 up to $r-1$.

Let $r-1$ be even (if $r-1$ is odd, one proceeds in the same way). Put $(\cdot) = (m+1) d_s^{1/s} |t|$. Then by (7), (8) and (64),

$$\mathbf{E}\left|\prod\nolimits'_{\nu}\hat{\xi}_j^{(l_\nu)}\prod\nolimits'_{\mu}\hat{\hat{\xi}}_j^{(l_\mu)}\right|^s \le (\cdot)^{s(r-1)/2}+8(r-3)2^{s(r-1)/2}\alpha(m+1)$$

and

$$\mathbf{E}\left|A_j\prod\nolimits''_{\nu}\hat{\xi}_j^{(l_\nu)}\prod\nolimits''_{\mu}\hat{\hat{\xi}}_j^{(l_\mu)}\right|^{s/(s-1)} \le d_s^{1/(s-1)}\big[(\cdot)^{s(r-1)/(2(s-1))}$$
$$+8(r-1)2^{s(r-1)/(2(s-1))}(\alpha(m+1))^{(s-2)/(s-1)}\big].$$

Therefore

$$\mathbf{E}\left|A_j\prod_{\nu=1}^{p}\hat{\xi}_j^{(l_\nu)}\prod_{\mu=p+1}^{r-1}\hat{\hat{\xi}}_j^{(l_\mu)}\right| \le 8d_s^{1/s}(\alpha(m+1))^{(s-1)/s}(r-1)2^{r-1}$$
$$+d_s^{1/s}(\cdot)^{r-1}+7d_s^{1/s}(2(\cdot))^{(r-1)/2}(\alpha(m+1))^{(s-2)/s}(r-1).$$

Since there are at most 2^{r-1} terms in (67), for $|t|\le(32(m+1)d_s^{1/s})^{-1}=T_3$ and $r=3,4,\dots,k$, we obtain

$$|a_j^{(r-1)}|\le C\left\{m^2d_s^{3/s}t^2\left(\frac{1}{2}\right)^{4r}\right.$$
$$\left.+d_s^{1/s}(\alpha(m+1))^{(s-2)/s}\left[r\left(\frac{1}{2}\right)^{r}+r4^r(\alpha(m+1))^{1/s}\right]\right\}=a^{(r-1)}. \quad (68)$$

It is now possible to estimate $|E_4|$. For $k\ge \ln n/(8\ln 2)\ge 3$ and $|t|\le T_3$, we find that

$$|E_4|\le na^{(k)}\le C[m^2n^{1/2}d_s^{3/s}t^2+d_s^{1/s}n(\alpha(m+1))^{(s-2)/s}). \quad (69)$$

We estimate the terms E_2, E_3 and E_5 subject to the additional condition

$$k^{3/2}4^k(\alpha(m+1))^{1/s}\le 1. \quad (70)$$

Substituting (66), (65) and (68) in (58), we find for $|t|\le T_3$ that

$$|E_2|\le C[m^2nd_s^{3/s}t^2+nd_s^{1/s}(\alpha(m+1))^{(s-2)/s}] \quad (71)$$

under condition (70).

Consider $\operatorname{cov}(\xi,\eta)=\mathbf{E}(\xi-\mathbf{E}\xi)\overline{(\eta-\mathbf{E}\eta)}$. By virtue of (6),

$$|E_3|\le\sum_{r=2}^{k}a^{(r-1)}\left(\sum_{j=1}^{n}\sum_{p=1}^{n}|\operatorname{cov}(\eta_j^{(r)},\eta_p^{(r)})|\right)^{1/2}$$
$$\le\sum_{r=2}^{k}a^{(r-1)}\left(\sum_{j=1}^{n}\sum_{|p-j|\le 2rm}\mathbf{E}^{1/2}|\eta_j^{(r)}|^2\mathbf{E}^{1/2}|\eta_p^{(r)}|^2\right.$$
$$\left.+\sum_{j=1}^{n}\sum_{|p-j|>2rm}24\mathbf{E}^{1/s}|\eta_j^{(r)}|^s\mathbf{E}^{1/s}|\eta_p^{(r)}|^s(\alpha(|p-j|-2rm))^{(s-2)/s}\right)^{1/2}.$$

Noting (66), (65) and (68), we obtain

$$|E_3| \le C[m^{1/2} + A_{n,s}^{1/2}][m^2 n^{1/2} d_s^{3/s} t^2 + n^{1/2} d_s^{1/s} (\alpha(m+1))^{(s-2)/s}] \tag{72}$$

for $|t| \le T_3$ under assumption (70).

The definition of $z_j^{(r)}$ shows that $\hat{z}_j^{(r)}$ and $\hat{\hat{z}}_j^{(r)}$ cannot vanish simultaneously. We shall consider that both do not vanish (otherwise the computations merely become simpler). Observe that

$$\begin{aligned}
&\left| \mathbf{E} \left[iA_j \prod_{l=1}^{r-1} \xi_j^{(l)} (e^{tiz_j^{(r)}} - \mathbf{E}e^{itz_j^{(r)}}) \right] \right| \\
&\le \left| \mathbf{E} \left(iA_j \prod_{l=1}^{r-1} \xi_j^{(l)} e^{itz_j^{(r)}} \right) - \mathbf{E} \left(iA_j \prod_{l=1}^{r-1} \xi_j^{(l)} e^{it\hat{z}_j^{(r)}} \right) \mathbf{E}e^{it\hat{\hat{z}}_j^{(r)}} \right| \\
&\quad + \left| \mathbf{E} \left[iA_j \prod_{l=1}^{r-1} \xi_j^{(l)} (e^{it\hat{z}_j^{(r)}} - \mathbf{E}e^{it\hat{z}_j^{(r)}}) \right] \right| \\
&\quad + \left| \mathbf{E}e^{it\hat{z}_j^{(r)}} \mathbf{E}e^{it\hat{\hat{z}}_j^{(r)}} - \mathbf{E}e^{itz_j^{(r)}} \right| \cdot \left| \mathbf{E} \left(iA_j \prod_{l=1}^{r-1} \xi_j^{(l)} \right) \right| .
\end{aligned}$$

The right-hand side of this last inequality can be estimated by means of inequalities (5) and (4) and one can show that it does not exceed $48 \cdot 2^{r-1} d_s^{1/s} (\alpha(m+1))^{(s-1)/s}$. Summing the resultant inequalities over all j and r, we conclude that

$$|E_5| \le 48 n d_s^{1/s} (\alpha(m+1))^{(s-2)/s} \tag{73}$$

under condition (70).

$|E_6|$ may be estimated in similar fashion with the result

$$|E_6| \le 32 n d_s^{1/s} (\alpha(m+1))^{(s-1)/s}. \tag{74}$$

Put $A_s = \sum_{\tau=1}^{\infty} (\alpha(\tau))^{(s-2)/s}$. The substitution of (63), (69), (71)–(74) in (49) yields this: If $A_s < \infty$, $k^{3/2} 4^k (\alpha(m+1))^{1/s} \le 1$, $3 \le \ln n/(8 \ln 2) \le k$ and $2k(m+1) \le n$, then

$$f_n'(t) = (-t + \theta a_n(t)) f_n(t) + \theta b_n(t) \tag{75}$$

for $|t| \le T_3$, where

$$\begin{aligned}
a_n(t) &= a_n^{(0)} + a_n^{(1)} |t| + a_n^{(2)} t^2 + a_n^{(3)} |t|^{s-1}, \\
b_n(t) &= b_n^{(0)} + b_n^{(2)} t^2,
\end{aligned}$$

$$\begin{aligned}
a_n^{(0)} &= C n d_s^{1/s} (\alpha(m+1))^{(s-2)/s}, & a_n^{(1)} &= C n^2 d_s^{2/s} (\alpha(m+1))^{(s-2)/s}, \\
a_n^{(2)} &= C m^2 n d_s^{3/s}, & a_n^{(3)} &= C m^{s-1} L_s^*, \\
b_n^{(0)} &= C(A_s) n d_s^{1/s} (\alpha(m+1))^{(s-2)/s}, & b_n^{(2)} &= C(A_s) m^{5/2} n^{1/2} d_s^{3/s}.
\end{aligned}$$

The differential equations (50) and (75) are of the same form and so (52) yields the following

Proposition 4. *Suppose that* (12) *is a s.m. sequence,* $\mathbf{E}|X_j|^s < \infty$, $2 < s \leq 3$ *and* $j = 1, 2, \ldots, n$, $A_s < \infty$ *and* $k^{3/2} 4^k (\alpha(m+1))^{1/s} \leq 1$, *with the positive integers* k *and* m *satisfying* $3 \leq \ln n/(8 \ln 2) \leq k$ *and* $2k(m+1) \leq n$. *Then*

$$\Delta_n \leq C(A_s)[m^{s-1} L_s^* + n^2 d_s^{2/s} (\alpha(m+1))^{(s-2)/s}).$$

It merely remains to select k and m depending on the rate of decay of the s.m. coefficient.

If $n \geq C(K, \lambda, s)$, part 1 of Theorem 10 follows from Proposition 4 with $m = \left[\dfrac{3s}{\lambda(s-2)} \ln(n+1)\right]$ and $k = \left[\dfrac{1}{(s-2)} \ln(n+1)\right]$.

If $n > C(K, \beta, s)$, part 2 of Theorem 10 follows from Proposition 4 with $m = [n^\varepsilon]$, $\varepsilon = s(s-2)/(2(s-1)(2\beta + s - 2))$ and $k = \left[\dfrac{5\beta}{8(2\beta+1)} \ln(n+1)\right]$.

For small n, Theorem 10 is a consequence of the estimate (17).

2.5. Heinrich's Method for m-Dependent Random Variables.

Heinrich's method (1982) is a fairly general way of deducing various limit theorems for the sums Z_n of m-dependent r.v.'s. It is based on the factorization of the c.f. $f_n(t) = \mathbf{E} e^{itZ_n}$ (or moment-generating function $\mathbf{E} e^{zZ_n}$, $z \in (C^1)$ in a neighborhood of $t = 0$ (or $z = 0$).

We shall only consider here the factorization of $f_n(t)$. By making use of the factorization of $f_n(t)$ once it has been found, one can obtain, for instance, these sorts of results for sums of m-dependent r.v.'s: convergence to unbounded distributions, uniform and non-uniform bounds for the rate of convergence in the CLT, asymptotic expansions, moderate and large deviations and so on.

The gist of Heinrich's method will be demonstrated by the proof of a single lemma.

Lemma 7. *Suppose that* (12) *is an* m*-dependent sequence and that* $\mathbf{E}|X_j|^s < \infty$, $2 < s \leq 3$ *and* $j = 1, 2, \ldots, n$. *Then the following inequalities hold in the interval*

$$|t| \leq \left(\frac{1-2c}{2C_1(s)} \cdot \frac{1}{(m+1)^{s-1} L_s^*}\right)^{1/(s-2)} = T_4$$

for $m \geq 0$*:*

1. $$\left|\ln f_n(t) + \frac{t^2}{2}\right| \leq C_1(s)(m+1)^{s-1} L_s^* |t|^s,$$
2. $$|f_n(t)| \leq e^{-ct^2},$$
3. $$|f_n(t) - e^{-t^2/2}| \leq C_1(s)(m+1)^{s-1} L_s^* |t|^s e^{-ct^2},$$

where

$$C_1(s) = \left(149 + \frac{1}{s}\right) \frac{2^{4-s}}{s-1}, \quad 0 < c < 1/2.$$

When $s = 3$, Lemma 7 follows (with different constants) from Corollary 3.2 in Heinrich (1982). Since $|e^z| \le e^{|z|}$ and $|e^z - 1| \le |z|e^{|z|}$ for any complex z, the second and third inequalities in Lemma 7 are consequence of the first one. To prove the first inequality, we need several preliminary assertions.

For any sequence of complex r.v.'s ξ_1, ξ_2, ... such that $\mathbf{E}|\xi_j|^k < \infty$, $j = 1, 2, \ldots, k$, the symbol $\hat{\mathbf{E}}\xi_1\xi_2\ldots\xi_k$ means that $\hat{\mathbf{E}}\xi_1 = \mathbf{E}\xi_1$ and for $k \ge 2$, that (see Heinrich (1982))

$$\hat{\mathbf{E}}\xi_1\xi_2\ldots\xi_k = \mathbf{E}\xi_1\xi_2\ldots\xi_k - \sum_{j=1}^{k-1}\hat{\mathbf{E}}\xi_1\ldots\xi_j\mathbf{E}\xi_{j+1}\ldots\xi_k. \tag{76}$$

This symbol was first introduced in another way by Statulevičius (1970b) and is known as the k-th centered moment. Among the many interesting properties of centered moments are the following.

Lemma 8 (Heinrich (1982)). *Let $\xi_1, \xi_2, \ldots, \xi_k$ be 1-dependent complex r.v.'s.*

1. *If $\mathbf{E}|\xi_j|^k < \infty$, $j = 1, 2, \ldots, k$, then*

$$\hat{\mathbf{E}}(\xi_1 + a_1)(\xi_2 + a_2)\ldots(\xi_k + a_k) = \hat{\mathbf{E}}\xi_1\xi_2\ldots\xi_k, \tag{77}$$

where $a_1, a_2, \ldots, a_k$ are any complex numbers.

2. *If $\mathbf{E}|\xi_j|^2 < \infty$, $j = 1, 2, \ldots, k$, then*

$$|\hat{\mathbf{E}}\xi_1\xi_2\ldots\xi_k| \le 2^{k-1}\prod_{j=1}^{k}\mathbf{E}^{1/2}|\xi_j|^2. \tag{78}$$

Corresponding to a sequence (12) of m-dependent r.v.'s, we form new 1-dependent r.v.'s

$$Y_j = \sum_{p=(j-1)(m+1)+1}^{j(m+1)} A_p, \quad j = 1, 2, \ldots, N = [n/(m+1)),$$

$$Y_{N+1} = \begin{cases} \sum_{p=N(m+1)+1}^{n} A_p & \text{if } N(m+1) < n, \\ 0 & \text{if } N(m+1) = n, \end{cases}$$

where $[x]$ is the integer part of x.

Put $U_j = \sum_{i=1}^{j} Y_i$, $j = 1, 2, \ldots, N+1$, $w = \max\limits_{1\le j\le N+1} \mathbf{E}^{1/2}|e^{itY_j} - 1|^2$ and $u_j(t) = 2\mathbf{E}^{1/2}|e^{itY_{j-1}} - 1|^2\mathbf{E}^{1/2}|e^{itY_j} - 1|^2$.

Then the following is true.

Lemma 9 (Heinrich (1982)). *If* $w \le 1/6$, *then*

1.
$$f_n(t) = \prod_{j=1}^{N+1} g_j(t), \tag{79}$$

where $g_1(t) = \mathbf{E}e^{itY_1}$, *and for* $j = 2, 3, \dots, N+1$,

$$g_j(t) = \frac{\mathbf{E}e^{itU_j}}{\mathbf{E}e^{itU_{j-1}}} = \mathbf{E}e^{itY_j} + \sum_{a=1}^{j-1} \frac{\hat{\mathbf{E}}(e^{itY_a} - 1)(e^{itY_{a+1}} - 1)\dots(e^{itY_j} - 1)}{\prod_{p=a}^{j-1} g_p(t)}; \tag{80}$$

2. *for* $j = 1, 2, \dots, N+1$,

$$\begin{aligned} |g_j(t) - 1| &\le |\mathbf{E}e^{itY_j} - 1| + 3u_j(t) && (81)\\ &\le 2w. && (82) \end{aligned}$$

Lemma 9 is proved by induction using relations (76)–(78).
By virtue of (80),

$$\begin{aligned} \sum_{j=1}^{N+1} [g_j(t) - 1] &= \sum_{j=1}^{N+1} (\mathbf{E}e^{itY_j} - 1) \\ &+ \sum_{j=2}^{N+1} \sum_{a=1}^{j-1} \left(\frac{1}{g_a(t)\dots g_{j-1}(t)} - 1 \right) \hat{\mathbf{E}}(e^{itY_a} - 1)\dots(e^{itY_j} - 1) \\ &+ \sum_{j=3}^{N+1} \sum_{a=1}^{j-2} \hat{\mathbf{E}}(e^{itY_a} - 1)\dots(e^{itY_j} - 1) \\ &+ \sum_{j=2}^{N+1} \hat{\mathbf{E}}(e^{itY_{j-1}} - 1)(e^{itY_j} - 1). \end{aligned}$$

Since $Y_1, Y_2, \dots, Y_{N+1}$ are 1-dependent r.v.'s with zero expectations,

$$\begin{aligned} \frac{t^2}{2} &= \sum_{j=1}^{N+1} \left[\mathbf{E}e^{itY_j} - 1 - \frac{(it)^2}{2}\mathbf{E}Y_j^2 \right] \\ &+ \sum_{j=2}^{N+1} [\hat{\mathbf{E}}(e^{itY_{j-1}} - 1)(e^{itY_j} - 1) - (it)^2\mathbf{E}(Y_{j-1}Y_j)] \\ &- \sum_{j=1}^{N+1} (\mathbf{E}e^{itY_j} - 1) - \sum_{j=2}^{N+1} \hat{\mathbf{E}}(e^{itY_{j-1}} - 1)(e^{itY_j} - 1). \end{aligned}$$

Adding the last two relations, we find that if (12) is a sequence of m-dependent r.v.'s and $\mathbf{E}|X_j|^s < \infty$, $2 < s \le 3$ and $j = 1, 2, \dots, n$, then

$$\ln f_n(t) + \frac{t^2}{2} = \Sigma_1 + \ldots + \Sigma_5, \tag{83}$$

where

$$\Sigma_1 = \sum_{j=1}^{N+1} [\ln g_j(t) - (g_j(t) - 1)],$$

$$\Sigma_2 = \sum_{j=2}^{N+1} \sum_{a=1}^{j-1} \left(\frac{1}{g_a(t) \ldots g_{j-1}(t)} - 1 \right) \hat{\mathbf{E}}(e^{itY_a} - 1) \ldots (e^{itY_j} - 1),$$

$$\Sigma_3 = \sum_{j=3}^{N+1} \sum_{a=1}^{j-2} \hat{\mathbf{E}}(e^{itY_a} - 1) \ldots (e^{itY_j} - 1),$$

$$\Sigma_4 = \sum_{j=1}^{N+1} \left[\mathbf{E}e^{itY_j} - 1 - \frac{(it)^2}{2} \mathbf{E}Y_j^2 \right],$$

$$\Sigma_5 = \sum_{j=2}^{N+1} \left[\hat{\mathbf{E}}(e^{itY_{j-1}} - 1)(e^{itY_j} - 1) - (it)^2 \mathbf{E}(Y_{j-1} Y_j) \right].$$

Everywhere below when estimating the right-hand side of (83), we are assuming that $w \leq 1/6$.

By the simple inequality $|\ln z - (z-1)| \leq |z-1|^2$, which is true for $|z-1| \leq 1/2$, and the estimates (81)–(82), it follows that

$$|E_1| \leq 2w \sum_{j=1}^{N+1} |\mathbf{E}e^{itY_j} - 1| + 6w \sum_{j=2}^{N+1} u_j(t). \tag{84}$$

According to (82),

$$\left| \frac{1}{g_a(t) \ldots g_{j-1}(t)} - 1 \right| \leq 3(j-a) 2^{j-a-1} w,$$

and the estimate (78) leads to

$$|\Sigma_2| \leq 27w \sum_{j=2}^{N+1} u_j(t). \tag{85}$$

The estimate (78) also assures that

$$|\Sigma_3| \leq 3w \sum_{j=3}^{N+1} u_j(t). \tag{86}$$

For $2 < s \leq 3$,

$$|e^{ix} - 1| \le 2^{3-s}|x|^{s-2},$$
$$|e^{ix} - 1 - ix| \le (2^{3-s}/(s-1))|x|^{s-1},$$
$$\left|e^{ix} - 1 - ix - \frac{(ix)^2}{2}\right| \le (2^{3-s}/(s-1)s)|x|^s,$$

and so

$$|\Sigma_4| \le (2^{3-s}/(s-1)s)|t|^s \sum_{j=1}^{N+1} \mathbf{E}|Y_j|^s, \tag{87}$$

$$|\Sigma_5| \le 3(2^{3-s}/(s-1))|t|^s \sum_{j=1}^{N+1} \mathbf{E}|Y_j|^s, \tag{88}$$

and

$$w \sum_{j=1}^{N+1} |\mathbf{E}e^{itY_j} - 1| \le (2^{3-s}/(s-1))(N+1) \max_{1\le j\le N+1} \mathbf{E}|Y_j|^s|t|^s. \tag{89}$$

Adding (84)–(88) and using (89) in conjunction with the fact that

$$\sum_{j=2}^{N+1} u_j(t) \le 4 \sum_{j=1}^{N+1} |\mathbf{E}e^{itY_j} - 1|,$$

we find for $w \le 1/6$ that

$$\left|\ln f_n(t) + \frac{t^2}{2}\right| \le C_1(s)(m+1)^{s-1}L_s^*|t|^s. \tag{90}$$

It remains to observe that $w \le 1/6$ when $|t| \le T_4$. Consequently, the first inequality of Lemma 7 has been proved and thereby all of Lemma 7.

From Lemma 9 it is seen that the functions $g_j(t)$ (whose product is the c.f. of Z_n) although not c.f.'s, behave primarily like the c.f.'s of the Y_i's. The next lemma, for example, underscores this fact.

Lemma 10 (Heinrich (1982)). *Let the sequence* (12) *be* m*-dependent and let* $\max_{1\le j\le n} \mathbf{E}|X_j|^p < \infty$ *for some* $p = 1, 2, \ldots$. *Then when* $w \le 1/6$,

1. $\max_{1\le j\le N+1} \left|\frac{d^p}{dt^p} g_j(t)\right| \le C(p) \max_{1\le j\le N+1} \mathbf{E}|Y_j|^p,$
2. $\max_{1\le j\le N+1} \left|\frac{d^p}{dt^p} \ln g_j(t)\right| \le C(p) \max_{1\le j\le N+1} \mathbf{E}|Y_j|^p,$

where the constants $C(p)$ *can be determined explicitly.*

Lemma 9 therefore plays a fundamental role in the study of the limiting law for the distribution of the normalized sum Z_n of m-dependent r.v.'s (see Heinrich (1982), (1984), (1985a,b,c,d)).

§3. Estimation of the Rate of Convergence in the Central Limit Theorem for Weakly Dependent Random Fields

Let $Z^d = \{a = (a_1, \ldots, a_d) : a_i \in \{0, \pm 1, \ldots\},\ i = 1, 2, \ldots, d\}$, $||a|| = \max_{1 \le i \le d} |a_i|$ and $\mathcal{V} = \{V \subset Z^d : |V| < \infty\}$, where $|V| = \#\{a : a \in V\}$ is the number of elements in V. The distance between $V_1, V_2 \in \mathcal{V}$ is defined as follows: $d(V_1, V_2) = \min\{||a - b|| : a \in V_1, b \in V_2\}$. $\mathcal{F}_V$ denotes the σ-algebra of events generated by the r.v.'s $\{X_a, a \in V\}$.

In what follows, we shall consider a real random field $\{X_a, a \in Z^d\}$, $d \ge 1$, satisfying one of the following weak dependence conditions:

1. *m-dependence*: $\mathcal{F}_{V_1}$ and $\mathcal{F}_{V_2}$ are independent for $\forall V_1, V_2 \in \mathcal{V}$ with $d(V_1, V_2) > m$;

2. *strong mixing* (s.m.): if there exist functions $M : Z_+^2 \to [1, \infty)$ and $\alpha : N \to [0, \infty)$ such that M is nondecreasing in each argument, $\alpha(r) \downarrow 0$ as $r \to \infty$ and for $\forall V_1, V_2 \in \mathcal{V}$,

$$\sup_{\substack{A \in \mathcal{F}_{V_1} \\ B \in \mathcal{F}_{V_2}}} |\mathbf{P}\{AB\} - \mathbf{P}\{A\}\mathbf{P}\{B\}| \le M(|V_1|, |V_2|)\alpha(d(V_1, V_2)).$$

The definition of these and other mixing coefficients for random fields as well as references on the subject may be found, for instance, in Dobrushin (1968), Bulinskii (1987), (1989), Takahata (1983), (1984), Sunklodas (1986) and Nakhapetyan (1987). Let

$$\{X_a,\ a \in Z^d\}, \quad d \ge 1, \tag{91}$$

be a real random field with $\mathbf{E}X_a = 0$ and $\mathbf{E}X_a^2 < \infty$ for $a \in V$. For $V \in \mathcal{V}$, $V \ne \emptyset$, put

$$\begin{aligned} S_V &= \sum_{a \in V} X_a, \quad B_V^2 = \mathbf{E}S_V^2, \quad Z_V = S_V / B_V, \\ F_V(x) &= \mathbf{P}\{Z_V < x\}, \quad A_a = X_a / B_V, \quad L_r = \sum_{a \in V} \mathbf{E}|A_a|^r, \\ L_s^* &= |V| d_s, \quad d_s = \max_{a \in V} \mathbf{E}|A_a|^s, \quad \Delta_V(x) = F_V(x) - \Phi(x), \\ \Delta_V &= \sup_x |\Delta_V(x)|, \quad ||\Delta_V(x)||_1 = \int_{-\infty}^{\infty} |\Delta_V(x)| dx, \end{aligned}$$

$$d_i^{(p)}(F_V, \Phi) = \sup_{h \in \mathcal{H}_i^{(p)}} |\mathbf{E}h(Z_V) - \mathbf{E}h(\mathcal{N})| / ||h||_{BH_i^{(p)}},$$

where $\mathcal{H}_i^{(p)}$ is the class of functions $h : R \to R$ with norm $||h||_{BH_i^{(p)}}$ defined in §2, 2.1, and $i = 1, 2$.

The rate of convergence in the CLT for weakly dependent random fields has been estimated by generalizing the methods developed for sequences of weakly depenent r.v.'s. The specific difficulties that have to be overcome in estimating Δ_V for multi-indexed terms, the distinctive features of mixing fields, the limits of applicability of Bernstein's method to random fields and other related question are discussed in detail in Bulinskii's book (1989). Leonenko (1975) found a bound for Δ_V in the case of m-dependent random fields for integer parallelepipeds $V \subset Z^d$. When $d = 1$, it reduces to Petrov's result (1960) cited above. By generalizing Maejima's results (1978) for r.v.'s, Rao (1981) found a nonuniform bound for $|\Delta_V(x)|$ for integer parallelepipeds for m-dependent random fields. He conjectured particularly that it was impossible to obtain an estimate for Δ_V of order $O(|V|^{-\gamma})$, $0 < \gamma \leq 1/2$, even for an m-dependent random field. This conjecture was disproved by Takahata (1983) and Guyon and Richardson (1984). A more precise uniform estimate (compared to those of Leonenko (1975) and Rao (1981)) for weakly dependent random additive functions (encompassing the class of m-dependent ones) was found by Bulinskii (1977). He subsequently strengthened this estimate (1987). The proofs by Rao, Leonenko and Bulinskii (1977) utilize Bernstein's method.

More exact estimates of the rate of convergence in the CLT for weakly dependent random fields have been found by means of the techniques of Stein and Tikhomirov.

Guyon and Richardson (1984) study the rate of convergence in the CLT for centered weakly dependent random fields $\{X_a, a \in Z^d\}$ (either m-dependent or s.m. with $M \equiv 1$) that satisfy $\sup_{a \in Z^d} \mathbf{E}|X_a|^{2+\delta} < \infty$, $\delta > 0$. The summation $S_{V_n} = \sum_{a \in V_n} X_a$ is over a strictly increasing sequence of sets $V_n \in \mathcal{V}$ such that $\liminf_{n \to \infty} B_{V_n}^2/|V_n| > 0$, where $B_{V_n}^2 = \mathbf{E}S_{V_n}^2$.

In particular, they show that

1. for m-dependent fields

$$\Delta_{V_n} = \begin{cases} O(B_{V_n}^{-\delta}) & \text{if } 0 < \delta < 1; \qquad (92) \\ O[B_{V_n}^{-1}(\log B_{V_n})^{(d-1)/2}) & \text{if } \delta \geq 1; \qquad (93) \end{cases}$$

2. for s.m. random fields with $M \equiv 1$ and α an exponentially decreasing function (here $a \wedge b = \min(a, b)$)

$$\Delta_{V_n} = O[B_{V_n}^{-(\delta \wedge 1)}(\log B_{V_n})^{d(1+\delta)\wedge 2}). \tag{94}$$

If $\sup_{a \in Z^d} \mathbf{E}|X_a|^{4+\delta} < \infty$, $\delta > 0$, the last estimate can be improved to

$$\Delta_{V_n} = O[B_{V_n}^{-1}(\log B_{V_n})^d). \tag{95}$$

Guyon and Richardson (1984) also investigated the case where $M \not\equiv 1$ and α decreases like a power function. The proofs are carried out by Tikhomirov's method.

Takahata (1983) found upper bounds for Δ_{V_n} and $||\Delta_{V_n}(x)||_1$ when the random field is m-dependent or s.m. with $M \not\equiv 1$ and α exponentially decreasing; the summation $S_{V_n} = \sum_{a \in V_n} X_a$ is over a sequence of sets $V_n \in \mathcal{V}$

such that $|V_n| \to \infty$ $(n \to \infty)$ and $\liminf_{n\to\infty} B_{V_n}^2/|V_n| > 0$. Under these conditions, the following was shown:

1. for m-dependent random fields

$$\Delta_{V_n} = O(|V_n|^{-1/2}) \tag{96}$$

if $\sup_{a\in Z^d} \mathbf{E}X_a^8 < \infty$, and

$$||\Delta_{V_n}(x)||_1 = O(|V_n|^{-1/2}) \tag{97}$$

if $\sup_{a\in Z^d} \mathbf{E}X_a^4 < \infty$;

2. for s.m. random fields with $M(n,m) \le B(n+m)^k$ for some $k > 1$ and $\alpha(\tau) \le Ke^{-\lambda\tau}$

$$\Delta_{V_n} = O[|V_n|^{-1/2}(\log|V_n|)^d] \tag{98}$$

if $\sup_{a\in Z^d} \mathbf{E}|X_a|^{8+\delta} < \infty$, $\delta > 0$, and

$$||\Delta_{V_n}(x)||_1 = O[|V_n|^{-1/2}(\log|V_n|)^d] \tag{99}$$

if $\sup_{a\in Z^d} \mathbf{E}|X_a|^{4+\delta} < \infty$, $\delta > 0$.

The estimate (96) established by Takahata (1983) for an m-dependent random field refines Riauba's paper (1980). Takahata (1983) used Stein's technique to prove his result.

The methods of Stein (Takahata (1983), Sunklodas (1986)) and Tikhomirov (Guyon and Richardson (1984), Sunklodas (1986), Bulinski (1986a), (1987)) may be extended to weakly dependent nonstationary random fields whose terms have finite absolute moments of order s, $2 < s \le 3$. Without any assumptions about the linear growth of the variance B_V^2 of the sum S_V, $V \in \mathcal{V}$, one may estimate $||\Delta_V(x)||_1$ (by Stein's method) and Δ_V (by Tikhomirov's method) in such a way that for $d = 1$ these estimates yield the best known estimates (or ones close to them) for a weakly dependent sequence of r.v.'s (Tikhomirov (1980), Erickson (1974), Sunklodas (1982)).

We state a number of results of this kind.

Theorem 13 (Sunklodas (1986)). *Suppose that the random field* (91) *is m-dependent and that* $\mathbf{E}|X_a|^s < \infty$, $2 < s \le 3$ *and* $a \in V$. *Then*

$$\Delta_V \le C(d)\{(m+1)^{d(s-1)}L_s^* + (m+1)^d d_s^{1/s}(\ln(|V|+1))^{(d-1)/2}\}$$

for $m+1 \le |V|^{1/d}/(C\ln(|V|+1))$, $C > 1$.

Theorem 14 (Sunklodas (1986)). *Suppose that* (91) *is a s.m. random field with* $M(n,m) \le B(n+m)^p$, $\alpha(\tau) \le Ke^{-\lambda\tau}$ *and that* $\mathbf{E}|X_a|^s < \infty$, $2 < s \le 3$, *with* $a \in V$. *If* $0 \le p < \infty$, *then*

$$\Delta_V \le C(B,K,\lambda,d,p,s)\{L_s^*(\ln(|V|+1))^{d(s-1)} + |V|^{1/2}d_s^{2/s}(\ln(|V|+1))^{1+(dp(s-2)/(2s))}\}.$$

Theorem 15 (Sunklodas (1986)). *Suppose that* (91) *is an m-dependent random field and* $\mathbf{E}|X_a|^s < \infty$, $2 < s \leq 3$, $a \in V$. *Then for* $m \geq 0$

$$||\Delta_V(x)||_1 \leq C(d)(m+1)^{d(s-1)} L_s.$$

The author has also found more precise estimates for $||\Delta_V(x)||_1$ when the terms have finite second moments.

It should be noted that for weakly dependent random fields (just as for a sequence of r.v.'s), Δ_V can be estimated in terms of L_s^* and d_s and $||\Delta_V(x)||_1$ can be estimated in terms of Lyapunov's quotient L_s.

Close results to Theorems 13 and 14 were obtained independently by Bulinskii (1986a), (1987). His mixing conditions are more general since they take into account a "geometric" aspect of selecting the sets used to define the mixing coefficient (see also Bulinskii (1989)).

Herrndorf (1983b) constructed an example of a s.m. strictly stationary sequence of r.v.'s whose s.m. coefficient decrease arbitrarily fast and whose partial sums have a variance increasing regularly. However, the CLT is not obeyed if just the second moments of the terms exist. It is therefore reasonable to estimate the rate of convergence in the CLT for weakly dependent random fields by imposing moment restrictions such as $\sup_{a \in Z^d} \mathbf{E}\mathcal{G}(|X_a|) < \infty$ on the terms, where $\mathcal{G}$ satisfies the condition $\lim_{x\to\infty} x^{-2}\mathcal{G}(x) = \infty$. Bulinskii and Doukhan (1990) obtained an estimate for the rate of convergence in the CLT for $\tilde{\alpha}$-mixing random fields (see (1990)) assuming the finiteness of the moments of the terms of "small" order (for example, of the type $\mathbf{E}X^2 \ln_+^\delta(x) < \infty$). These estimates were derived by means of truncation as applied to Bulinskii's results (1986a). The author (1990) found estimates for $d_i^{(p)}$, $i = 1, 2$, for various types of mixing random fields. The problem is reduced to estimating an absolute moment of Z_V whose order depends on p and the type of mixing.

Here we state just one estimate which generalizes Theorem 3 to random fields.

Theorem 16 (Sunklodas (1990)). *Suppose that the function* $h : R \to R$ *satisfies the condition* $H_1^{(p)} < \infty$ *or* $H_2^{(p)} < \infty$ (see §2, 2.1) *and that* (91) *is an m-dependent random field with* $\mathbf{E}|X_a|^{2+p+\alpha} < \infty$ *and* $a \in V$. *If* $(6m+1)^d < |V|$, *then*

$$\begin{aligned}|\mathbf{E}h(Z_n) - \mathbf{E}h(\mathcal{N})| \leq C(d,p,\alpha)H_i^{(p)}\{&(m+1)^{d(1+\alpha)}L_{2+\alpha}(1+\mathbf{E}|Z_V|^p) \\ &+ (m+1)^{d(1+p+\alpha)}L_{2+p+\alpha}\}.\end{aligned}$$

The boundedness of h is not required in Theorem 16.

Zuev (1989) found an estimate for Δ_V for $m(d)$-dependent random fields (for an exact definition, see that paper).

Shergin (1988), (1990) found estimates for Δ_V, $\Delta_V(x)$ and $\int_{-\infty}^{\infty} |x|^l |\Delta_V(x)| dx$ for finitely dependent r.v.'s (see Chen (1978)).

Mukhamedov (1987) obtained estimates for $\int_{-\infty}^{\infty} |x|^l |\Delta_V(x)| dx$ using Stein's technique for s.m. and u.s.m. random fields.

For weakly dependent stationary r.v.'s and random fields with a very slowly increasing mixing coefficient, Nakhapetyan (1987), (1989) applied his small blocking method to find estimates of the rate of convergence in the CLT that are sufficient for the law of the iterated logarithm.

Tikhomirov (1983) found a nonuniform estimate for $|\Delta_V(x)|$ for strongly mixing strictly stationary random fields defined on the integer lattice Z_+^d and taking values in a finite-dimensional Euclidean space R^k. Tikhomirov (1983) extended his method to the estimation of the rate of convergence in the CLT for strictly stationary Hilbert-valued r.v.'s.

Asymptotic expansions for weakly dependent r.v.'s and random fields may be found in the papers by Heinrich (1985a), (1986), (1990a), Rhee (1985) and Götze and Hipp (1983), (1989). Stein's method was investigated by Barbour (1990) in the context of functional approximation of Wiener and other Gaussian processes.

References*

Aminev, F.A., and Dubrovin, V.T. (1984): Estimation of the rate of convergence in the central limit theorem for homogeneous random fields. Issled. Prikl. Mat., Kazan Univ. Press, Kazan **10**, 3–12. English transl.: J. Sov. Math. **44**, No. 5, 559–567 (1989). Zbl. 594.60027

Babu, G.J. (1980): An inequality for moments of sums of truncated φ-mixing random variables and its applications. Sankhya, Ser. A **42**, No. 1–2, 1–8. Zbl. 486.60030

Babu, G.J., Ghosh, H., and Singh, K. (1978): On rates of convergence to normality for φ-mixing processes. Sankhya, Ser. A **40**, No. 3, 278–293. Zbl. 414.60025

Bakirov, N.K. (1987): The central limit theorem for weakly dependent variables. Mat. Zametki **41**, No. 1, 104–109. English transl.: Math. Notes. **41**, 63–67 (1987). Zbl. 645.60026

Barbour, A.D. (1986): Asymptotic expansions based on smooth functions in the central limit theorem. Probab. Theor. Relat. Fields **72**, No. 2, 289–303. Zbl. 572.60029

Barbour, A.D. (1990): Stein's method for diffusion approximation. Probab. Theor. Relat. Fields **84**, No. 3, 297–322. Zbl. 685.60009

Barbour, A.D., and Eagleson, G.K. (1985): Multiple comparisons and sums of dissociated random variables. Adv. Appl. Probab. **17**, No. 1, 147–162. Zbl. 559.60027

Basalykas, A. (1986): Estimation of the rate of convergence of some estimators of weakly dependent observations. Liet. Mat. Rink. **26**, No. 4, 607–615 (Russian). Zbl. 647.62028

Bergström, H. (1970): A comparison method for distribution functions of sums of independent and dependent random variables. Teoriya Veroyat. Primen. **15**, No. 3, 442–468. Also: Theory Probab. Appl. **15**, 430–457. Zbl. 209.20005

* For the convenience of the reader, references to reviews in Zentralblatt für Mathematik (Zbl.), compiled by means of the MATH database, and Jahrbuch über die Fortschritte der Mathematik (Jbuch) have, as far as possible, been included in this bibliography.

Bergström, H. (1971): Reduction of the limit problem for sums of random variables under a mixing condition. Proc. 4th Conf. Probab. Theory, Romania, pp. 107–120. Zbl. 305.60006

Bergström, H. (1972): On the convergence of sums of random variables in distribution under a mixing condition. Period. Math. Hung. **2**, No. 1–4, 173–190. Zbl. 252.60009

Berk, K.N. (1973): A central limit theorem for m-dependent random variables with unbounded m. Ann. Probab. **1**, No. 2, 352–354. Zbl. 263.60006

Bernstein, S.N. (1926): Sur l'extension du théorème limite du calcul des probabilités aux sommes de quantités dépendantes. Math. Ann. **97**, 1–59. Jbuch 52.0517.03

Bernstein, S.N. (1964): Collected Works, Vol. 4. Nauka, Moscow (Russian). Zbl. 198.50902

Bhattacharya, R.N., and Rao R. Ranga (1976): Normal Approximation and Asymptotic Expansions. John Wiley, New York. Zbl. 331.41023

Billingsley, P. (1968): Convergence of Probability Measures. John Wiley, New York. Zbl. 172.21201

Blum, J.R., Hanson, D.L., and Koopmans, L.H. (1963): On the strong law of large numbers for a class of stochastic processes. Z. Wahrscheinlichkeitstheorie Verw. Geb. **2**, No. 1, 1–11. Zbl. 117.35603

Bolthausen, E. (1982): On the central limit theorem for stationary mixing random fields. Ann. Probab. **10**, No. 4, 1047–1050. Zbl. 496.60020

Bolthausen, E. (1984): An estimate of the remainder in a combinatorial central limit theorem. Z. Wahrscheinlichkeitstheorie Verw. Geb. **66**, No. 3, 379–386. Zbl. 563.60026

Bradley, R.C. (1980a): A remark on the central limit question for dependent random variables. J. Appl. Probab. **17**, No. 1, 94–101. Zbl. 424.60025

Bradley, R.C. (1980b): On the φ-mixing condition for stationary random sequences. Duke Math. J. **47**, No. 2, 421–433. Zbl. 442.60035

Bradley, R.C. (1981): Central limit theorems under weak dependence. J. Multivariate Anal. **11**, No. 1, 1–16. Zbl. 453.60028

Bradley, R.C. (1983): Equivalent measures of dependence. J. Multivariate Anal. **13**, No. 1, 167–176. Zbl. 508.60004

Bradley, R.C. (1986): Basic properties of strong mixing conditions. Dependence in Probab. and Statistics. (Eds. E. Eberlein and M.S. Taqqu). Progress in Probability and Statistics, Birkhäuser, Boston **2**, 165–192. Zbl. 603.60034

Bradley, R.C. (1988): A central limit theorem for stationary ρ-mixing sequences with infinite variance. Ann. Probab. **16**, No. 1, 313–332. Zbl. 643.60018

Bradley, R.C. (1988): A caution on mixing conditions for random fields. Statist. Probab. Lett. **8**, No. 5, 489–491. Zbl. 697.60054

Bradley, R.C. (1990): On ρ-mixing except on small sets. Pac. J. Math. **146**, No. 2, 217–226. Zbl. 733.60059

Bradley, R.C., and Bryc, W. (1985): Multilinear forms and measures of dependence between random variables. J. Multivariate Anal. **16**, No. 3, 335–367. Zbl. 586.62086

Bradley, R.C., Bryc, W., and Janson, S. (1985): Remarks on the foundation of measures of dependence. [Center for Stoch. Proc. Univ. of North Carolina. Tech. Rep. 105.] New perspectives in theoretical and applied statistics, Selected Papers of the 3rd International Meeting of Statistics, Bilbao/Spain 1986, 421–437 (1987). Zbl. 619.60011

Bradley, R.C., Bryc, W., and Janson, S. (1987): On dominations between measures of dependence. J. Multivariate Anal. **23**, No. 2, 312–329. Zbl. 627.60009

Bradley, R.C., and Peligrad, M. (1986): Invariance principles under a two-part mixing assumption. Stoch. Proc. Appl. **22**, No. 2, 271–289. Zbl. 609.60048

Bulinskii, A.V. (1977): On the speed of convergence in the central limit theorem for additive random functions. Dokl. Akad. Nauk. SSSR **235**, No. 1, 741–744. English transl.: Sov. Math., Dokl. **18**, No. 4, 1009–1013. Zbl. 391.60025

Bulinskii, A.V. (1979): The central limit theorem for random fields with strong and weak dependence. Dokl. Akad. Nauk, SSSR **248**, No. 1, 17–19. English transl.: Sov. Math., Dokl. **20**, No. 5, 929–931. Zbl. 435.60047

Bulinskii, A.V. (1984): On measures of dependence close to the maximum correlation coefficient. Dokl. Adad. Nauk SSSR **277**, No. 6, 1296–1298. English transl.: Sov. Math., Dokl. **30**, No. 1, 249–252. Zbl. 584.62084

Bulinskii, A.V. (1985a): Asymptotic normality of mixing random fields. Dokl. Akad. Nauk. SSSR **284**, No. 5, 1044–1048. English transl.: Sov. Math., Dokl. **32**, No. 2, 523–527. Zbl. 597.60049

Bulinskii, A.V. (1985b): On mixing conditions of random fields. Teor. Veroyatn. Primen. **30**, No. 1, 200–201. English transl.: Theory Probab. Appl. **30**, No. 1, 219–220.

Bulinskii, A.V. (1986a): An estimate of the rate of convergence in the central limit theorem for random fields. Dokl. Akad. Nauk. SSSR **291**, No. 1, 22–25. English transl.: Soviet. Math. Dokl. **34**, No. 3 (1987), 416–419. Zbl. 665.60028

Bulinskii, A.V. (1986b): The central limit theorem and invariance principle for mixing random fields. First Int. Congr. of the Bernoulli Statist. Soc., Tashkent, Vol. 1, 105–107.

Bulinskii, A.V. (1987): Limit theorems under weak dependence conditions. Probab. Theory and Math. Statistic., Proceedings of the Fourth Vilnius Conf., VNU Sc. Press, Utrecht, Vol. 1, 307–326. Zbl. 648.60025

Bulinskii, A.V. (1988): On various mixing conditions and asymptotic normality of random fields. Dokl. Akad. Nauk. SSSR **229**, No. 4, 785–789. English transl.: Soviet Math. Dokl. **37**, No. 2, 443–448. Zbl. 689.60049

Bulinskii, A.V. (1989): Limit Theorems under Conditions of Weak Dependency. Moscow Univ. Press, Moscow (Russian).

Bulinskii, A.V., and Doukhan, P. (1990): Vitesse de convergence dans le théorème de limite centrale pour les champs mélangeants satisfaisant des hypothèses de moments faibles, C.R. Acad. Sci., Paris, Sér I **311**, No. 12, 801–805. Zbl. 719.60020

Bulinskii, A.V., and Zhurbenko, I.G. (1976a): The central limit theorem for random fields. Dokl. Akad. Nauk. SSSR **226**, No. 1, 23–25. English transl.: Sov. Math., Dokl. **17**, No. 1, 14–17.

Bulinskii, A.V., and Zhurbenko, I.G. (1976b): The central limit theorem for additive functions. Teoriya Veroyat. Primen. **21**, No. 4, 707–717. English transl.: Theory Probab. Appl. **21**, No. 4, 687–697. Zbl. 382.60025

Chen, L.H.Y. (1972): An elementary proof of the central limit theorem. Bull. Singapore Math. Soc. 1972, 1–12.

Chen, L.H.Y. (1978): Two central limit problems for dependent random variables. Z. Wahrscheinlichkeitstheorie Verw. Geb. **43**, No. 3, 223–243. Zbl. 372.60029

Chen, L.H.Y. (1979): Stein's method in limit theorems for dependent variables. SEA Bull. Math. (Special Issue), 36–50. Zbl. 419.60019

Cohn, H. (1965): On a class of dependent random variables. Rev. Roumaine Math. Pures Appl. **10**, No. 10, 1593–1606. Zbl. 203.19403

Dasgupta, R. (1988): Nonuniform rates of convergence to normality for strong mixing processes. Sankhya, Ser. A, **50**, No. 3, 436–451. Zbl. 679.60052

Davis, R. (1983): Stable limits for partial sums of dependent random variables. Ann. Probab. **11**, No. 2, 262–269. Zbl. 511.60021

Davydov, Yu.A. (1968): Convergence of distributions generated by stationary stochastic processes. Teor. Veroyatn. Primen. **13**, No. 4, 730–737. English transl.: Theory Probab. Appl. **13**, No. 4, 691–696. Zbl. 174.49201

Davydov, Yu.A. (1970): The invariance principle for stationary processes. Teor. Veroyatn. Primen. **15**, No. 3, 498–509. English transl.: Theory Probab. Appl. **15**, No. 3, 487–498. Zbl. 209.48904

Davydov, Yu.A. (1973): Mixing conditions for Markov chains. Teor. Veroyatn. Primen. **18**, No. 2, 321–338. English transl.: Theory Probab. Appl. **18**, No. 2, 312–328. Zbl. 297.60031

Dehling, H., Denker, M., and Philipp, W. (1986): Central limit theorem for mixing sequences of random variables under minimal conditions. Ann. Probab. **14**, No. 4, 1359–1370. Zbl. 605.60027

Denker, M. (1986): Uniform integrability and the central limit theorem. Dependence in Probab. and Statist (Eds. E. Eberlein and M.S. Taqqu). Birkhäuser, Boston **2**, 269–274. Zbl. 612.60028

Denker, M., and Jakubowski, A. (1989): Stable limit distributions for strongly mixing sequences. Statist. Probab. Lett. **8**, No. 5, 477–483. Zbl. 694.60017

Deo, C.M. (1973): A note on empirical processes of strong mixing conditions. Ann. Probab. **1**, No. 5, 870–875. Zbl. 281.60034

Deo, C.M., and Wong, H.S.-F. (1980): On Berry–Esseen approximation and a functional LIL for a class of dependent random fields. Pac. J. Math. **91**, No. 2, 269–275. Zbl. 476.60030

Diananda, P.H. (1955): The central limit theorem for m-dependent variables. Proc. Cambridge Philos. Soc. **51**, No. 1, 92–95. Zbl. 064.13104

Dobrushin, R.L. (1956a): Central limit theorem for nonstationary Markov chains, I. Teor. Veroyatn. Primen. **1**, No. 1, 72–89. English transl.: Theory Probab. Appl. **1**, No. 1, 65–80. Zbl. 093.15001

Dobrushin, R.L. (1956b): Central limit theorem for nonstationary Markov chains, II. Teor. Veroyatn. Primen. **1**, No. 4, 365–425. English transl.: Theory Probab. Appl. **1**, No. 4, 329–383. Zbl. 093.15001

Dubrovin, V.T. (1971): A central limit theorem for sums of functions of weakly dependent random variables. Probability Methods and Cybernetics, Kazan Univ. Press, No. 9, 21–33 (Russian). Zbl. 276.60025

Dubrovin, V.T. (1974): A multidimensional central limit theorem for number-theoretic endomorphisms. Probability Methods and Cybernetics, Kazan Univ. Press **10**, 17–29 (Russian). Zbl. 321.60017

Dubrovin, V.T., and Moskvin, D.A. (1979): The central limit theorem for sums of functions of mixing sequences. Teor. Veroyatn. Primen. **24**, No. 3, 553–564. English transl.: Theory Probab. Appl. **24**, No. 3, 560–571. Zbl. 408.60026

Dvoretzky, A. (1972): Asymptotic normality for sums of dependent random variables. Proc. Sixth Berkeley Symp. Math. Statist. Probab. **2**, 513–535. Zbl. 256.60009

Eberlein, E. (1979): An invariance principle for lattices of dependent random variables. Z. Wahrscheinlichkeitstheorie Verw. Geb. **50**, No. 2, 119–133. Zbl. 414.60036

Eberlein, E. (1984): Weak convergence of partial sums of absolutely regular sequences. Statist. Probab. Lett. **2**, No. 5, 291–293. Zbl. 564.60025

Egorov, V.A. (1970): Some limit theorems for m-dependent random variables. Liet. Mat. Rink. **10**, No. 1, 51–59 (Russian). Zbl. 213.20201

Erickson, R.V. (1973): On an L_p version of Berry–Esseen theorem for independent and m-dependent random variables. Ann. Probab. **1**, No. 3, 497–503. Zbl. 292.60040

Erickson, R.V. (1974): L_1 bounds for asymptotic normality of m-dependent sums using Stein's technique. Ann. Probab. **2**, No. 3, 522–529.

Erickson, R.V. (1975): Truncation of dependent random variables. Teor. Veroyatn. Primen. **20**, No. 4, 892–900. Also: Theory Probab. Appl. **20**, No. 4, 873–880. Zbl. 362.60005

Gabbasov, F.G. (1977): A multidimensional central limit theorem for sums of functions of mixing sequences. Liet. Mat. Rink. **17**, No. 4, 83–98. English transl.: Lith. Math. J. **17**, No. 4, 494–505. Zbl. 373.60021

Gebelein, H. (1941): Das statistische Problem der Korrelation als Variations- und Eigenwertproblem und sein Zusammenhang mit der Ausgleichungsrechnung. Z. Angew. Math. Mech. **21**, 364–379. Zbl. 026.33402

Gnedenko, B.V., and Kolmogorov, A.N. (1949): Limit Distributions for Sums of Independent Random Variables. Gostekhizdat, Moscow. English transl.: Addison-Wesley, Reading (MA), 1954. Zbl. 056.36001

Goldie, C.M., and Greenwood, P.E. (1986a): Variance of set-indexed sums of mixing random variables under weak convergence of set-indexed processes. Ann. Probab. **14**, No. 3, 817–839. Zbl. 604.60032

Goldie, C.M., and Greenwood, P.E. (1986b): Central limit results for random fields. Proc. First Int. Congress Bernoulli Soc., Tashkent, 345–352. Zbl. 677.60027

Goldie, C.M., and Morrow, G.J. (1986): Central limit questions for random fields. Dependence in Probab. and Statist., Birkhäuser, Boston **2**, 275–289. Zbl. 605.60029

Gordin, M.I. (1969): The central limit theorem for stationary processes. Dokl. Akad. Nauk. SSSR **188**, No. 4, 739–741. English transl.: Sov. Math., Dokl. **10**, No. 4, 1174–1176. Zbl. 212.50005

Gorodetskii, V.V. (1982): The invariance principle for strongly mixing stationary random fields. Teor. Veroyat. Primen. **27**, No. 2, 358–364. English transl.: Theory Probab. Appl. **27**, No. 2, 380–385.

Gorodetskii, V.V. (1984): The central limit theorem and an invariance principle for weakly dependent random fields. Dokl. Akad. Nauk. SSSR **276**, No. 3, 528–531. English transl.: Sov. Math., Dokl. **29**, No. 3, 529–532. Zbl. 595.60025

Götze, F., and Hipp, C. (1983): Asymptotic expansions for sums of weakly dependent random vectors. Z. Wahrscheinlichkeitstheorie Verw. Geb. **64**, No. 2, 211–239. Zbl. 514.60027

Götze, F., and Hipp, C. (1989): Asymptotic expansions for potential functions of i.i.d. random fields. Probab. Theory Relat. Fields **82**, 3, 349–370. Zbl. 687.60022

Grin', A.G. (1982): On a condition for regularity of stationary sequences. Teor. Veroyatn. Primen. **27**, No. 4, 789–795. English transl.: Theory Probab. Appl. **27**, No. 4, 850–855. Zbl. 504.60043

Grin', A.G. (1990a): Domains of attraction of mixing sequences. Sib. Mat. Zh. **31**, 1. 53–65. English transl.: Sib. Math. J. **31**, No. 1, 43–52. Zbl. 714.60017

Grin', A.G. (1990b): On Domains of attraction for sums of dependent variables. Teor. Veroyatn. Primen. **35**, No. 2, 255–270. English transl.: Theory Probab. Appl. **35**, No. 2, 241–257. Zbl. 273.60024

Gudynas, P. (1977): The invariance principle for nonhomogeneous Markov chains. Liet. Mat. Rink. **17**, No. 2, 63–73. English transl.: Lith. Math. J. **17**, No. 4, 490–505. Zbl. 398.60033

Gudynas, P. (1983): On approximating distributions of sums of dependent Banach-valued random variables. Liet. Mat. Rink. **23**, No. 3, 3–21. English transl.: Lith. Math. J. **23**, No. 3, 251–263. Zbl. 539.60013

Gudynas, P. (1989): A generalization of an approximation inequality. Liet. Mat. Rink. **29**, No. 1, 27–34. English transl.: Lith. Math. J. **29**, No. 1, 17–22. Zbl. 682.60002

Guyon, X., and Richardson, S. (1984): Vitesse de convergence du théorème de la limite centrale pour des champs faiblement dépendants. Z. Wahrscheinlichkeitstheorie Verw. Geb. **66**, No. 2, 297–314. Zbl. 544.60031

Hall, P., and Heyde, C.C. (1980): Martingale Limit Theory and Its Application. Academic Press, New York. Zbl. 462.60045

Hegerfeldt, G.C., and Nappi, C. (1977): Mixing properties in lattice systems. Commun. Math. Phys. **53**, No. 1, 1–7. Zbl. 349.60105

Heinrich, L. (1982): A method of the derivation of limit theorems for sums of m-dependent random variables. Z. Wahrscheinlichkeitstheorie Verw. Geb. **60**, No. 4, 501–515. Zbl. 492.60028

Heinrich, L. (1984): Non-uniform estimates and asymptotic expansions of the remainder in the central limit theorem for m-dependent random variables. Math. Nachr. **115**, 7–20. Zbl. 558.60026

Heinrich, L. (1985a): Some remarks on asymptotic expansions in the central limit theorem for m-dependent random variables. Math. Nachr. **122**, 151–155. Zbl. 576.60019

Heinrich, L. (1985b): Stable limits for sums of m-dependent random variables. Serdica Bulgaricae Math. I Publ. **11**, No. 2, 189–199. Zbl. 582.60031

Heinrich, L. (1985c): Some estimates of the cumulant-generating function of a sum of m-dependent random vectors and their application to large deviations. Math. Nachr. **120**, 91–101. Zbl. 375.60028

Heinrich, L. (1985d): Non-uniform estimates, moderate and large deviation, in the central limit theorem for m-dependent random variables. Math. Nachr. **121**, 107–121. Zbl. 572.60031

Heinrich, L. (1986): Stable limit theorems for sums of multiply indexed m-dependent random variables. Math. Nachr. **127**, 193–210. Zbl. 609.60033

Heinrich, L. (1987): Asymptotic expansions in the central limit theorem for a special class of m-dependent random fields, I. Math. Nachr. **134**, 83–106. Zbl. 668.60026

Heinrich, L. (1990a): Asymptotic expansions in the central limit theorem for a special class of m-dependent random fields, II. Math. Nachr. **145**, 309–327. Zbl. 705.60025

Heinrich, L. (1990b): Non-uniform bounds for the error in the central limit theorem for random fields generated by functions of independent random variables. Math. Nachr. **145**, 345–364. Zbl. 705.60026

Heinrich, L., and Richter, W.-D. (1984): On moderate deviations of sums of m-dependent random vectors. Math. Nachr. **118**, 253–264. Zbl. 558.60027

Herrndorf, N. (1983a): The invariance principle for φ-mixing sequences. Z. Wahrscheinlichkeitstheorie Verw. Geb. **63**, No. 1, 97–108. Zbl. 506.60029

Herrndorf, N. (1983b): Stationary strongly mixing sequences not satisfying the central limit theorem. Ann. Probab. **11**, No. 3, 809–813. Zbl. 513.60033

Herrndorf, N. (1985): A functional central limit theorem for strongly mixing sequences of random variables. Z. Wahrscheinlichkeitstheorie Verw. Geb. **69**, No. 4, 541–550. Zbl. 558.60032

Hipp, C. (1979a): Convergence rates of the strong law for stationary mixing sequences. Z. Wahrscheinlichkeitstheorie Verw. Geb. **49**, No. 1, 49–62. Zbl. 398.60029

Hipp, C. (1979b): Convergence rates in the central limit theorem for stationary mixing sequences of random vectors. J. Multiv. Anal. **9**, 560–578. Zbl. 431.60021

Ho, S.-T., and Chen, L.H.Y. (1978): An L_p bound for the remainder in a combinatorial central limit theorem. Ann. Probab. **6**, No. 2, 231–249. Zbl. 375.60028

Hoeffding, W., and Robbins, H. (1948): The central limit theorem for dependent random variables. Duke Math. J. **15**, 773–780. Zbl. 031.36701

Ibragimov, I.A. (1962): Some limit theorems for stationary processes. Teor. Veroyatn. Primen. **7**, No. 4, 361–392. English transl.: Theory Probab. Appl. **7**, No. 4, 349–382. Zbl. 119.14204

Ibragimov, I.A. (1967): The central limit theorem for sums of functions of dependent variables and sums of the form $\sum f(2^k t)$. Teor. Veroyatn. Primen. **12**, No. 4, 655–665. English transl.: Theory Probab. Appl. **12**, No. 4, 596–607. Zbl. 217.49803

Ibragimov, I.A. (1975): A note on the central limit theorem for dependent random variables. Teor. Veroyatn. Primen. **20**, No. 1, 134–140. English transl.: Theory Probab. Appl. **20**, No. 2, 135–141. Zbl. 335.60023

Ibragimov, I.A., and Linnik, Yu.V. (1965): Independent and Stationary Sequences of Random Variables. Nauka, Moscow. English transl.: Wolters-Noordhoff, Groningen, 1971. Zbl. 219.60027

Ibragimov, I.A., and Rozanov, Yu.A. (1970): Gaussian Random Processes. Nauka, Moscow. English transl.: Springer-Verlag, New York (1978). Zbl. 392.60037

Iosifescu, M. (1968): La loi logarithme itéré pour une classe de variables aléatoires dépendantes. Teor. Veroyat. Primen. **13**, No. 2, 315–325. English transl.: Theory Probab. Appl. **13**, No. 2, 304–313. Zbl. 159.47302

Iosifescu, M. (1974): Limit theorems for φ-mixing sequences: a survey. Proc. Fifth Conf. on Probab. Theory, Sept. 1–6, 1974, Brasov, Romania. Editura Academiei Republicii Socialiste Romania, Bucharest, 51–57. Zbl. 376.60025

Iosifescu, M. (1980): Recent advances in mixing sequences of random variables. Third Int'l Summer School on Probab. Theory and Math. Statist., Varna, 1978. Bulgarian Academy of Sciences Publishing House, Sofia, 111–138. Zbl. 435.60024

Iosifescu, M., and Teodorescu, R. (1969): Random Processes and Learning. Springer-Verlag, Berlin. Zbl. 194.51101

Jakimavičius, D., and Statulevičius, V.A. (1987): Estimates of Cumulants and Centered Moments of Mixing Random Processes. Preprint No. 3, Acad. Sci. Lith. SSR, Inst. Math. and Cyb., Vilnius (Russian). Zbl. 661.60024

Janson, S. (1988): Normal convergence of higher semi-invariants with applications to sums of dependent random variables and random graphs. Ann. Probab. **16**, No. 1, 305–312. Zbl. 639.60029

Kallianpur, G. (1955): On a limit theorem for dependent random variables. Dokl. Akad. Nauk SSSR **101**, No. 1, 13–16 (Russian). Zbl. 067.10702

Kanagawa, S. (1981): Rates of convergence of the invariance principle for weakly dependent random variables. Keio Math. Seminar Rep. **6**, 23–25. Zbl. 464.60032

Kanagawa, S. (1982): On the rate of convergence of the invariance principle for stationary sequences. Keio Sci. Techn. Reports **35**, No. 3, 53–61. Zbl. 503.60045

Kesten, H., and O'Brien, G.L. (1976): Examples of mixing sequences. Duke Math. J. **43**, No. 2, 405–415. Zbl. 337.60035

Kolmogorov, A.N., and Rozanov, Yu.A. (1960): On strong mixing conditions for stationary Gaussian processes. Teor. Veroyatn. Primen. **5**, No. 2, 222–227. English transl.: Theory Probab. Appl. **5**, No. 2, 204–208. Zbl. 091.30001

Kolmogorov, A.N., and Sarmanov, O.V. (1960): S.N. Bernstein's work in probability theory. Teor. Veroyatn. Primen. **5**, No. 2, 215–221. English transl.: Theory Probab. Appl. **5**, No. 2, 197–203. Zbl. 087.33002

Kornfel'd, I.P., Sinai, Ya.G., and Fomin, S.V. (1980): Ergodic Theory. Nauka, Moscow. English transl.: Springer-Verlag, New York, 1982. Zbl. 493.28007

Krieger, H.A. (1984): A new look of Bergström's theorem on convergence in distribution for sums of dependent random variables. Israel J. Math. **47**, 32–64. Zbl. 536.60032

Lapinskas, R. (1976): On the rate of convergence for sums of infinite-dimensional random variables connected in a Markov chain. Liet. Mat. Rink. **16**, No. 4, 125–132. English transl.: Lith. Math. J. **16**, No. 4, 559–664. Zbl. 366.60093

Lapinskas, R. (1980): Limit theorems for weakly dependent random variables. Liet. Mat. Rink. **20**, No. 3, 91–97. English transl.: Lith. Math. J. **20**, No. 3, 244–249. Zbl. 468.60001

Lappo, P.M. (1986): Rate of convergence in the central limit theorem for a mixing sequence. Mat. Zametki **39**, No. 2, 295–299. English transl.: Math. Notes **39**, 160–162. Zbl. 619.60027

Leonenko, N.N. (1975): On estimating the rate of convergence in the central limit theorem for m-dependent random fields. Mat. Zametki **17**, No. 1, 129–132. English transl.: Math. Notes **17**, 76–78. Zbl. 347.60021

Leonenko, N.N. (1976): Limit theorems for additive random functions. Studies in Theory of Stochastic Processes. Math. Inst. Akad. Nauk Ukraine SSR, 94–105 (Russian).

Leonenko, N.N., Ivanov, A.V. (1986): Statistical Analysis of Random Fields. Vishcha Shkola, Kiev (Russian). English transl.: Kluwer, Dordrecht, 1989. Zbl. 713.62094

Leonenko, N.N., and Yadrenko, M.I. (1979): On the invariance principle for homogeneous random fields. Teor. Veroyatn. Primen. **24**, No. 1, 175–181. English transl.: Theory Probab. Appl. **24**, No. 1, 175–181. Zbl. 402.60049

Lifshits, B.A. (1978): On the central limit theorem for Markov chains. Teor. Veroyatn. Primen. **23**, No. 2, 295–312. English transl.: Theory Probab. Appl. **23**, No. 2, 279–296. Zbl. 389.60047

Lifshits, B.A. (1984): The invariance principle for weakly dependent variables. Teor. Veroyatn. Primen. **29**, No. 1, 33–40. English transl.: Theory Probab. Appl. **29**, No. 1, 33–40. Zbl. 535.60026

Lifshits, M.A. (1985): Partitioning of multidimensional sets. Rings and Modules. Limit Theorems of Probability Theory. Leningrad Univ. Press, Leningrad. No. 1, 175–178 (Russian). Zbl. 701.00008

Liptser, R.Sh., and Shiryaev, A.N. (1986): Theory of Martingales. Nauka, Moscow. English transl.: Kluwer, Dordrecht, 1989. Zbl. 728.60048

Maejima, M. (1978): A non-uniform estimate in the central limit theorem for m-dependent random variables. Keio Sci. Techn. Rep. **31**, No. 2, 15–20. Zbl. 451.60032

Malevich, T.L. (1980): On the rate of convergence in the central limit theorem for sums of m-dependent variables. Izv. Akad. Nauk. Uzb. SSR, Ser. Fiz.-Mat. Nauk. **2**, 25–29 (Russian). Zbl. 445.60016

Malyshev, V.A. (1975): The central limit theorem for Gibbsian random fields. Dokl. Adad. Nauk. SSSR **224**, No. 1, 35–38. English transl.: Sov. Math., Dokl. **16**, No. 5, 1141–1145.

Malyshev, V.A., and Minlos, R.A. (1985): Gibbs Random Fields. Nauka, Moscow. English transl.: Kluwer, Dordrecht-New York, 1991. Zbl. 731.60099

McLeish, D.L. (1974): Dependent central limit theorems and invariance principles. Ann. Probab. **2**, No. 4, 620–628. Zbl. 287.60025

McLeish, D.L. (1975): Invariance principles for dependent variables. Z. Wahrscheinlichkeitstheorie Verw. Geb. **32**, No. 3, 165–178. Zbl. 305.60010

Mukhamedov, A.K. (1987): On a global form of the central limit theorem for weakly dependent random fields. Math. Analysis, Algebra and Probability Theory. Tashkent Univ., Tashkent, 77–81. Zbl. 738.60045

Nagaev, S.V. (1957): Some limit theorems for stationary Markov chains. Teor. Veroyatn. Primen. **2**, No. 4, 389–416. English transl.: Theory Probab. Appl. **2**, No. 4, 378–406. Zbl. 078.03804

Nagaev, S.V. (1962): The central limit theorem for time-discrete Markov processes. Izv. Akad. Nauk. Uzb. SSR, Ser. Fiz.-Mat. Nauk **2**, 12–20 (Russian). Zbl. 122.13603

Nakhapetyan, B.S. (1978): The central limit theorem for strongly mixing random fields. Multidimensional Random Systems, Nauka, Moscow, 276–288. English transl.: Advances in Probability and Related Topics **6**, 531–547 (1980) Marcel Dekker. Zbl. 442.60053

Nakhapetyan, B.S. (1987): An approach to proving limit theorems for dependent random variables. Teor. Veroyatn. Primen. **32**, No. 3, 589–594. English transl.: Theory Probab. Appl. **32**, No. 3, 535–539. Zbl. 629.60045

Nakhapetyan, B.S. (1989): Weakly Dependent Random Fields and the Summation Problem for Gibbsian Fields in Large Volumes. Doctoral Dissertation, Erevan (Russian).

Neaderhouser, C.C. (1978a): Some limit theorems for random fields. Commun. Math. Phys. **61**, No. 3, 293–305. Zbl. 393.60019

Neaderhouser, C.C. (1978b): Limit theorems for multiply-indexed mixing random variables with applications to Gibbs random fields. Ann. Probab. **6**, No. 2, 207–215. Zbl. 374.60033

Negishi, H. (1977): The rate of convergence to normality for strong mixing sequences of random variables. Sci. Repts. Yokohama National Univ., Sec. **1**, 14, 17–25. Zbl. 431.60022

Oodaira, H., and Yoshihara, K. (1972): Functional central limit theorem for strictly stationary processes satisfying the strong mixing conditions. Kodai Math. Sem. Rep. **24**, No. 3, 259–269. Zbl. 245.60006

Orey, S. (1958): A central limit theorem for m-dependent random variables. Duke Math. J. **25**, No. 4, 543–546. Zbl. 107.13403

Paulauskas, V., and Račkauskas, A. (1987): Approximation Theory in the Central Limit Theorem. Exact Results in Banach Spaces, Mokslas, Vilnius. English transl.: Kluwer, Dordrecht (1989). Zbl. 715.60023

Peligrad, M. (1982): Invariance principles for mixing sequences of random variables. Ann. Probab. **10**, No. 4, 968–981. Zbl. 503.60044

Peligrad, M. (1983): A note on two measures of dependence and mixing sequences. Adv. Appl. Probab. **15**, 461–464. Zbl. 508.60033

Peligrad, M. (1985): An invariance principle for φ-mixing sequences. Ann. Probab. **13**, No. 4, 1304–1313. Zbl. 597.60018

Peligrad, M. (1986a): Recent advances in the central limit theorem and its weak invariance principle for mixing sequences of random variables (a survey). Dependence in Probab. and Statist. Birkhäuser, Basel-Stuttgart, 193–223. Zbl. 603.60022

Peligrad, M. (1986b): Invariance principles under weak dependence. J. Multivariate Anal. **19**, No. 2, 299–310. Zbl. 603.60023

Peligrad, M. (1987): On the central limit theorem for ρ-mixing sequences of random variables. Ann. Probab. **15**, No. 4, 1387–1394. Zbl. 638.60032

Peligrad, M. (1990): On Ibragimov-Iosifescu conjecture for φ-mixing sequences. Stoch. Proc. Appl. **35**, No. 2, 293–308. Zbl. 712.60020

Petrov, V.V. (1960): On the central limit theorem for m-dependent variables. Proc. All-Union Conf. on Probab. and Statist., 1958. Akad. Nauk. Armen. SSR Publ., Erevan (Russian). 38–44. Zbl. 100.34503

Petrov, V.V. (1987): Limit Theorems for Sums of Independent Random Variables. Nauka, Moscow. English transl.: Oxford, Clarendon Press 1995. Zbl. 621.60022 (826.60001)

Petrov, V.V. (1988): Sequences of m-orthogonal random variables. Zap. Nauchn. Semin. Leningrad. Otd. Mat. Inst. **119**, No. 7, 198–202. English transl.: J. Sov. Math. **27**, 3136–3139 (1984). Zbl. 494.60032

Philipp, W. (1969a): The central limit theorem for mixing sequences of random variables. Z. Wahrscheinlichkeitstheorie Verw. Geb. **12**, No. 2, 155–171. Zbl. 174.49904

Philipp, W. (1969b): The remainder in the central limit theorem for mixing stochastic processes. Ann. Math. Statist. **40**, No. 2, 601–609. Zbl. 179.23503

Philipp, W. (1980): Weak and L^p-invariance principles for sums of B-valued random variables. Ann. Probab. **8**, No. 1, 68–82. Zbl. 426.60033

Philipp, W., and Stout, W.F. (1975): Almost sure invariance principles for partial sums of weakly dependent random variables. Mem. Am. Math. Soc. **161**. Zbl. 361.60007

Philipp, W., and Webb, G.R. (1973): An invariance principle for mixing sequences of random variables. Z. Wahrscheinlichkeitstheorie Verw. Geb. **25**, No. 3, 223–237. Zbl. 259.60007

Prokhorov, Yu.V. (1956): Convergence of random processes and limit theorems in probability theory. Teor. Veroyatn. Primen. **1**, No. 2, 177–238. English transl.: Theory Probab. Appl. **1**, No. 2, 157–214.

Prokhorov, Yu.V. (1972): Multidimensional distributions: inequalities and limit theorems. Itogi Nauki Tekhn. Ser. Teor. Veroyatn., Mat. Statist., Teor. Kibern. **10**, 5–24. English transl.: J. Sov. Math. **2**, 475–488 (1974). Zbl. 295.60013

Rao, B.L.S.P. (1975): Remark on the rate of convergence in the random central limit theorem for mixing sequences. Z. Wahrscheinlichkeitstheorie Verw. Gew. **31**, No. 2, 157–160. Zbl. 306.60011

Rao, B.L.S.P. (1981): A non-uniform estimate of the rate of convergence in the central limit theorem for m-dependent random fields. Z. Wahrscheinlichkeitstheorie Verw. Geb. **58**, No. 2, 247–256. Zbl. 465.60028

Renyi, A. (1959): On measures of dependence. Acta Math. Acad. Sci. Hung. **10**, 441–451. Zbl. 091.14403

Rhee Wan Soo (1985): An Edgeworth expansion for a sum of m-dependent random variables. Int. J. Math., Math. Sci. **8**, No. 3, 563–569. Zbl. 585.60032

Rhee Wan Soo (1986): On the characteristic function of a sum of m-dependent random variables valued in certain Banach spaces. Int. J. Math., Math. Sci. **9**, No. 2, 397–404. Zbl. 612.60026

Rhee Wan Soo, and Talagrand, M. (1981): On Berry–Esseen type bounds for m-dependent random variables valued in certain Banach spaces. Z. Wahrscheinlichkeitstheorie Verw. Geb. **58**, No. 4, 433–451. Zbl. 474.60006

Rhee Wan Soo, and Talagrand, M. (1986): Uniform bound in the central limit theorem for Banach-space-valued dependent random variables. J. Multivariate Anal. **20**, No. 2, 303–320. Zbl. 606.60010

Rio, E. (1996): Sur le theorème de Berry–Esseen pour les suites faiblement dépendantes. Probab. Theory Relat. Fields **104**, 255–282. Zbl. 838.60017

Riauba, B. (1980): On the rate of convergence in the central limit theorem for m-dependent random fields. Liet. Mat. Rink. **20**, No. 1, 157–163. English transl.: Lith. Math. J. **20**, No. 1, 71–75. Zbl. 428.60030

Riauba, B. (1988): Central limit theorem for stationary random fields. Liet. Mat. Rink. **28**, No. 4, 758–769. English transl.: Lith. Math. J. **28**, No. 4, 375–382. Zbl. 674.60055

Rosén, B. (1967): On the central limit theorem for sums of dependent random variables. Z. Wahrscheinlichkeitstheorie Verw. Geb. **7**, No. 1, 48–82. Zbl. 147.17001

Rosenblatt, M. (1956): A central limit theorem and a strong mixing condition. Proc. Nat. Acad. Sci. USA **42**, No. 1, 43–47. Zbl. 070.13804

Rosenblatt, M. (1971): Markov Processes. Structure and Asymptotic Behavior. Springer-Verlag, New York. Zbl. 236.60002

Rosenblatt, M. (1972): Central limit theorem for stationary processes. Proc. Sixth Berkeley Symp. Math. Statist. and Probab. **2**, 551–561. Zbl. 255.60002

Rosenblatt, M. (1984): Asymptotic normality, strong mixing and spectral density estimates. Ann. Probab. **12**, No. 4, 1167–1180. Zbl. 545.62058

Roussas, G.G., and Ionnides, D. (1987): Moment inequalities for mixing sequences of random variables. Stoch. Anal. Appl. **5**, No. 1, 61–120. Zbl. 619.60022

Rozanov, Yu.A. (1960): A central limit theorem for additive random functions. Teor. Veroyatn. Primen. **5**, No. 2, 243–246. English transl.: Theory Probab. Appl. **5**, No. 2, 221–223. Zbl. 091.30503

Rozanov, Yu.A. (1963): Stationary Random Processes. Fizmatgiz, Moscow. English transl.: Springer-Holden-Day, San Francisco (1967). Zbl. 152.16302

Rozanov, Yu.A. (1981): Markov Random Fields. Nauka, Moscow. English transl.: Springer-Verlag, New York, 1982. Zbl. 498.60057

Rychlik, Z., and Szyszkowski, I. (1987): The invariance principle for φ-mixing sequences. Teor. Veroyatn. Primen. **32**, No. 3, 616–619. Also: Theory Probab. Appl. **32**, No. 3, 559–562. Zbl. 626.60031

Samur, J. (1984): Convergence of sums of mixing triangular arrays of random vectors with stationary rows. Ann. Probab. **12**, No. 2, 390–426. Zbl. 542.60012

Sarmanov, I.O. (1964): On Lyapunov's theorem for sums of weakly dependent random variables. Izv. Vysh. Uch. Zaved. Mat., No. 3, 123–130 (Russian). Zbl. 123.36201

Sarmanov, O.V., and Zakharov, V.K. (1960): Measures of dependence between random variables and the spectra of stochastic matrices. Mat. Sb. **52**, No. 4, 953–990 (Russian). Zbl. 143.19903

Saulis, L., and Statulevičius, V. (1989): Limit Theorems on Large Deviations. Mokslas, Vilnius. English transl.: Kluwer Academic, Dordrecht-Boston, 1991. Zbl. 744.60028

Schneider, E. (1981): On the speed of convergence in the random central limit theorem for φ-mixing processes. Z. Wahrscheinlichkeitstheorie Verw. Geb. **58**, No. 1, 125–138. Zbl. 465.60027

Serfling, R.J. (1968): Contributions to central limit theorem for dependent variables. Ann. Math. Statist. **39**, No. 4, 1158–1175. Zbl. 176.48004

Serfling, R.J. (1980): Approximation Theorems of Mathematical Statistics. John Wiley, New York. Zbl. 538.62002

Shergin, V. (1976): Estimation of the remainder in the central limit theorem for m-dependent random variables. Liet. Mat. Rink. **16**, No. 4, 245–250. English transl.: Lith. Math. J. **16**, No. 4, 637–641. Zbl. 371.60028

Shergin, V. (1979): On the rate of convergence in the central limit theorem for m-dependent random variables. Teor. Veroyatn. Primen. **24**, No. 4, 781–794. English transl.: Theory Probab. Appl. **24**, No. 4, 782–796. Zbl. 437.60018

Shergin, V. (1983): On a global form of the central limit theorem for m-dependent random variables. Teor. Veroyatn. Mat. Statist. **29**, 122–128. English transl.: Theory Probab. Appl. Statist. **29** (1984), 149–156. Zbl. 577.60024

Shergin, V. (1988): On the central limit theorem for finitely-dependent random variables. Teor. Sluchajnykh Protsessov **16**, 93–97. English transl.: J. Sov. Math. **67**, No. 4, 3244–3248 (1993). Zbl. 793.60029

Shergin, V. (1990): The central limit theorem for finitely dependent random variables. Probab. Theory and Math. Statist. Proc. of the Fifth Vilnius Conf, 1989. VSP BV, Utrecht; Mokslas, Vilnius. 424–431. Zbl. 732.60028

Shiryaev, A.N. (1960): Some problems in the spectral theory of higher-order moments. Teor. Veroyatn. Primen. **5**, No. 3, 193–213. English transl.: Theory Probab. Appl. **5**, No. 3, 265–284. Zbl. 109.36001

Sirazhdinov, S.Kh. (1955): Limit Theorems for Homogeneous Markov Chains. Izd-vo Akad. Nauk Uzbek. SSSR, Tashkent (Russian).

Sirazhdinov, S.Kh., and Formanov, Sh.K. (1979): Limit Theorems for Sums of Random Vectors in a Markov Chain. Fan, Tashkent (Russian). Zbl. 482.60019

Statulevičius, V. (1961): Local limit theorems and asymptotic expansions for nonhomogeneous Markov chains. Liet. Mat. Rink. **1**, Nos 1–2, 231–314 (Russian). Zbl. 126.14401

Statulevičius, V. (1962): On refinements of limit theorems for weakly dependent random variables. Proc. Sixth All-Union Conf. on Probab. Theory and Math. Statist., Vilnius, 113–119 (Russian). Zbl. 128.38202

Statulevičius, V. (1969a): Limit theorems for sums of random variables related in a Markov chain, I. Liet. Mat. Rink. **9**, No. 2, 345–362 (Russian). Zbl. 203.50206

Statulevičius, V. (1969b): Limit theorems for sums of random variables related in a Markov chain, II. Liet. Mat. Rink. **9**, No. 3, 635–672 (Russian). Zbl. 203.50206

Statulevičius, V. (1970a): Limit theorems for sums of random variables related in a Markov chain, III. Liet. Mat. Rink. **10**, No. 1, 161–169 (Russian). Zbl. 203.50206

Statulevičius, V. (1970b): On limit theorems for random functions, I. Liet. Mat. Rink. **10**, No. 3, 582–592 (Russian). Zbl. 266.60014

Statulevičius, V. (1974): Limit theorems for dependent random variables under various regularity conditions. Proc. Int. Congr. Math. Vancouver **2**, 173–181. Zbl. 366.60027

Statulevičius, V. (1977a): Application of semi-invariants to asymptotic analysis of distributions of random processes. J. Multivar. Anal. **4**, 325–337. Zbl. 445.60019

Statulevičius, V. (1977b): On limit theorems for dependent random variables. Abstracts of Commun. in Second Vilnius Conf. on Probab. Theory and Math. Statist., Vilnius **3**, 212–215.

Statulevičius, V. (1983): On conditions for almost Markov regularity. Teor. Veroyatn. Primen. **28**, No. 2, 358–361. English transl.: Theory Probab. Appl. **28**, No. 2, 379–383. Zbl. 513.60064

Stein, C. (1972): A bound for the error in the normal approximation to the distribution of a sum of dependent random variables. Proc. Sixth Berkeley Symp. Math. Statist. and Probab. **2**, 583–602. Zbl. 278.60026

Stein, C. (1981): Estimation of the mean of a multivariate normal distribution. Ann. Statist. **9**, No. 6, 1135–1151. Zbl. 476.62035

Sunklodas, J. (1977): Estimation of the rate of convergence in the central limit theorem for weakly dependent random variables. Liet. Mat. Rink. **17**, No. 3, 41–51. English transl.: Lith. Math. J. **17**, No. 3, 313–320. Zbl. 404.60028

Sunklodas, J. (1978): Estimation of the rate of convergence in the central limit theorem for m-dependent random vectors. Liet. Mat. Rink. **18**, No. 4, 175–186. English transl.: Lith. Math. J. **18**, No. 4, 566–574. Zbl. 388.60027

Sunklodas, J. (1982): Distance in the L_1 metric between the distribution of a sum of weakly dependent random variables and the normal distribution function. Liet. Mat. Rink. **22**, No. 2, 171–188. English transl.: Lith. Math. J. **22**, No. 2, 177–189. Zbl. 495.60039

Sunklodas, J. (1984): On the rate of convergence in the central limit theorem for strongly mixing random variables. Liet. Mat. Rink. **24**, No. 2, 174–185. English transl.: Lith. Math. J. **24**, No. 2, 182–190. Zbl. 558.60023

Sunklodas, J. (1986): Estimation of the rate of convergence in the central limit theorem for weakly dependent random fields. Liet. Mat. Rink. **26**, No. 3, 541–559. English transl.: Lith. Math. J. **26**, No. 3, 272–287. Zbl. 628.60032

Sunklodas, J. (1989): Estimation of the bounded Lipschitz metric for sums of weakly dependent random variables. Liet. Mat. Rink. **29**, No. 2, 385–393. English transl.: Lith. Math. J. **29**, No. 2, 187–193. Zbl. 696.60027

Sunklodas, J. (1990): Approximation by the normal distribution. Liet. Mat. Rink. **30**, No. 2, 382–391 (Russian). Zbl. 709.60012

Sunklodas, J. (1998): On a lower bound of the rate of convergence in the central limit theorem for m-dependent random fields. Teor. Veroyatn. Primen. **43**, No. 1, 171–179. English transl.: Theory Probab. Appl. **43**, No. 1, 162–169. Zbl. 990.28431

Szweczak, Z.S. (1988): On a central limit theorem for m-dependent sequences. Bull. Polish Acad. Sci. Math. **36**, No. 5–6, 327–331. Zbl. 759.60022

Szyszkowski, I. (1990): An invariance principle for dependent random variables. Acta Math. Hung. **56**, No. 1–2, 45–51. Zbl. 737.60023

Takahata, H. (1981): L_∞-bound for asymptotic normality of weakly dependent summands using Stein's result. Ann. Probab. **9**, No. 4, 676–683. Zbl. 465.60033

Takahata, H. (1983): On the rates in the central limit theorem for weakly dependent random fields. Z. Wahrscheinlichkeitstheorie Verw. Geb. **64**, No. 4, 445–456. Zbl. 514.60028

Takahata, H. (1984): The central limit theorem problems for energy in the Gibbs random fields. A short survey. Bull. Tokyo Gakugei Univ., Sec. 4 **36**, 1–15. Zbl. 552.60017

Tikhomirov, A.N. (1976): On the rate of convergence in the central limit theorem for weakly dependent variables. Vestn. Leningr. Univ., No. 7, Mat. Mekh. Astron. No. 2, 158–159 (Russian). Zbl. 346.60016

Tikhomirov, A.N. (1980): On the convergence rate in the central limit theorem for weakly dependent variables. Teor. Veroyatn. Primen. **25**, No. 4, 800–818. English transl.: Theory Probab. Appl. **25**, No. 4, 790–809 (1981). Zbl. 448.60019

Tikhomirov, A.N. (1983): On normal approximation of sums of mixing random fields. Dokl. Akad. Nauk. SSSR **272**, No. 2, 312–314. English transl.: Sov. Math., Dokl. **28**, 396–397. Zbl. 546.60052

Tikhomirov, A.N. (1986): On the distribution of the maximum sum of weakly dependent variables. Teor. Veroyatn. Primen. **31**, No. 4, 829–834. English transl.: Theory Probab. Appl. **31**, No. 4, 733–738. Zbl. 658.60045

Tikhomirov, A.N. (1990): On the normal approximation of sums of weakly dependent Hilbert-valued random variables. Probab. Theory and Math. Statist. Proc. of the Fifth Vilnius Conf., 1989. VSP BV, Utrecht; Mokslas, Vilnius, 482–494. Zbl. 726.60007

Utev, S.A. (1984): Inequalities for sums of weakly dependent random variables and estimates of rate of convergence in the invariance principle. Limit Theorems for Sums of Random Variables. Nauka, Novosibirsk, Tr. Inst. Mat. **3**, 50–77. English transl.: Transl. Ser. Math. Eng., 73–114 (1986). Zbl. 541.60016

Utev, S.A. (1989): Sums of φ-mixing random variables. Asymptotic Analysis of Distributions of Random Processes. Nauka, Novosibirsk. 78–100 (Russian). Zbl. 711.60032

Utev, S.A. (1990a): Central limit theorem for dependent random variables. Probab. Theory and Math. Statist. Proc. of Fifth Vilnius Conference, 1989. VSP BV, Utrecht; Mokslas, Vilnius, 519–528. Zbl. 732.60029

Utev, S.A. (1990b): On the central limit theorem for a double array of φ-mixing random variables. Teor. Veroyatn. Primen. **35**, No. 1, 110–117. English transl.: Theory Probab. Appl. **35**, No. 1, 131–139. Zbl. 724.60028

Utev, S.A. (1991): On a method of studying sums of weakly dependent random variables. Sib. Mat. Zh. **32**, No. 4, 165–183. English transl.: Sib. Math. J. No. 4, 675–690 (1991). Zbl. 778.60014, Zbl. 769.60023

Volkonskii, V.A., and Rozanov, Yu.A. (1959): Some limit theorems for random functions, I. Teor. Veroyatn. Primen. **4**, No. 2, 186–207. English transl.: Theory Probab. Appl. **4**, No. 2, 178–197. Zbl. 092.33502

Volkonskii, V.A., and Rozanov, Yu.A. (1961): Some limit theorems for random functions, II. Teor. Veroyatn. Primen. **6**, No. 2, 202–215. English transl.: Theory Probab. Appl. **6**, No. 2, 186–198. Zbl. 108.31301

Withers, C.S. (1981): Central limit theorems for dependent random variables, I. Z. Wahrscheinlichkeitstheorie Verw. Geb. **57**, No. 4, 509–534. Zbl. 461.60036

Withers, C.S. (1987): Central limit theorems for dependent variables, II. Probab. Theory Related Fields **76**, No. 1, 1–13. Zbl. 608.60019

Yokoyama, R. (1979): Convergence of moments in the central limit theorem for stationary φ-mixing sequences. Tsukuba J. Math. **3**, No. 2, 1–6. Zbl. 434.60026

Yokoyama, R. (1980): Moment bounds for stationary mixing sequences. Z. Wahrscheinlichkeitstheorie Verw. Geb. **52**, No. 1, 45–57. Zbl. 416.60004

Yokoyama, R. (1983): The convergence of moments in the central limit theorem for weakly dependent random variables. Tsukuba Math. J. **7**, No. 1, 147–156. Zbl. 521.60025

Yoshihara, K. (1976): Limiting behavior of U-statistics for stationary absolutely regular processes. Z. Wahrscheinlichkeitstheorie Verw. Geb. **35**, No. 3, 237–252. Zbl. 336.60037

Yoshihara, K. (1977): Convergence rates for integral-type functionals of absolutely regular processes. Yokohama Math. J. **25**, No. 2, 145–153. Zbl. 373.60032

Yoshihara, K. (1978a): Moment inequalities for mixing sequences. Kodai Math. J. **1**, No. 2, 316–328. Zbl. 392.60020

Yoshihara, K. (1978b): Probability inequalities for sums of absolutely regular processes and their applications. Z. Wahrscheinlichkeitstheorie Verw. Geb. **43**, No. 4, 319–329. Zbl. 372.60035

Yoshihara, K. (1979a): Summability of random variables satisfying the strong mixing condition. Sci. Rep. Yokohama Nat. Univ. Sect. 1 **26**, 9–15. Zbl. 429.60023

Yoshihara, K. (1979b): Convergence rates of the invariance principle for absolutely regular sequences. Yokohama Math. J. **27**, No. 1, 49–55. Zbl. 429.60027

Yoshihara, K. (1985): Central limit theorems for stationary mixing sequences. Yokohama Math. J. **33**, No. 1–2, 131–137. Zbl. 601.60022

Yudin, M.D. (1981): On the rate of convergence of the distribution of a sum of $f(n)$-dependent random variables to the normal law. Vestsi Akad. Nauk BSSR, Ser. Fiz.-Mat. Navuk **3**, 57–60 (Russian). Zbl. 466.60050

Yudin, M.D. (1984): On the rate of convergence of the distribution of a sum of weakly dependent random variables. Vestsi Akad. Nauk BSSR, Ser. Fiz.-Mat. Navuk **2**, 44–52 (Russian). Zbl. 546.60024

Yudin, M.D. (1987): A note on approximating distributions of sums of dependent variables with infinitely divisible distributions. Vestsi Akad. Nauk BSSR, Ser. Fiz.-Mat. Navuk **2**, 38–41 (Russian). Zbl. 624.60027

Yudin, M.D. (1989a): On approximating distributions of sums of m_n-dependent variables by distributions in class L. Izv. VUZ. Mat. **4**, 83–88 (Russian). English transl.: Sov. Math. **33**, No. 4, 103–109. Zbl. 618.60027

Yudin, M.D. (1989b): Approximation of distributions of sums of mixing random variables by distributions in class L. Teor. Veroyatn. Mat. Statist. **41**, 120–125. English transl.: Theory Probab. Math. Statist. **41** (1990), 143–147. Zbl. 60024

Yudin, M.D. (1990): Convergence of Distributions of Sums of Random Variables. Universitetskoe, Minsk (Russian). Zbl. 745.60021

Yudin, M.D. (1991): On the approximation of distributions of sums of dependent random variables by distributions from class L. Vestsi Akad. Nauk BSSR, Ser. Fiz.-Mat. Navuk, 1991, No. 1, 35–41 (Russian).

Zaremba, S.K. (1958): Note on the central limit theorem. Math. Z. **69**, 295–298. Zbl. 081.35201

Zhurbenko, I.G. (1972): On strong estimate of mixed cumulants of stochastic processes. Sib. Mat. Zh. **13**, No. 2, 293–308. English transl.: Sib. Math. J. **13**, No. 2, 202–213. Zbl. 246.60040

Zolotarev, V.M. (1986): Modern Theory of Summation of Independent Random Variables. Nauka, Moscow. English transl.: VSP, Utrecht (1997). Zbl. 649.60016

Zolotukhina, L.A. (1978): The central limit theorem for discrete random fields. Izv. Akad. Nauk Uzb. SSR, Ser. Fiz.-Mat. Nauk, **3**, 15–19 (Russian). Zbl. 405.60023

Zolotukhina, L.A., and Chugueva, V.N. (1978): Sufficient conditions for asymptotic normality of sums of discrete random fields dependent in strips. Mat. Zametki **23**, No. 5, 725–732. English transl.: Math. Notes **23**, No. 5, 400–403. Zbl. 404.60056

Zuev, N.M. (1986): On the rate of convergence in the central limit theorem for $m(d)$-dependent random fields. Vestsi Akad. Nauk BSSR, Ser. Fiz.-Mat. Navuk **4**, 28–32 (Russian). Zbl. 614.60019

Zuev, N.M. (1989): Estimation of the rate of convergence in the central limit theorem for $m(d)$-dependent random fields. Vestsi Akad. Nauk BSSR, Ser. Fiz.-Mat. Navuk **3**, 17–22 (Russian). Zbl. 687.60053

Zuparov, T.M. (1981): Moment inequalities and estimation of the remainder in the central limit theorem for a sequence of weakly dependent random variables. Limit Theorems for Random Processes and Statistical Inferences. Fan, Tashkent, 69–87 (Russian). Zbl. 543.60029

Zuparov, T.M. (1983): Estimates of the rate of convergence in the central limit theorem for absolutely regular Banach-valued random variable. Dokl. Akad. Nauk. SSSR **272**, No. 5, 1042–1045. English transl.: Soviet Math. Dokl. **28**, No. 2, 475–478. Zbl. 553.41027

Zuparov, T.M. (1984): Estimates of the rate of convergence in the central limit theorem for absolutely regular Banach-valued random variables. Asymptotic Problems for Probability Distributions, Fan, Tashkent, 78–87 (Russian).

Zuparov, T.M. (1985): On estimates of the rate of convergence in the invariance principle for weakly dependent random variables. Limit Theorems for Probability Distributions. Fan, Tashkent, 32–52 (Russian). Zbl. 616.60036

IV. Refinements of the Central Limit Theorem for Homogeneous Markov Chains

P. Gudynas

Contents

Besides independent random variables (r.v.'s), modern probability also studies so-called weakly dependent r.v.'s. This is related both to the logical growth of probability theory and to its specific applications. Among the various weak dependency schemes, a special place is occupied by the mixing homogeneous Markov chains. They serve like an intermediary link between independent and identically distributed r.v.'s and the more complicated weak dependency schemes. In addition, homogeneous Markov chains have a fairly simple structure and are therefore comparatively easy to investigate. Finally, they are good models of many processes observed in practice.

The first goal of this article is to give the reader an idea of the two most highly developed ways of proving limit theorems for sums of r.v.'s connected in a homogeneous Markov chain. The second goal is to familiarize the reader with some results obtained by using these methods. Because of its short length, this article does not pretend to be a complete presentation. Only the most typical procedures and results for homogeneous chains are touched upon. But (and this should be underscored) the general-purpose methods for weakly dependent r.v.'s work fairly well also for homogeneous Markov chains.

§1. Introduction

We begin with some basic notation and definitions. Let $(\mathcal{X}, \mathcal{F})$ be a measurable space and let $P(x, A)$ be a *transition probability function* (t.p.f.). For fixed x, $P(x, \cdot)$ is by definition a probability measure on $(\mathcal{X}, \mathcal{F})$ and for fixed A, $P(\cdot, A)$ is measurable with respect to the σ-algebra $\mathcal{F}$ ($\mathcal{F}$-measurable).

Consider a *homogeneous Markov chain* $\{X_i,\ i = 0, 1, 2, \ldots\}$ with state space $(\mathcal{X}, \mathcal{F})$, initial distribution $G(\cdot)$, and transition probability $P(\cdot, \cdot)$. By the Ionescu-Tulcea theorem,

$$\begin{aligned}&\mathbf{P}_G(X_0 \in A_0,\ X_1 \in A_1, \ldots, X_n \in A_n)\\ &\quad = \int_{A_0} G(dx_0) \int_{A_1} P(x_0, dx_1) \ldots \int_{A_n} P(x_{n-1}, dx_n)\end{aligned} \tag{1}$$

for any sets A_i, $i = 0, 1, \ldots, n$, in $\mathcal{F}$. The subscript G in this emphasizes that $\mathbf{P}_G$ is the distribution of a chain with initial distribution G. Similarly, we shall add the subscript G (or some other one) to the expectation symbol and we shall take $\mathbf{E}_G f$ to mean an integral with respect to the measure $\mathbf{P}_G$. $P^n(x, A)$ is the n-step transition probability of the chain X_i from state x to set A. $\mathcal{A}_n$ is the σ-algebra generated by $\{X_i,\ i \leq n\}$ and $\mathcal{B}_m$ is the σ-algebra generated by $\{X_j,\ j \geq m\}$. Let $f(x)$ be a measurable real function on $(\mathcal{X}, \mathcal{F})$. Consider the sequence of r.v.'s

$$Y_i := f(X_i), \quad i = 1, 2, \ldots . \tag{2}$$

The r.v.'s $\{Y_i\}$ so defined are said to be *connected in a Markov chain.* Let

$$S_n := Y_1 + \ldots + Y_n, \tag{3}$$

and

$$\mu_n := \mathbf{E}_G S_n, \quad \sigma_n^2 := \mathbf{E}_G (S_n - \mu_n)^2 . \tag{4}$$

In what follows, we shall assume that there exists a probability measure Q which is invariant under P. In other words, for any $A \in \mathcal{F}$

$$Q(A) = \int Q(dx) P(x, A).$$

Many methods have been developed as of now for proving limit theorems for Markov chains. But the most definitive results have been deduced for chains defined on an arbitrary phase space and satisfying Doeblin's condition by a method based on the spectral theory of perturbed linear operators. This technique, which we shall call the spectral method, was later extended to broader classes of Markov chains. It is precisely with the gist of this method that we intend to begin. But before we do this, we shall make a few additional remarks. The first concerns Doeblin's condition D_0 (see Doob (1953)). It is equivalent to (see Rosenblatt (1971), p. 209)

$$\sup_{x\in\mathcal{X}}\ \sup_{A\in\mathcal{F}} |P^n(x,A) - Q(A)| \leq \alpha(n) \to 0. \tag{5}$$

In other words, condition D_0 of Markov chain theory is essentially equivalent to the *uniformly strong mixing condition* for weakly dependent r.v.'s. Furthermore, it is known that (5) holds if and only if there exist constants $\gamma > 0$ and $0 < \rho < 1$ such that

$$|P^n(x,A) - Q(A)| < \gamma\rho^n \tag{6}$$

for all $x \in \mathcal{X}$ and $A \in \mathcal{F}$. We see that the conditions equivalent to D_0 impose rather stringent constraints on the ergodic properties of a chain. This made it compelling to seek other conditions under which the spectral method is applicable. One of these is for a chain to be asymptotically uncorrelated. Let us state it here.

Let $L_2(\mathcal{A}_n)$ be the set of $\mathcal{A}_n$-measurable functions with zero means and finite second moments. Define $L_2(\mathcal{B}_m)$ in similar fashion. Let

$$c(n) :=$$
$$\sup\left\{\frac{\mathbf{E}_Q(fg)}{(\mathbf{E}_Q f^2 \mathbf{E}_Q g^2)^{1/2}} : f \in L_2(\mathcal{A}_m),\ g \in L_2(\mathcal{B}_{m+n}),\ m = 0, 1, \dots\right\}.$$

$c(n)$ is called the *maximal correlation coefficient.* If

$$c(n) \to 0 \tag{7}$$

as $n \to \infty$, then it is customary to say that the chain X_i is *asymptotically uncorrelated.*

We now attempt to clarify what conditions (5) and (7) have in common. This will help us understand the gist of the spectral method. Let $L^p = L^p(\mathcal{X}, \mathcal{F}, Q)$, $1 \leq p < \infty$, denote the Banach space of complex $\mathcal{F}$-measurable functions on $\mathcal{X}$ with norm $||g||_{L^p} = (\int |g(x)|^p Q(dx))^{1/p}$. Let $L^\infty = L^\infty(\mathcal{X}, \mathcal{F}, Q)$ be the Banach space of complex $\mathcal{F}$-measurable and essentially bounded functions on $\mathcal{X}$ with norm $||g||_{L^\infty} = \operatorname{ess\,sup} |g(x)|$. Consider operators $\mathbf{P}$ and $\mathbf{Q}$ defined on L^1 by the formulas

$$(\mathbf{P}g)(x) = \int_{\mathcal{X}} P(x, dy) g(y)$$

and

$$(\mathbf{Q}g)(x) = \int_{\mathcal{X}} Q(dy) g(y).$$

The first one is frequently called the *chain transition operator.* The second one is the projector on constants. Notice that $\mathbf{PQ} = \mathbf{QP} = \mathbf{Q}$. In other words, $\lambda = 1$ is an eigenvalue of $\mathbf{P}$. It is not hard to see also that (6) is equivalent to the condition

$$||\mathbf{P}^n - \mathbf{Q}||_{L^\infty \to L^\infty} < \gamma_0 \rho^n, \tag{8}$$

where $||\cdot||_{L^\infty \to L^\infty}$ is the operator norm in L^∞ and γ_0 is a positive constant. But $(\mathbf{P} - \mathbf{Q})^n = \mathbf{P}^n - \mathbf{Q}$. Therefore (8) implies that the spectral radius of

$\mathbf{P}-\mathbf{Q}$ in L^{∞} does not exceed ρ. Consequently, (6) (and hence also (5)) holds if and only if the entire spectrum of $\mathbf{P}$ lies in the disc $|\lambda| < \rho < 1$, except for the eigenvalue $\lambda = 1$.

It turns out that condition (7) has a similar spectral interpretation. But in this case, the operator $\mathbf{P}$ has to be considered in the more restricted Banach space $L^2(Q)$ of Q-square-integrable functions. Namely, (7) holds (see Lifshits (1978)) if and only if the spectrum of $\mathbf{P}$ in L^2 lies in some disc $|\lambda| < \rho_1 < 1$ except for the eigenvalue $\lambda = 1$. For this to happen, if is clearly necessary and sufficient that there exist a positive γ_1 such that

$$||\mathbf{P}^n - \mathbf{Q}||_{L^2 \to L^2} < \gamma_1 \rho_1^n. \tag{9}$$

We introduce one additional object that is very important to the study of Markov chains. Let B be a Banach space whose elements are complex-valued measurable functions on $(\mathcal{X}, \mathcal{F})$. For those complex numbers z for which it is meaningful, we define $\mathbf{P}(z)$ to be the operator acting according to the formula

$$(\mathbf{P}(z)g)(x) = \int P(x, dy) e^{z f(y)} g(y) \tag{10}$$

with $g \in B$.

By analogy with independent r.v.'s, $\mathbf{P}(z)$ is sometimes called the *operator-valued generating function* of the moments of the chain. Clearly, $\mathbf{P}(0) = \mathbf{P}$. Let $\psi = \psi(x)$ be a function identically equal to one (that is, $\psi(x) \equiv 1$). Let $f_n(t)$ be the characteristic function of S_n and $\mathcal{G}$ the functional defined by

$$\mathcal{G}g = \int_{\mathcal{X}} G(dx) g(x). \tag{11}$$

Then it is easy to see that $f_n(t)$ is expressible as

$$f_n(t) = \mathbf{E}_G e^{itS_n} = \mathcal{G}\mathbf{P}^n(it)\psi, \tag{12}$$

that is, the characteristic function of S_n can be expressed in terms of the n-th power of the operator-valued generating function. This demonstrates a certain analogy to independent r.v.'s. If (5) or (7) holds, then it is possible to deduce from (12), for example, the following asymptotic relation. In a sufficiently small neighborhood of zero

$$f_n(t) = \lambda^n(it)(1 + t\theta_1(t)) + \rho_2^n t\theta_2(n, t), \tag{13}$$

where $\lambda(it)$ is the largest (in absolute value) eigenvalue of $\mathbf{P}(it)$, $\theta_1(t)$ and $\theta_2(n, t)$ are bounded functions and ρ_2 is a positive number less than one. In addition, it turns out that the properties of $\lambda(it)$ in a neighborhood of zero are similar to those of an ordinary characteristic function. Consequently, if (5) or (7) holds, then the characteristic function of S_n behaves almost the same as in the case of independent r.v.'s. By applying classical methods, one can now easily obtain from relations such as (13) various refinements of the central limit theorem for Markov chains.

From what we have said, it becomes clear that the basic difficulties in proving limit theorems for S_n are associated with going from the identity (12) to expressions for $f_n(t)$ similar to (13). It is exactly for this reason that one also uses the *spectral method* whose basic steps we shall endeavor to demonstrate in §3. In the next section, we shall state several results which may be proved using this method.

§2. Results for B-Regular Chains

We first introduce a weak dependence condition that generalizes (5) and (7). Let B be a Banach space of complex functions on $\mathcal{X}$ such that

(a) if $g \in B$, then g is $\mathcal{F}$-measurable;
(b) if g is constant on $\mathcal{X}$, then $g \in B$;
(c) the operators $\mathbf{P}$ and $\mathbf{Q}$ defined above are bounded endomorphisms of B.

Definition. A Markov chain $\{X_i\}$ is *B-regular* (see Gudynas (1982c)) if there exists a number k_0 such that

$$||\mathbf{P}^{k_0} - \mathbf{Q}||_{B\to B} = \delta < 1. \tag{14}$$

Here $\mathbf{P}^{k_0}$ is the k_0-th iterate of $\mathbf{P}$ and $||\cdot||_{B\to B}$ is the operator norm in B. It is not hard to find another equivalent form of this condition: There exist positive numbers γ_3 and $\rho_3 < 1$ such that

$$||\mathbf{P}^n - \mathbf{Q}||_{B\to B} < \gamma_3 \rho_3^n. \tag{15}$$

Observe that when (14) holds, one may take $\gamma_3 = \delta^{-1}(||\mathbf{P}||_{B\to B}+||\mathbf{Q}||_{B\to B})^{k_0-1}$ and $\rho_3 = \delta^{1/k_0}$ in (15). Finally, B-regularity may be defined in one further way in terms of the spectral theory of linear operators. Condition (14) holds if an only if there is a ρ_4, $0 < \rho_4 < 1$, for which the set of points of the complex plane $\{\xi : |\xi| \geq \rho_4\}$ has a single point in common with the spectrum of the transition operator $\mathbf{P}$ (acting in B), namely, the simple eigenvalues $\lambda = 1$ to which the projector $\mathbf{Q}$ corresponds. It will be recalled that an eigenvalue is simple if it is bounded away from the rest of the spectrum and it has algebraic multiplicity one. Immediate consequences of this spectral interpretation of B-regularity and the spectral interpretation of (5) and (7) are the following assertions: Condition (5) is equivalent to L^∞-regularity and condition (7) is equivalent to L^2-regularity of a homogeneous Markov chain. In other words, the above two conditions may be viewed as special cases of the general scheme of L^p-regularity (where p runs over the interval $[1,\infty]$). A natural question is the interconnection among the conditions of L^p-regularity for different values of p.

Proposition 1. *If L^p-regularity is satisfied for at least one value of q in the interval $[1,\infty]$, then it is satisfied for all $p \in (1,\infty)$.*

Proof. This is an immediate consequence of the Riesz–Thorin theorem on interpolating linear operators in L^p (see Krein (1972), p. 150). It is merely necessary to use that L^p-regularity is equivalent to condition (15) and to take into account that $||\mathbf{P}^k - \mathbf{Q}||_{L^1 \to L^1} \le 2$ and $||\mathbf{P}^k - \mathbf{Q}||_{L^\infty \to L^\infty} < 2$. Let $r \in (1, q)$. One can find a θ, $0 < \theta < 1$, such that $1/r = (1-\theta) + \theta/q$. Then by the Riesz–Thorin theorem,

$$\begin{aligned} ||\mathbf{P}^n - \mathbf{Q}||_{L^r \to L^r} &\le ||\mathbf{P}^n - \mathbf{Q}||_{L^1 \to L^1}^{1-\theta} ||\mathbf{P}^n - \mathbf{Q}||_{L^q \to L^q}^{\theta} \\ &\le 2^{1-\theta} (\gamma_4 \rho_4^n)^\theta = \gamma_5 \rho_5^n, \ 0 < \rho_5 < 1, \end{aligned}$$

and, consequently, the chain $\{X_i\}$ is L^r-regular. The case $r \in (q, \infty)$ is proved in similar fashion.

It should be noted that the class of B-regular chains is not exhausted by the L^p-regular chains. L^p-regularity plays a central role only for chains with arbitrary state spaces. But for chains with topological state spaces, the interesting possibilities of B-regularity are considerably greater. We shall not make this last statement more precise. Instead, we give an example.

Example. Let $\{Z_i, i = 0, 1, \ldots\}$ be a Markov chain with phase space $\mathcal{X} = [0, 1]$ and transition probability

$$P(x, A) := \begin{cases} p(x) & \text{if } A = \{x/2\}, \\ 1 - p(x) & \text{if } A = \{x/2 + 1/2\}, \end{cases} \tag{16}$$

where $p(x)$ is a continuously differentiable function on $[0, 1]$ such that $0 < p(x) < 1$. The transition operator for this chain acts according to

$$(\mathbf{P}g)(x) = p(x)g(x/2) + (1 - p(x))g(x/2 + 1/2).$$

Let $C = C[0, 1]$ be the Banach space of all continuous functions on $[0, 1]$ with norm $||g||_0 := \max\{|g(x)| : x \in [0, 1]\}$ and let $C^{(1)} = C^{(1)}[0, 1]$ be the Banach space of differentiable functions on $[0, 1]$ with norm $||g|| := ||g||_0 + ||g||_1$, where $||g||_1 := \max\{|g'(x)| : x \in [0, 1]\}$. It has been shown (Gudynas (1982c)) that to any $q \in (0, 1)$, one may select $\varepsilon(q) > 0$ so that $||p(x) - q|| < \varepsilon(q)$ implies the existence of a measure Q invariant under the t.p.f. (16) and the $C^{(1)}$-regularity of the Markov chain with initial distribution Q and t.p.f. (16).

We mention right away that $\{Z_i\}$ is not a uniformly strong mixing Markov chain. In addition, even in the simple case where $p(x) \equiv 1/2$, the chain $\{Z_i\}$ is $L^q[0, 1]$-regular for no value of $q \in [1, \infty)$.

We now proceed to state various limit theorems for B-regular chains. Henceforth, we assume that

$$\int f(x) Q(dx) = 0. \tag{17}$$

Let $\Phi(x)$ be the standard normal distribution function and let

$$\sigma^2 := \mathbf{E}_Q(f^2(X_1)) + 2\sum_{k=1}^{\infty} \mathbf{E}_Q(f(X_1)f(X_{k-1})), \tag{18}$$
$$F_{n,G}(x) := \mathbf{P}_G(S_n < \sigma\sqrt{n}x).$$

Define operators $\mathbf{P}^{(k)}$, $k = 1, 2, \ldots$, by the relation

$$(\mathbf{P}^{(k)}g)(x) = (1/k!)\int_{\mathcal{X}} P(x, dy) f^k(y) g(y). \tag{19}$$

A function $f(x)$ will be said to satisfy condition (M_B) if

(a) $\mathbf{P}^{(1)}$, $\mathbf{P}^{(2)}$ and $\mathbf{P}(it)$, $t \in (-\infty, \infty)$, are bounded linear operators mapping B into B;
(b) in the operator norm,

$$\mathbf{P}(it) = \mathbf{P} + it\mathbf{P}^{(1)} - t^2\mathbf{P}^{(2)} + O(t^3).$$

The reader might question whether σ^2 is well defined. The next statement answers the question.

Proposition 2 (Gudynas (1982c)). *Let the Markov chain $\{X_i\}$ be B-regular and let $f(x)$ satisfy condition (M_B). Then σ^2 is well defined by relation (18) and $\sigma^2 < \infty$.*

Theorem 1 (Gudynas (1982c)). *Suppose that*

(a) *$\{X_i\}$ is a B-regular Markov chain;*
(b) *$f(x)$ satisfies condition (M_B);*
(c) *the linear functional $\mathcal{G}$ corresponding to the initial distribution G is bounded on B;*
(d) *$\sigma^2 > 0$.*

Then

$$\sup_x |F_{n,G}(x) - \Phi(x)| = O(n^{-1/2}). \tag{20}$$

One must recognize that condition (M_B) has been stated in very abstract terms and this complicates the application of Theorem 1. But it is doubtful whether this shortcoming can be avoided without narrowing the class of admissible spaces B. One of the reasonable ways of narrowing this class is to confine oneself to the spaces L^p, $p \in [1, \infty]$. Let us state a consequence of Theorem 1.

Theorem 2 (see Gudynas (1982b)). *For a fixed $p \in [1, \infty]$, suppose that*

(a) *$\{X_i\}$ is an L^p-regular chain;*
(b) *the operator $\mathbf{P}^{(3)}$ is a bounded endomorphism of L^p;*
(c) *the functional $\mathcal{G}$ belongs to $(L^p)^*$, the dual space of L^p;*
(d) *$\sigma^2 > 0$.*

Then relation (20) *holds.*

Remark 1. It is not hard to see that condition (b) of Theorem 2 with $p \in (1, \infty)$ may be replaced by the simpler though more restrictive condition

$$\operatorname*{ess\,sup}_x \int_{\mathcal{X}} P(x, dy)|f(y)|^{3q} < \infty, \tag{21}$$

where $1/p + 1/q = 1$.

It should be noted that the spectral method yields more precise results than Theorem 1. In particular, it permits one to describe in detail the structure of the remainder term in relation (20). A good illustration of this remark is Nagaev's classical theorem on L^∞-regular chains (in our terminology). We state a version of it here with a few insignificant changes.

Theorem 3 (see Nagaev (1961), Sirazhdinov (1979)). *Suppose that* (6) *holds and a positive function of a real variable* $H(x)$ *exists such that* $\lim H(x) = \infty$ *and*

$$\sup_x \int P(x, dy)|f(y)|^3 H(|f(y)|) < \infty.$$

Then σ^2 *is well defined and* $\sigma^2 < \infty$. *If* $F_{n,G}^{(0)}$ *denotes the distribution function of the normalized sum*

$$Z_n^{(0)} := \frac{1}{\sigma\sqrt{n}} \sum_{k=0}^{n} f(X_k),$$

$m_{G,1} := \int |f(x)|G(dx)$ *and* $\sigma^2 > 0$, *then*

$$\sup_x |F_{n,G}^{(0)}(x) - \Phi(x)| \le c_1 M^3(\rho) \frac{m_3}{\sigma^3\sqrt{n}} + c_2 \frac{m_{G,1}}{\sigma\sqrt{n}} + c_3 \left(\frac{1}{3} + \frac{2}{3}\rho\right)^n,$$

where c_1, c_2, *and* c_3 *are absolute constants and*

$$m_3 := \sup_x \int P(x, dy)|f(y)|^3,$$

$$M(\rho) := \frac{6(1 + 2\rho) + 3}{1 - \rho}.$$

An n-dimensional analogue of this limit theorem was proved by Formanov (see Sirazhdinov (1979), p. 108). Nagaev (1961) also studied asymptotic expansions for L^∞-regular chains. The structure of the remainder term in the CLT has actually not been investigated for other types of B-regularity.

The spectral method can be used to derive a number of results on large deviations. Classical large deviations have only been studied for L^p-regular chains (see Nagaev (1961) and Gudynas (1982b)).

Theorem 4 (Gudynas (1982b)). *For some* $p \in [1, \infty]$, *suppose that*

(a) $\{X_i\}$ *is an* L^p*-regular Markov chain;*
(b) $\mathcal{G}$ *is a bounded linear functional on* L^p*;*
(c) *there exists a positive* α *such that*

$$\sup\left\{\left\|\int P(\cdot, dy)\exp(\alpha|f(y)|)g(y)\right\|_{L^p} : ||g||_{L^p} \leq 1\right\} < \infty;$$

(d) $\sigma^2 > 0$.

If $x > 1$ *and* $x = o(\sqrt{n})$, *then*

$$\frac{1 - F_{n,G}(x)}{1 - \Phi(x)} = \exp(x^3 r(x/\sqrt{n})/\sqrt{n})(1 + O(x/\sqrt{n}))$$

and

$$\frac{F_{n,G}(-x)}{\Phi(-x)} = \exp(x^3 r(-x/\sqrt{n})/\sqrt{n})(1 + O(x/\sqrt{n})),$$

where $r(y)$ *is a real holomorphic function at* $y = 0$.

The spectral method yields especially interesting results when studying large deviations of order n. True, in this case one has to require more than B-regularity. Figuratively speaking, one has to have "B-regularity of rare events". We do not strive here to present the most general results for a broad class of spaces B. The reader will be able to find them in Gudynas (1982a) and (1986). We limit ourselves to a special case of a theorem in the latter paper where $B = L^s$, $s \in (1, \infty)$.

Theorem 5. *Let* $f(x)$ *be bounded and as before* $\int f(x)Q(dx) = 0$. *Let* $\nu(\cdot)$ *be a probability measure on* $(\mathcal{X}, \mathcal{F})$ *such that* $Q(dx) = q(x)\nu(dx)$ *and* $P(x, dy) = p(x, y)\nu(dy)$ *with* $q(x) > 0$ ν-a.s. *and* $p(x, y) > 0$ $(\nu \otimes \nu)$-a.s. *Let* $1 < s < \infty$, $1/s + 1/t = 1$, $q \in L^t(\mathcal{X}, \nu)$ *and*

$$\int \left(\int |p(x, y)|^s \nu(dx)\right)^{t/s} \nu(dy) < \infty. \tag{22}$$

Then $\lim_{n\to\infty} \mu_n/n = 0$ *and* $\lim_{n\to\infty} \sigma_n^2/n = \sigma^2$. *If* $r(h)$ *is the spectral radius of the operator* $\mathbf{P}(h)$ *in* $L^s(\mathcal{X}, \nu)$ *with* h *a real number,*

$$M_1 := \sup\left\{\frac{d}{dh}\ln r(h) : h > 0\right\},$$

and if, in addition, $\sigma^2 > 0$, *then for any* a, $0 < a < M_1$,

$$\lim_{n\to\infty} n^{-1}\ln \mathbf{P}_G(S_n > na) = \inf\{-ah + \ln r(h) : h > 0\} \tag{23}$$

and for all $a > M_1$,

$$\lim_{n\to\infty} n^{-1}\ln \mathbf{P}_G(S_n > na) = -\infty. \tag{24}$$

Corollary 1 (Gudynas (1986)). *Suppose that* $\{X_i\}$ *is a nondegenerate regular stationary Gaussian Markov sequence with values in* R^1 *and* $f(x)$ *is bounded and centered in the sense of* (17). *Then Theorem* 5 *is true, where* $r(h)$ *is the spectral radius of* $\mathbf{P}(h)$ *in* $L^2(R^1, \Phi)$ *and* Φ *is the standard normal distribution function.*

Remark 2. In contrast to the independent case for a homogeneous Markov chain, sometimes it is difficult to say something definite about the validity of (23) when $a = M_1$. And this difficulty can be essential. But if $\inf\{-ah + \ln r(h) : \ h \in [0, M_1)\} = -\infty$, then relation (23) also evidently holds for $a = M_1$.

It should be emphasized that under the hypotheses of Theorem 5, $\mathbf{P}(z)$ has for any complex z a simple eigenvalue $\lambda(z)$ which exceeds all other eigenvalues of this operator in absolute value. Furthermore, $\lambda(th)$ is real and positive for real values of h and so $r(h) = \lambda(h)$. Consequently, Theorem 5 provides a very important probabilistic interpretation of the largest eigenvalue of the operator-valued conditional moment-generating function $\mathbf{P}(z)$. In addition, the theorem once more shows that $\lambda(z)$ plays the same role in limit theorems for chains as the moment-generating function in the independent case. Therefore the gist of the spectral method for chains just consists in studying and applying the analytic properties of $\mathbf{P}(z)$ and $\lambda(z)$.

§3. Proof of Theorem 1

To illustrate the application of the spectral method, we present the principal steps in the proof of Theorem 1. Denote be $\mathbf{R}(\xi)$ the resolvent of the operator $\mathbf{P}$, by $\mathbf{R}(\xi, it)$ the resolvent of $\mathbf{P}(it)$, by $\lambda(it)$ the perturbed eigenvalue $\lambda = 1$ of the operator $\mathbf{P}(0) = \mathbf{P}$ (see Kato (1966)), by $\mathbf{Q}(it)$ the perturbed eigenprojector $\mathbf{Q}$ and by $\mathbf{S}$ the value of the resolvent of $\mathbf{P}$ at $\xi = 1$. Then (see Kato (1966), p. 229)

$$(\mathbf{P} - 1)\mathbf{S} = 1 - \mathbf{Q}, \quad \mathbf{SQ} = \mathbf{QS} = 0 \tag{25}$$

and in a neighborhood of $\xi = 1$,

$$\mathbf{R}(\xi) = -\mathbf{Q}/(\xi - 1) + \sum_{n=0}^{\infty} (\xi - 1)^n \mathbf{S}^{n+1}. \tag{26}$$

Below we need a specific expression for $\mathbf{S}$. By definition, the resolvent of $\mathbf{P}$ at $\xi = 1$ can be expressed as

$$\mathbf{R}_1(\xi) = \mathbf{R}(\xi) + \mathbf{Q}/(\xi - 1).$$

The spectrum of $\mathbf{P} - \mathbf{Q}$ lies in the disc $|\xi| \leq \rho_3 < 1$. Therefore, the following relation holds for $|\xi| > \rho_3$ (see Kato (1966), p. 53):

$$\mathbf{R}_1(\xi) = -\sum_{n=0}^{\infty} \xi^{-n-1}\mathbf{P}^n + \mathbf{Q}/(\xi - 1) = -\sum_{n=0}^{\infty} \xi^{-n-1}(\mathbf{P}^n - \mathbf{Q}).$$

Substituting $\xi = 1$, we obtain

$$\mathbf{S} = \mathbf{R}_1(1) = -\sum_{n=0}^{\infty}(\mathbf{P}^n - \mathbf{Q}). \tag{27}$$

The techniques of perturbation theory of linear operators (see, for example, Kato (1966)) may be used to prove the next assertion.

Proposition 3 (Gudynas (1982b)). *Suppose that a Markov chain is B-regular and satisfies condition* (M_B). *Let* Γ *be a closed contour isolating* $\lambda = 1$ *from the rest of the spectrum of* $\mathbf{P}$. *Then* $\lambda(it)$ *is still a simple eigenvalue in a some neighborhood of the origin and* Γ *once more isolates it from the rest of the spectrum of* $\mathbf{P}(it)$. *In addition,*

$$\mathbf{Q}(it) = \mathbf{Q} + it\mathbf{Q}^{(1)} - t^2\mathbf{Q}^{(2)} + O(t^3) \tag{28}$$

and

$$\lambda(it) = 1 + it\lambda^{(1)} - t^2\lambda^{(2)} + O(t^3), \tag{29}$$

where

$$\begin{aligned}\mathbf{Q}^{(1)} &= \mathbf{QP}^{(1)}\mathbf{S} - \mathbf{SP}^{(1)}\mathbf{Q}; \quad \mathbf{Q}^{(2)} = -\mathbf{QP}^{(2)}\mathbf{S} - \mathbf{SP}^{(2)}\mathbf{Q}\\ &\quad + \mathbf{QP}^{(1)}\mathbf{SP}^{(1)}\mathbf{S} + \mathbf{SP}^{(1)}\mathbf{QP}^{(1)}\mathbf{S} + \mathbf{SP}^{(1)}\mathbf{SP}^{(1)}\mathbf{Q}\\ &\quad - \mathbf{QP}^{(1)}\mathbf{QP}^{(1)}\mathbf{S}^2 - \mathbf{QP}^{(1)}\mathbf{S}^2\mathbf{P}^{(1)}\mathbf{Q} - \mathbf{S}^{(2)}\mathbf{P}^{(1)}\mathbf{QP}^{(1)}\mathbf{Q}\end{aligned}$$

and

$$\lambda^{(1)} = \mathrm{tr}\mathbf{P}^{(1)}\mathbf{Q}; \quad \lambda^{(2)} = \mathrm{tr}(\mathbf{P}^{(2)}\mathbf{Q} - \mathbf{P}^{(1)}\mathbf{SP}^{(1)}\mathbf{Q}). \tag{30}$$

The substitution of the specific expressions for $\mathbf{Q}$, $\mathbf{S}$, $\mathbf{P}^{(1)}$, and $\mathbf{P}^{(2)}$ in (30) yields

$$\lambda^{(1)} = \mathbf{E}_Q f(X_1) = 0, \quad \lambda^{(2)} = \sigma^2/2. \tag{31}$$

We proceed now to the proof of Theorem 1. Because of Proposition 3.

$$\mathbf{P}^n(it) = \lambda^n(it)\mathbf{Q}(it) + \mathbf{T}^n(it) \tag{32}$$

in a small enough neighborhood of zero, where $\mathbf{T}(it) = \mathbf{P}(it)(1 - \mathbf{Q}(it))$. Substituting (32) in (12), we obtain

$$f_n(t) = \mathbf{E}_G \exp(itS_n) = \lambda^n(it)\mathcal{G}\mathbf{Q}(it)\psi + \mathcal{G}\mathbf{T}(it)\psi. \tag{33}$$

In view of Proposition 3 and (31), as $t \to 0$,

$$\lambda(it) = 1 - t^2\sigma^2/2 + O(t^3), \tag{34}$$

$$\mathcal{G}\mathbf{Q}(it)\psi = 1 + O(t) \tag{35}$$

and

$$||(\mathbf{1}-\mathbf{Q}(it))\psi||_B = ||(\mathbf{1}-\mathbf{Q}(it))\mathbf{Q}\psi||_B = ||(\mathbf{Q}-\mathbf{Q}(it))\psi||_B = O(t). \quad (36)$$

Furthermore, relation (14) and the continuity of $\mathbf{P}(it)$, $\lambda(it)$ and $\mathbf{Q}(it)$ at $t=0$ imply that to any δ_1, $\delta < \delta_1 < 1$, there is a neighborhood of zero in which

$$\begin{aligned} ||\mathbf{T}^{k_0}(it)||_{B\to B} &= ||\mathbf{P}^{k_0}(it)(\mathbf{1}-\mathbf{Q}(it))||_{B\to B} \\ &= ||\mathbf{P}^{k_0}(it) - \lambda^{k_0}(it)\mathbf{Q}(it))||_{B\to B} < \delta_1. \end{aligned}$$

Therefore, taking $\rho_4 = \delta_1^{1/k_0}$, one can select a positive constant γ_4 so that

$$||\mathbf{T}^n(it)||_{B\to B} < \gamma_4\rho_4^n \quad (37)$$

in the same neighborhood of zero. By (36), (37) and condition (M_B),

$$\begin{aligned} |\mathcal{G}\mathbf{T}^n(it)\psi| &= |\mathcal{G}\mathbf{T}^{n-1}(it)\mathbf{P}(it)(\mathbf{1}-\mathbf{Q}(it))\psi| \\ &\le ||\mathcal{G}||_{B^*}||\mathbf{T}^{n-1}(it)||_{B\to B}||\mathbf{P}(it)||_{B\to B}||\mathbf{1}-\mathbf{Q}(it))\psi||_B \le C_4\rho_4^n t \quad (38) \end{aligned}$$

for sufficiently small t, where C_4 is a constant not depending on n and t. Now inserting (34), (35) and (38) in (33), we obtain

$$f_n(t) = \lambda^n(it)(1+t\theta_1(t)) + \rho_4^n t\theta_2(n,t) \quad (39)$$

for t sufficiently small, where $\theta_1(t)$ and $\theta_2(t)$ are bounded functions.

Thus, we have proved our basic relation (13). The balance of the proof of the theorem actually does not differ from that of analogous statements in the independent case. As usual, we apply Esseen's inequality. Noting (39), we find that

$$\begin{aligned} \sup_x |F_{n,G} - \Phi(x)| &\le A/\Delta + \pi^{-1}\int_{-\Delta}^{\Delta}\left|\frac{f_n(t/\sigma\sqrt{n}) - \exp(-t^2/2)}{t}\right| dt \\ &= A/\Delta + \pi^{-1}\int_{-\Delta}^{\Delta}\left|t^{-1}\Big(\lambda^n(it/\sigma\sqrt{n})\Big(1+\frac{t}{\sigma\sqrt{n}}\theta_1(t/\sigma\sqrt{n})\Big)\right. \\ &\quad \left.-\exp(-t^2/2)\Big)\right| dt + \pi^{-1}\int_{-\Delta}^{\Delta}\left|\frac{\rho_4^n\theta_2(n,t/\sigma\sqrt{n})}{\sigma\sqrt{n}}\right| dt \\ &= J_1 + J_2 + J_3, \end{aligned} \quad (40)$$

where $\Delta > 0$ and A is an absolute constant. Taking $\Delta = \varepsilon\sqrt{n}$ and estimating J_2 and J_3 with the help of (34), with no especial difficulty, we deduce from (40) that

$$\sup_x |F_{n,G} - \Phi(x)| = O(n^{-1/2}).$$

This completes the proof. The proofs of the remaining theorems of §2 are in many respects similar to that of Theorem 1.

§4. Harris-Recurrent Markov Chains

It will be recalled that a renewal sequence is one that can be split into mutually independent segments (renewal cycles) by a sequence of random times (renewal times). The classical example of such a sequence is a homogeneous Markov chain with a recurrent state. Its renewal times are obviously the times it hits this state. In this connection, a method has been created for proving limit theorems whose basic idea is the following: a limit theorem for Markov chains is reduced to a statement about a sequence of independent random variables – the increments in the renewal cycles; the powerful techniques of the summation of independent random variables can then be applied to it. The source of this method may be found in Kolmogorov (1937) and Doeblin (1938), (1940). It was improved considerably later on. This method has been used to deduce various limit theorems for homogeneous chains (see Chung (1960)) including bounds for the rate of convergence in the CLT (see Landers (1976), Bolthausen (1980), (1982)).

For a long time, it seemed that a Markov chain could only be studied by the *renewal method* providing its phase space contained a positively recurrent state. This appeared to be an essential shortcoming of the method. However subsequent studies of the ergodic properties specified on any phase space of Markov chains showed that their renewal in the broader sense of existence of a renewal increment is assured by Harris's recurrency condition (see Nummelin (1978), Athreya & Ney (1978)). This condition, of which Doeblin's is a special case, also imposes certain restrictions on the ergodic properties of the chain. But these restrictions are quite reasonable and are satisfied in many practically interesting situations. Results found by the earlier renewal method for discrete chains can be carried over readily to chains defined on an arbitrary phase space and satisfying Harris's reccurency condition (see, for instance, Bolthausen (1982), Malinovskii (1986), (1989), Nummelin (1984), and Levental (1988)). The renewal method has thus proved to be a suitable way of studying a broad class of Markov chains.

We now discuss the notion of Harris recurrency. We shall not be able to do this in an entirely simple fashion because the authors who have used it in limit theorems for chains have modified it somewhat. We state here the definition given by Levental (1988).

Definition. A homogeneous Markov chain $\{X_i\}$ is (C, λ, β, m)-*recurrent* if there exist a set $C \in \mathcal{F}$, a number λ, $0 < \lambda < 1$, a natural number m, and a probability measure β such that

(a) $P^m(x, K) > \lambda\beta(K)$ for all $x \in C$ and $K \in \mathcal{F}$;
(b) $P_x(X_n \in C$ for some $n \geq 1) = 1$ for all $x \in \mathcal{X}$.

The chain $\{X_i\}$ is said to be positively recurrent if in addition to (a) and (b) one has: (c) $\sup\{\mathbf{E}_x(\tau_c) : \; x \in C\} < \infty$, where τ_c is the first time that the chain falls in C by step m (in other words, $\tau_c := \min\{i \geq 1 : \; X_{i \cdot m} \in C\}$) and

$\mathbf{E}_x$ is the mean evaluated with respect to the distribution of the chain with initial distribution G concentrated at x.

Remark 3 (see Levental (1988)). If $\mathcal{F}$ is a countably generated σ-algebra, then the definition is equivalent to Harris recurrency (see Orey (1971) and Nummelin (1984)). When we speak here of *Harris recurrency*, we shall mean (C, λ, β, m)-recurrency.

As was already mentioned above, the renewal method was devised to study Markov chains having recurrent atoms. We shall not discuss here in detail the initial stage of its development and the corresponding results that have now become traditional. We are interested mostly in everything that allows one to compare the method with the spectral method and, particularly, to compare their generality. From that standpoint, an important step on the development of the renewal method was the construction of *renewal extensions* of Harris-recurrent chains (Athreya and Ney (1978), Nummelin (1978)). We now give a brief description of Nummelin's construction. In so doing, we confine ourselves to the simplest case of a $(C, \lambda, \beta, 1)$-recurrent chain $\{X_i\}$. Our aim is to form a Markov chain $\{X_i^*\}$ with $\{X_i\}$ embedded in $\{X_i^*\}$ such that there exists an atom which $\{X_i^*\}$ visits infinitely often with probability one.

For any $x \in X$ and $A \in \mathcal{F}$, put

$$\begin{aligned} x_0 &:= (x,0), \quad x_1 := (x,1), \quad \mathcal{X}^* := \{x_i :\ x \in \mathcal{X},\ i = 0,1\}, \\ A_0 &:= A \times \{0\}, \quad A_1 := A \times \{1\}, \quad A^* = A \times \{0,1\}. \end{aligned}$$

Let $\mathcal{F}^*$ be the σ-algebra generated by the sets A_i $(A \in \mathcal{F},\ i = 0,1)$. Below we shall identify any $A \in \mathcal{F}$ with $A^* \in \mathcal{F}^*$. In particular, we write $\mathcal{F} \subset \mathcal{F}^*$. Any probability measure μ on $\mathcal{F}$ can be extended automatically to a measure μ^* on $\mathcal{F}^*$ by defining its values on the sets A_i $(A \in \mathcal{F},\ I = 0,1)$ as

$$\mu^*(A_0) = \int_A (1 - \lambda I_C(x))\mu(dx), \quad \mu^*(A_1) = \lambda\mu(A \cap C), \tag{41}$$

where $I_C(\cdot)$ is the indicator of C and $\lambda \in (0,1)$. We shall call this new measure μ^* the extension of μ. We define the transition probability P^* from $(\mathcal{X}^*, \mathcal{F}^*)$ to $(\mathcal{X}, \mathcal{F})$ as follows. For any $x \in \mathcal{X}$ and $A \in \mathcal{F}$,

$$\begin{aligned} P^*(x_0, A) &:= (1 - \lambda I_C(x))^{-1}(P(x,A) - \lambda I_C(x)\beta(A)), \\ P^*(x_1, A) &:= \beta(A). \end{aligned}$$

We extend P^* to be a t.p.f. from $(\mathcal{X}^*, \mathcal{F}^*)$ to $(\mathcal{X}^*, \mathcal{F}^*)$ as follows: for any $z \in \mathcal{X}^*$, the measure $P^*(z, \cdot)$ on $\mathcal{F}^*$ is defined to be the extension (in the sense of (41) of the measure $P^*(z, \cdot)$ on $\mathcal{F}$.

Let $\{X_1^*\} = \{(V_i, W_i)\}$, $(V_i \in \mathcal{X},\ W_i \in \{0,1\})$, be the Markov chain with state space $(\mathcal{X}^*, \mathcal{F}^*)$, t.p.f. $P^*(\cdot, \cdot)$ and initial distribution G^*, with G^* the extension of G. We see at once that $\mathcal{X}_1 := \mathcal{X} \times \{1\}$ is an atom of P^* and that $Q^*(\mathcal{X}_1) > 0$. More precisely, the next assertion is true.

Proposition 4 (Nummelin (1978)). *Let G be any initial distribution. The marginal distribution of the first component process $\{V_i\}$ of the chain $\{X_i^*\}$ is identical to the distribution of $\{X_i\}$. Moreover, the set $B := \mathcal{X}_1$ is a recurrent atom of $\{X_i^*\}$.*

Finally, we state a result which was deduced by means of the renewal method. Recall in this connection that the strong mixing coefficient is given by

$$\gamma(k) = \sup_n \sup_{A \in \mathcal{A}_n} \sup_{B \in \mathcal{B}_{n+k}} |\mathbf{P}_Q(A \cap B) - \mathbf{P}_Q(A)\mathbf{P}_Q(B)|. \tag{42}$$

Theorem 6 (Bolthausen (1982)). *Suppose that $\{X_i\}$ is a stationary Markov chain (that is, $G = Q$) and that $3 < p \leq \infty$. If*

$$\sum_{m=1}^{\infty} m^{(p+3)/(p-3)} \gamma(m) < \infty, \tag{43}$$

$$||f||_{L^p} < \infty \tag{44}$$

and (17) *holds, then the series in* (18) *is convergent. Moreover, if $\sigma^2 > 0$ and $\{X_i\}$ is $(C, \lambda, \beta, 1)$-recurrent, then*

$$\sup_x |F_{n,Q}(x) - \Phi(x)| = O(n^{-1/2}).$$

In conclusion, it should be noted that the zones of applicability of the spectral and renewal methods only coincide partially. Thus, (43) does not imply B-regularity of the chain. On the other hand, the chains discussed in the above example evidently do not have (C, λ, β, m)-regularity. It should also be mentioned that there is a significant difference in the terminology used to formulate results for B-regular chains and Harris-recurrent chains.

References*

Athreya, K.B., and Ney, P. (1978): A new approach to the limit theory of recurrent Markov chains. Trans. Am. Math. Soc. **245**, 493–501. Zbl. 397.60053

Bolthausen, E. (1980): The Berry–Esseen theorem for functionals of discrete Markov chains. Z. Wahrscheinlichkeitstheorie Verw. Geb. **54**, No. 1, 59–73. Zbl. 438.60026

Bolthausen, E. (1982): The Berry–Essen theorem for strongly mixing Harris-recurrent Markov chains. Z. Wahrscheinlichkeitstheorie Verw. Geb. **60**, No. 3, 283–289. Zbl. 492.60026

* For the convenience of the reader, references to reviews in Zentralblatt für Mathematik (Zbl.), compiled by means of the MATH database, and Jahrbuch über die Fortschritte der Mathematik (Jbuch) have, as far as possible, been included in this bibliography.

Chung, K.L. (1960): Markov Chains with Stationary Transition Probabilities. Springer-Verlag Berlin. Zbl. 092.34304

Doeblin, W. (1938): Sur deux problèmes de Kolmogoroff concernant les chaines denombrables. Bull. Math. de France **66**, 210–220. Zbl. 020.14604

Doeblin, W. (1940): Elements d'une théorie générale des chaines simples constantes de Markoff. Ann. Sci. École Norm. Sup. **37**, No. 3, 61–111. Zbl. 024.26503

Doob, J.L. (1953): Stochastic Processes. John Wiley, New York. Zbl. 053.26802

Gudynas, P. (1982a): Large deviations of order n for sums of random variables related to a Markov chain. Liet. Mat. Rink. **21**, No. 4, 65–74. English transl.: Lith. Math. J. **21**, No. 4, 306–312. Zbl. 496.60025

Gudynas, P. (1982b): Refinements of the central limit theorem for a homogeneous Markov chain. Liet. Mat. Rink. **22**, No. 1, 66–78. English transl.: Lith. Math. J. **22**, No. 1, 36–45. Zbl. 501.60032

Gudynas, P. (1982c): B-regularity of a homogeneous Markov chain. Liet. Mat. Rink. **22**, No. 3, 67–80. English transl.: Lith. Math. J. **22**, No. 3, 265–275. Zbl. 531.60064

Kaplan, E.I., and Sil'vestrov, D.S. (1979): Theorems of the invariance principle type for recurrent semi-Markov processes with arbitrary phase space. Teor. Veroyatn. Primen. **24**, No. 3, 529–541. English transl.: Theory Probab. Appl. **24**, No. 3, 536–547. Zbl. 408.60031

Kato, T. (1966): Perturbation Theory for Linear Operators. Springer-Verlag, New York. Zbl. 148.12601

Kolmogorov, A.N. (1937): Markov chains with a denumerably many possible states. Bull. Mosk. Gos. Univ. **1**, No. 3, 1–16 (Russian). Zbl. 018.41302

Krein, S.G., ed. (1972): Functional Analysis. Nauka, Moscow (Russian). Zbl. 261.47001

Landers, D., and Rogge, L. (1976): On the rate of convergence in the central limit theorem for Markov chains. Z. Wahrscheinlichkeitstheorie Verw. Geb. **35**, No. 1, 57–63. Zbl. 322.60025

Levental, S. (1988): Uniform limit theorems for Harris-recurrent Markov chains. Probab. Theory Related Fields **80**, No. 2, 101–118. Zbl. 638.60030

Lifshits, B.A. (1978): On the central limit theorem for Markov chains, Teor. Veroyatn. Primen. **23**, No. 2, 295–312. English transl.: Theory Probab. Appl. **23**, No. 2, 279–296. Zbl. 389.60047

Malinovskii, V.K. (1986): On limit theorems for Harris-recurrent Markov chains, I. Teor. Veroyatn. Primen. **31**, No. 2, 315–332. English transl.: Theory Probab. Appl. **31**, No. 2, 269–285. Zbl. 657.60087

Malinvoskii, V.K. (1989): On limit theorems for Harris-recurrent Markov chains, II. Teor. Veroyatn. Primen. **34**, No. 2, 289–303. English transl.: Theory Probab. Appl. **34**, No. 2, 252–265. Zbl. 711.60066

Nagaev, S.V. (1958): Some limit theorems for stationary Markov chains. Teor. Veroyatn. Primen. **2**, No. 4, 389–416. English transl.: Theory Probab. Appl. **2**, 378–406. Zbl. 078.31804

Nagaev, S.V. (1961): More exact statements of limit theorems for homogeneous Markov chains. Teor. Veroyatn. Primen. **6**, No. 1, 67–86. English transl.: Theory Probab. Appl. **6**, No. 2, 62–81. Zbl. 116.10602

Nummelin, E. (1978): A splitting technique for Harris-recurrent Markov chains. Z. Wahrscheinlichkeitstheorie Verw. Geb. **43**, No. 4, 309–318. Zbl. 372.60090

Nummelin, E. (1984): General Irreducible Markov Chains and Non-Negative Operators. Cambridge Univ. Press, Cambridge. Zbl. 551.60066

Orey, S. (1971): Lecture Notes on Limit Theorems for Markov Chains Transition Probabilities. Van Nostrand, New York. Zbl. 295.60054

Revuz, D. (1975): Markov Chains. North Holland, New York, etc. Zbl. 322.60045

Rosenblatt, M. (1971): Markov Processes. Structure and Asymptotic Behavior. Springer-Verlag, New York. Zbl. 236.60002

Sirazhdinov, S.Kh., and Formanov, Sh.K. (1979): Limit Theorems for Sums of Random Vectors Connected in a Markov Chain. FAN, Tashkent (Russian). Zbl. 482.60019

V. Limit Theorems on Large Deviations

L. Saulis and V. Statulevičius

Contents

Preface

Most papers dealing with limit theorems on large deviations study independent random variables. There are powerful analytic tools that are applicable to this case that allow one to understand the general behavior of the probabilities of large deviations.

Most frequently investigated have been the following situations: Cramér's condition is satisfied, that is, the characteristic functions of the terms are analytic in a neighborhood of zero; all of the moments of the terms are finite but their growth does not assure the analyticity of the respective characteristic functions in a neighborhood of zero, considered by Linnik; the so-called moderate deviations first studied by Rubin and Sethuraman (1965) where the terms have only finitely many moments; and finally neither Cramér's nor Linnik's condition holds but the tail distributions of the terms behave sufficiently regularly (the case where the terms belong to the zone of attraction of a stable law with exponent $\alpha < 2$ was considered by Fortus (1957), Heyde (1968), and Tkachuk (1975); ultralarge deviations were studied by S.V. Nagaev (1961), (1963) and A.V. Nagaev (1967)).

Many of the basic ideas and results have been presented fairly completely in the books by Ibragimov and Linnik (1965) and Petrov (1972).

Another area that has attracted the attention of researchers are large deviations in functional limit theorems. The most complete results appear in theorems establishing the behavior of the large deviation probabilities up to logarithmic equivalence. Mention should be made here primarily of the papers by Sazonov (1974) and Bahadur (1960) on sample distribution functions, by Borovkov (1964a,b) and Borovkov and Mogul'skii (1978), (1989) on large deviations in the invariance principle for sums of independent random variables and for processes with independent increments, by Donsker and Varadhan (1975a,b) on large deviations in ergodic theory of Markov processes, by

Ventsel' (1976) on large deviations for broad classes of families of Markov processes, and by Liptser and Shiryaev (1989) for martingales and semimartingales. Most of the results in this area are presented in the books by Ventsel' (1986), Stroock (1984) and Liptser and Shiryaev (1989).

This article is devoted to applications of the cumulant method to limit theorems on large deviations. The reader will be able to see that this method is an effective way of studying large deviation probabilities for sums of independent and dependent random variables, polynomial forms, multiple stochastic integrals of random processes and polynomial statistics. Only the case of normal approximation will be considered here.

To obtain a more thorough acquaintance with the cumulant method and its usefulness in asymptotic analysis taking large deviations into account, the reader is referred to the book: Saulis and Statulevičius: "Limit Theorems on Large Deviations", Moscow and Vilnius, 1989 (English translation: Kluwer, Dordrecht-Boston, 1991).

Chapter 1
Basic Lemmas

Introduction

Let ξ be a random variable (r.v.) with distribution function (d.f.) $F_\xi = \mathbf{P}\{\xi < x\}$ and characteristic function (c.f.) $f_\xi(t) = \mathbf{E}\exp(it\xi)$ whose absolute moment $\beta_l = \mathbf{E}|\xi|^l$ exists. The existence of β_l implies the existence of the *cumulants* (semi-invariants) γ_k of any order not exceeding l ($k \le l$). They are given by the formula

$$\gamma_k = \frac{1}{i^k}\frac{d^k}{dt^k}(\ln f_\xi(t))\Big|_{t=0}. \tag{1.1}$$

The cumulant γ_k of ξ will also be denoted by $\Gamma_k(\xi)$. It is generally simpler to use cumulants than moments. Thus, for example, if $\xi_1, \dots, \xi_n$ are independent r.v.'s and if $S_n = \xi_1 + \dots + \xi_n$, then

$$\Gamma_k(S_n) = \sum_{j=1}^{n} \Gamma_k(\xi_j). \tag{1.2}$$

The present chapter is intended to acquaint the reader with the potentialities of the cumulant method in studying the asymptotic behavior of distributions. It deals mainly with large deviation probabilities for various functionals of random processes and fields and primarily for sums

$$S_n = \sum_{j=1}^{n} \xi_j \tag{1.3}$$

of independent r.v.'s ξ_j, sums

$$S_n = \sum_{t=1}^{n} X_t \tag{1.4}$$

of dependent r.v.'s X_t, *polynomial forms*

$$\zeta_n^{(p)} = \sum_{1 \le t_1 \le \ldots \le t_p \le n} a_{t_1,\ldots,t_p} X_{t_1} \cdot \ldots \cdot X_{t_p}, \tag{1.5}$$

multiple stochastic integrals

$$Y_n^{(p)} = \underbrace{\int \cdots \int}_{p} a(t_1, \ldots, t_p) dX(t_1) \ldots dX(t_p) \tag{1.6}$$

with respect to a Wiener or Poisson process, *Pitman polynomial estimators*, *U-statistics* and others.

In studying random processes, we shall make use of their finite-dimensional distributions or, in other words, the distributions of the vector $X = (X_{t_1}, \ldots, X_{t_k})$, $t_1, \ldots, t_k \in T$. If $\mathbf{E}|X_t^m| < \infty$, $t \in T$, then the functions

$$m_X(t_1, \ldots, t_k) := \mathbf{E} X_{t_1} \ldots X_{t_k} \tag{1.7}$$

are defined for all $k \le m$ (the symbol := means "by definition"); $m_X(t_1, \ldots, t_k)$ is the *k-th moment function* or *simple moment* of k-th order of X_t. Let

$$f_X(u_1, \ldots, u_k) := \mathbf{E} \exp\{i\langle u, X\rangle\} \tag{1.8}$$

be the characteristic function of the random vector X, where $\langle a, b\rangle = \sum_{j=1}^{k} a_j b_j$ is the inner product of $a = (a_1, \ldots, a_k) \in R^k$ and $b = (b_1, \ldots, b_k) \in R^k$. Then similarly to the one-dimensional case, when $\mathbf{E}|X_t^m| < \infty$, the *mixed cumulants* of X

$$\Gamma_\nu(X) := \frac{1}{i^{|\nu|}} \frac{\partial^{\nu_1 + \cdots + \nu_k}}{\partial u_1^{\nu_1} \ldots \partial u_k^{\nu_k}} \ln f_X(u_1, \ldots, u_k) \bigg|_{\substack{u_1 = 0 \\ \vdots \\ u_k = 0}} \tag{1.9}$$

exist for all k and $\nu = (\nu_1, \ldots, \nu_k)$, $\nu_i \ge 0$, with $k|\nu| \le m$, where $|\nu| := |\nu_1| + \ldots + |\nu_k|$. Sometimes we shall write $\Gamma_\nu(X_{t_1}, \ldots, X_{t_k})$ instead of $\Gamma_\nu(X)$.

If $\nu = (1, \ldots, 1) \in R^k$, the corresponding cumulant $\Gamma_\nu(X_{t_1}, \ldots, X_{t_k})$ will be denoted by $\Gamma_k(X)$, $\Gamma(X_{t_1}, \ldots, X_{t_k})$ or $c_X(t_1, \ldots, t_k)$. $c_X(t_1, \ldots, t_k)$ is called the *correlation function* or *simple cumulant* of k-th order of the random process X_t. If $S_n = \sum_{t=1}^{n} X_t$ or $S(T) = \int_0^T X_t dt$ (where X_t is measurable and $\int_0^T X_t(\omega) dt$ exists for almost all ω), then the definition implies that

$$\Gamma_k(S_n) = \sum_{1\le t_1,\dots,t_k\le n} \Gamma(X_{t_1},\dots,X_{t_k}) \tag{1.10}$$

and

$$\Gamma_k(S(T)) = \underbrace{\int_0^T \cdots \int_0^T}_{k} \Gamma(X_{t_1},\dots,X_{t_k})dt_1\dots dt_k. \tag{1.11}$$

To shorten the notation, we shall makes use of the following: If $\nu = (\nu_1,\dots,\nu_k)$ is an integer-valued non-negative vector and $a = (a_1,\dots,a_k)$ is a real vector, then

$$a^\nu := a_1^{\nu_1}\cdots a_k^{\nu_k}, \quad \nu! := \nu_1!\cdots\nu_k!, \quad |\nu| := \nu_1+\cdots+\nu_k. \tag{1.12}$$

Put

$$\mathbf{E}_\nu(X) = \mathbf{E}X_{t_1}^{\nu_1}\cdots X_{t_k}^{\nu_k}. \tag{1.13}$$

As in the one-dimensional case, one can derive the following formulas connecting $\mathbf{E}_\nu(X)$ and $\Gamma_\nu(X)$:

$$\mathbf{E}_\nu(X) = \sum_{\lambda^{(1)}+\cdots+\lambda^{(q)}=\nu} \frac{1}{q!}\frac{\nu!}{\lambda^{(1)}!\dots\lambda^{(q)}!}\prod_{p=1}^{q}\Gamma_{\lambda^{(p)}}(X), \tag{1.14}$$

$$\mathbf{\Gamma}_\nu(X) = \sum_{\lambda^{(1)}+\cdots+\lambda^{(q)}=\nu} \frac{(-1)^{q-1}}{q}\frac{\nu!}{\lambda^{(1)}!\dots\lambda^{(q)}!}\prod_{p=1}^{q}\mathbf{E}_{\lambda^{(p)}}(X), \tag{1.15}$$

where the two summations are over all ordered collections of non-negative integral vectors $\lambda^{(p)}$, $|\lambda^{(p)}|>0$, adding up to ν.

Let $I = \{t_i,\dots,t_k\}$ be an index set for the vector X. A partition of I is an unordered collection of nonempty disjoint sets I_p such that $\bigcup_p I_p = I$. In this notation, a relationship can be established between the simple moments and cumulants on the basis of (1.14) and (1.15):

$$\mathbf{E}X_{t_1}\cdots X_{t_k} = \sum_{\bigcup_{p=1}^{q} I_p = I}\ \prod_{p=1}^{q}\Gamma(X_{I_p}), \tag{1.16}$$

$$\mathbf{\Gamma}(X_{t_1},\dots,X_{t_k}) = \sum_{\bigcup_{p=1}^{q} I_P = I} (-1)^{q-1}(q-1)!\prod_{p=1}^{q}\mathbf{E}(X_{I_p}). \tag{1.17}$$

Here and throughout, the following notation has been adopted: If $I' = \{t_1',\dots,t_m'\}\subset I$, then

$$X_{I'} := (X_{t'_1}, \dots, X_{t'_m}),$$
$$\mathbf{E}(X_{I'}) := \mathbf{E}X_{t'_1}, \dots, X_{t'_m},$$
$$\Gamma(X_{I'}) := \Gamma(X_{t'_1}, \dots, X_{t'_m}).$$

Formulas (1.14)–(1.17) are due to Leonov and Shiryaev (1959).

To study and estimate the cumulants $\Gamma_k(S_n)$, $\Gamma_k(S(T))$ and $\Gamma_k(\xi_n^p)$, it is considerably more advantageous to express $\Gamma(X_{t_1}, \dots, X_{t_k})$ in terms of the *centered moments*

$$\hat{\mathbf{E}}(X_I) := \mathbf{E}X_{t_1}\widehat{X_{t_2} \dots \widehat{X_{t_{k-1}}\widehat{X_{t_k}}}}, \tag{1.18}$$

in which the symbol "^" over a r.v. stands for centering:

$$\hat{\xi} := \xi - \mathbf{E}\xi. \tag{1.19}$$

Instead of $\hat{\mathbf{E}}(X_I)$ we shall often also make use of the notation $\hat{\mathbf{E}}X_{t_1} \dots X_{t_k}$. We have

$$\begin{aligned} \hat{\mathbf{E}}X_t &= \mathbf{E}X_t, \quad \hat{\mathbf{E}}X_sX_t = \mathbf{E}X_sX_t - \mathbf{E}X_s\mathbf{E}X_t, \\ \hat{\mathbf{E}}X_{t_1}X_{t_2}X_{t_3} &= \mathbf{E}X_{t_1}X_{t_2}X_{t_3} - \mathbf{E}X_{t_1}\mathbf{E}X_{t_2}X_{t_3} \\ &\quad - \mathbf{E}X_{t_1}X_{t_2}\mathbf{E}X_{t_3} + \mathbf{E}X_{t_1}\mathbf{E}X_{t_2}\mathbf{E}X_{t_3}. \end{aligned} \tag{1.20}$$

An explicit expression for a centered moment in terms of moments is

$$\hat{\mathbf{E}}X_{t_1} \dots X_{t_k} = \sum_{\nu=1}^{k} (-1)^{\nu-1} \sum_{\bigcup_{p=1}^{\nu} I_p = I}^{*} \prod_{p=1}^{\nu} \mathbf{E}(X_{I_p}), \tag{1.21}$$

where $\mathbf{E}(X_{I_p}) = \mathbf{E}X_{t_1^{(p)}} \dots X_{t_p^{(p)}}$ and the summation $\sum^*$ is over ν-block partitions $\{I_1, \dots, I_\nu\}$ of I such that $\max I_p \le \min I_{p+1}$, $1 \le p \le \nu - 1$.

When $X_{t_1}, \dots, X_{t_k}$ are independent r.v.'s, $\hat{\mathbf{E}}X_{t_1} \dots X_{t_k}$ is nonvanishing only for $t_1 = t_2 = \dots = t_k$. The same thing is also true for $\Gamma(X_{t_1}, \dots, X_{t_k})$.

We now state a formula that expresses $\Gamma(X_{t_1}, \dots, X_{t_k})$ in terms of the centered moments (Statulevičius (1965), (1969a,b), (1970a,b), (1983)).

Lemma 1.1 (Saulis and Statulevičius (1989)). *The following representation holds:*

$$\Gamma(X_{t_1}, \dots, X_{t_k}) = \sum_{\nu=1}^{k} (-1)^{\nu-1} \sum_{\bigcup_{p=1}^{\nu} I_p = I} N_\nu(I_1, \dots, I_\nu) \prod_{p=1}^{\nu} \hat{\mathbf{E}}(X_{I_p}), \tag{1.22}$$

where $\sum_{\bigcup_{p=1}^{\nu} I_p = I}$ *means summation over all* ν*-block partitions* $\{I_1, \dots, I_\nu\}$ *of* I*. The integers* $N_\nu(I_1, \dots, I_\nu)$ *satisfy*

$$0 \leq N_\nu(I_1, \ldots, I_\nu) \leq (\nu - 1)!, \tag{1.23}$$

and depend only on the sets $\{I_1, \ldots, I_\nu\}$ *and if* $N_\nu\{I_1, \ldots, I_\nu\} > 0$, *then*

$$\sum_{p=1}^{\nu} \max_{t_i, t_j \in I_p} (t_j - t_i) \geq \max_{1 \leq i,j \leq k} (t_j - t_i). \tag{1.24}$$

Thus, for example,

$$\begin{aligned}
&\mathbf{\Gamma}(X_t) = \hat{\mathbf{E}} X_t = \mathbf{E} X_t,\\
&\mathbf{\Gamma}(X_s, X_t) = \hat{\mathbf{E}} X_s X_t,\\
&\mathbf{\Gamma}(X_{t_1}, X_{t_2}, X_{t_3}) = \hat{\mathbf{E}} X_{t_1} X_{t_2} X_{t_3} - \hat{\mathbf{E}} X_{t_2} \hat{\mathbf{E}} X_{t_1} X_{t_3},\\
&\mathbf{\Gamma}(X_{t_1}, X_{t_2}, X_{t_3}, X_{t_4}) = \hat{\mathbf{E}} X_{t_1} X_{t_2} X_{t_3} X_{t_4} - \mathbf{E} X_{t_2} \hat{\mathbf{E}} X_{t_1} X_{t_3} X_{t_4}\\
&- \mathbf{E} X_{t_3} \hat{\mathbf{E}} X_{t_1} X_{t_2} X_{t_4} - \hat{\mathbf{E}} X_{t_1} X_{t_3} \hat{\mathbf{E}} X_{t_2} X_{t_4} - \hat{\mathbf{E}} X_{t_1} X_{t_4} \hat{\mathbf{E}} X_{t_2} X_{t_3}\\
&+ \mathbf{E} X_{t_2} \mathbf{E} X_{t_3} \hat{\mathbf{E}} X_{t_1} X_{t_4}, \ldots .
\end{aligned}$$

Let Z_T be a r.v. depending on a parameter T all of whose moments exist: $\mathbf{E}|Z_T^k| < \infty$, $k \geq 1$. Let

$$\Gamma_k(Z_T) \to \Gamma_k(Z), \quad T \to \infty,$$

for any fixed k, where Z is a r.v. meeting *Carleman's test* for the uniqueness of the moment problem:

$$\sum_{k=1}^{\infty} (\mathbf{E} Z^{2k})^{-\frac{1}{2k}} = \infty. \tag{1.25}$$

Then since the moments of a r.v. are expressible in terms of its cumulants, we obtain

$$Z_T \xrightarrow{D} Z, \quad T \to \infty.$$

In other words, Z_T converges to a r.v. Z in distribution:

$$F_{Z_T}(x) \to F_Z(x), \quad T \to \infty,$$

at each point of continuity of the limiting function $F_Z(x)$.

In the case of a $(0,1)$-normal limiting distribution, when $\mathbf{E} Z_T = 0$ and $\mathbf{E} Z_T^2 = 1$, it is sufficient that

$$\Gamma_k(Z_T) \to 0, \quad T \to \infty,$$

for each fixed integer $k \geq 3$.

If one is not only interested in convergence to the normal distribution but a more precise asymptotic analysis of the distribution $\mathbf{P}_{Z_T}$ (rate of convergence, asymptotic expansions, large deviation probabilities; Z_T may be taken to be, for example, any of the normalized r.v.'s defined by one of the relations (1.3)–(1.6)), then one has to find accurate upper bounds for $\Gamma_k(Z_t)$ relative to both

T and k. After that one can make use of the general assertions about the behavior of $\mathbf{P}_{Z_T}$ if estimates for $\Gamma_k(Z_T)$ are known.

The next section is devoted to establishing precisely such general assertions about the behavior of $\mathbf{P}_{Z_T}$ knowing information about the cumulants of the r.v. Z_T.

§1.1. General Lemmas on Approximating the Distribution Function of an Arbitrary Random Variable by the Normal Distribution

Consider a r.v. $\xi = \xi_\Delta$ depending on a parameter Δ with distribution function $F_\xi(x) = \mathbf{P}\{\xi < x\}$, mean $\mathbf{E}\xi = 0$ and variance $\mathbf{V}\xi = 1$.[1] Let $\varphi_\xi(z) = \mathbf{E}\exp(z\xi)$ be the moment-generating function of ξ. Suppose that there exist quantities $\gamma \geq 0$ and $\Delta > 0$ such that

$$|\Gamma_k(\xi)| \leq (k!)^{1+\gamma}/\Delta^{k-2}, \quad k = 3, 4, \ldots . \qquad (S_\gamma)$$

The condition (S_γ) for $\gamma = 0$ assures the analyticity of the moment-generating function $\varphi_\xi(z)$ in the domain $|z| < \Delta$. More precisely, if

$$|\Gamma_k(\xi)| \leq k!H/\Delta^{k-2}$$

for some positive H, then it is easy to see that

$$|\ln\varphi_\xi(z)| \leq \frac{Hz^2}{(1-\rho)}, \quad |z| \leq \rho\Delta, \quad 0 < \rho < 1.$$

Conversely, if

$$|\ln\varphi_\xi(z)|_{|z|=\Delta^*} \leq H_1\Delta^{*2},$$

that is,

$$\left|\frac{\ln\varphi_\xi(z)}{\ln\varphi_\eta(z)}\right|_{|z|=\Delta^*} \leq \frac{1}{2}H_1,$$

where η is a $(0,1)$-normal r.v., then using Cauchy's formula, we obtain

$$|\Gamma_k(\xi)| \leq k!H_1/\Delta^{*k-2}, \quad k = 3, 4, \ldots .$$

Write

$$\Phi(x) = \frac{1}{\sqrt{2\pi}}\int_{-\infty}^{x}\exp\left(-\frac{t^2}{2}\right)dt, \qquad (1.26)$$

$$\Delta_\gamma = c_\gamma\Delta^{1/(1+2\gamma)}, \quad c_\gamma = \frac{1}{6}\left(\frac{\sqrt{2}}{6}\right)^{1/(1+2\gamma)}.$$

Let θ (with or without a subscript) denote a quantity (not always the same one) that does not exceed one in absolute value and let $[m]$ be the integral part of m.

[1] For simplicity, instead of $\operatorname{Var}\xi$ we shall write $\mathbf{V}\xi$ throughout.

Lemma 1.2 (Saulis (1981)). *If* (S_γ) *holds for a r.v.* ξ *with* $\mathbf{E}\xi = 0$ *and* $\mathbf{E}\xi^2 = 1$, *then* $\forall T$, $T \geq \Delta_\gamma$,

$$\sup_x |F_\xi(x) - \Phi(x)| \leq \frac{3}{\sqrt{2\pi}}\left\{\frac{3 \cdot 6^\gamma}{\Delta} + 100\Delta_\gamma \exp\left(-\frac{3}{2}\Delta_\gamma\right) + \frac{1.5}{T} + \frac{1}{\sqrt{2\pi}}\int_{\Delta_\gamma}^T \left|f_\xi(t) - \exp\left(-\frac{t^2}{2}\right)\right|\frac{dt}{t}\right\}. \tag{1.27}$$

Lemma 1.3 (Saulis (1981)). *If* (S_γ) *holds for a r.v.* ξ *with* $\mathbf{E}\xi = 0$ *and* $\mathbf{E}\xi^2 = 1$, *then*

$$\sup_x |F_\xi(x) - \Phi(x)| \leq \frac{18}{\Delta_\gamma}. \tag{1.28}$$

Lemma 1.4 (Rudzkis, Saulis and Statulevičius (1978)). *Let* ξ, *with* $\mathbf{E}\xi = 0$ *and* $\mathbf{E}\xi^2 = 1$, *satisfy the condition*

$$|\Gamma_k(\xi)| \leq (k-2)!/\Delta^{k-2}, \quad k = 3, 4, \ldots, s+2, \tag{S^*}$$

where s *is even and satisfies* $s \leq 2\Delta^2$. *Then the large deviation relations*

$$\begin{aligned}\frac{1 - F_\xi(x)}{1 - \Phi(x)} &= \exp\left(\tilde{L}(x)\right)\left(1 + \theta_1 \tilde{f}(x)\frac{x+1}{\sqrt{s}}\right), \\ \frac{F_\xi(-x)}{\Phi(-x)} &= \exp\left(\tilde{L}(x)\right)\left(1 + \theta_2 \tilde{f}(x)\frac{x+1}{\sqrt{s}}\right)\end{aligned} \tag{1.29}$$

hold in the interval $0 \leq x < \sqrt{s}/(3\sqrt{e})$. *Here*

$$\tilde{f}(x) = \frac{117 + 96s\exp\left(-\frac{1}{2}(1 - 3\sqrt{e}x/\sqrt{s})s^{1/4}\right)}{(1 - 3\sqrt{e}x/\sqrt{s})}, \tag{1.30}$$

and $\tilde{L}(x) = \sum_{k=0}^{\infty} \tilde{l}_k x^{k+3}$ *is a power series converging for* $|x| < \sqrt{2}\Delta/(3\sqrt{e})$. *In this range,* $|\tilde{L}(x)| \leq 5|x|^3/(4\Delta)$. *The coefficients* $\tilde{l}_k$, $k = 0, 1, \ldots$, *are expressible in terms of the first* $r_k = \min(k+3, s)$ *cumulants of* ξ. *For* $k \leq s - 3$, *the expressions for* $\tilde{l}_k$ *are the same as the coefficients in the classical Cramér-Petrov series.*

Lemma 1.5 (Rudzkis, Saulis and Statulevičius (1978)). *If* ξ *is an arbitrary r.v. with* $\mathbf{E}\xi = 0$ *and* $\mathbf{E}\xi^2 = 1$ *satisfying condition* (S_γ), *then the large deviation relations*

$$\begin{aligned}\frac{1 - F_\xi(x)}{1 - \Phi(x)} &= \exp(L_\gamma(x))\left(1 + \theta_1 f(x)\frac{x+1}{\Delta_\gamma}\right), \\ \frac{F_\xi(-x)}{\Phi(-x)} &= \exp(L_\gamma(-x))\left(1 + \theta_2 f(x)\frac{x+1}{\Delta_\gamma}\right)\end{aligned} \tag{1.31}$$

hold in the interval $0 \leq x < \Delta_\gamma$. *Here*

$$f(x) = \frac{60(1+10\Delta_\gamma^2 \exp\big(-(1-x/\Delta_\gamma)\sqrt{\Delta_\gamma}\big)}{(1-x/\Delta_\gamma)} \tag{1.32}$$

and

$$L_\gamma(x) = \sum_{3\le k<p} \lambda_k x^k + \theta(x/\Delta_\gamma)^3, \quad p = \begin{cases} 2+\frac{1}{\gamma}, & \gamma>0, \\ \infty, & \gamma=0. \end{cases} \tag{1.33}$$

The expressions for the coefficients λ_k *in terms of the cumulants of* ξ *coincide with the coefficients of the Cramér-Petrov series* (Petrov (1972)) *and may be found according to the formula*

$$\lambda_k = -\frac{b_{k-1}}{k}, \tag{1.34}$$

in which the b_k*'s are determined in succession from the equations*

$$\sum_{r=1}^{j} \frac{1}{r!}\Gamma_{r+1}(\xi) \sum_{\substack{j_1+\cdots+j_r=j \\ j_i\ge 1}} \prod_{i=1}^{r} b_{j_i} = \begin{cases} 1, & j=1 \\ 0, & j=2,3,\ldots. \end{cases} \tag{1.35}$$

In particular,

$$\begin{aligned} b_1 &= 1, \\ b_2 &= -\frac{1}{2}\Gamma_3(\xi), \\ b_3 &= -\frac{1}{6}(\Gamma_4(\xi) - 3\Gamma_3^2(\xi)), \\ b_4 &= -\frac{1}{24}(\Gamma_5(\xi) - 10\Gamma_3(\xi)\Gamma_4(\xi) + 15\Gamma_3^3(\xi)), \ldots. \end{aligned}$$

A bound for the coefficient λ_k *is*

$$|\lambda_k| \le \frac{2}{k}\left(\frac{16}{\Delta}\right)^{k-2}((k+1)!)^\gamma, \tag{1.36}$$

and therefore

$$L_\gamma(x) \le \frac{x^2}{2}\cdot\frac{x}{x+8\Delta_\gamma}, \quad L_\gamma(-x) \ge -\frac{x^3}{3\Delta_\gamma}.$$

Lemma 1.6 (Akhmedov (1990)). *Suppose that condition* (S_γ) *holds for* ξ *with* $\mathbf{E}\xi = 0$ *and* $\mathbf{E}\xi^2 = 1$ *for* $k = 2, 3, \ldots, l$. *Then*

$$|F(x) - \Phi(x)| \le \frac{c(k,\gamma)(\ln\Delta)^{k/2}}{(1+|x|^k)\Delta_\gamma} \tag{1.37}$$

for all x *and* $k \le l$, *where*

$$c(k,\gamma) = \left(7c(k) \vee \frac{2\sqrt{2}\cdot 4^{\gamma}}{7}\right)\left(\frac{1}{1+2\gamma}\right)^{k/2}$$

and

$$c(k) \le 2^{k/2}\left(5 + \sqrt{e/\pi}2^{k/2}\Gamma\left(\frac{k+1}{2}\right)\right), \quad k > 1.$$

Here and elsewhere $a \vee b = \max(a,b)$ and $a \wedge b = \min(a,b)$.

Lemma 1.7 (Bentkus and Rudzkis (1980)). *Let ξ be an arbitrary r.v. with $\mathbf{E}\xi = 0$ for which there exist quantities $\gamma \ge 0$, $H > 0$ and $\tilde{\Delta} > 0$ such that*

$$|\Gamma_k(\xi)| \le \left(\frac{k!}{2}\right)^{1+\gamma}\frac{H}{\tilde{\Delta}^{k-2}}, \quad k = 2,3,\dots. \tag{1.38}$$

Then for all $x \ge 0$.

$$\mathbf{P}\{\pm\xi \ge x\} \le \exp\left(-\frac{x^2}{2(H + (x/\tilde{\Delta}^{1/(1+2\gamma)}))^{(1+2\gamma)/(1+\gamma)}}\right). \tag{1.39}$$

Corollary 1.1. *The following holds under the hypotheses of Lemma 1.7:*

$$\mathbf{P}\{\pm\xi \ge x\} \le \begin{cases} \exp\left(-\dfrac{x^2}{4H}\right) & 0 \le x \le (H^{1+\gamma}\tilde{\Delta})^{1/(1+2\gamma)}, \\ \exp\left(-\dfrac{1}{4}(x\tilde{\Delta})^{1/(1+\gamma)}\right), & x \ge (H^{1+\gamma}\tilde{\Delta})^{1/(1+2\gamma)}. \end{cases} \tag{1.40}$$

If $x \le (H^{1+\gamma}\tilde{\Delta})^{1/(1+2\gamma)}$, then $H \ge (x/\tilde{\Delta}^{1/(1+2\gamma)}$ and so the right-hand side of (1.39) does not exceed $\exp(-x^2/4H)$. The second line of (1.40) is derived in similar fashion.

Lemma 1.8 (Rudzkis, Saulis and Statulevičius (1978)). *Let ξ be such that $\mathbf{E}\xi = 0$ and $\sigma^2 = \mathbf{E}\xi^2 > 0$ and let there exist $\gamma \ge 0$ and $K > 0$ such that*

$$|\mathbf{E}\xi^k| \le (k!)^{1+\gamma}K^{k-2}\sigma^2, \quad k = 3,4,\dots. \tag{$\tilde{B}_\gamma$}$$

Then the k-th order cumulant of ξ has the following bound:

$$|\Gamma_k(\xi)| \le (k!)^{1+\gamma}\big(2(K \vee \sigma)\big)^{k-2}\sigma^2, \quad k = 3,4,\dots. \tag{1.41}$$

Condition $(\tilde{B}_\gamma)$ is a generalization of Bernstein's familiar condition

$$\mathbf{E}|\xi|^k \le \frac{k!}{2}K^{k-2}\sigma^2, \quad k = 2,3,\dots.$$

Let us clarify the connection between the classical *Linnik condition* and (S_γ). A r.v. ξ is subject to Linnik's condition if there exists a constant C_γ such that

$$\mathbf{E}\exp(|\xi|^{1/(1+\gamma)}) \le C_\gamma < \infty, \quad \gamma > 0. \tag{L}$$

Proposition 1.1. *If ξ satisfies condition (L), then*

$$|\Gamma_k(\xi)| \le (k!)^{1+\gamma}(C_\gamma^{(1)})^{k-2}, \quad k = 3, 4, \dots, \tag{1.42}$$

where

$$C_\gamma^{(1)} = 4C_\gamma e^\beta \beta^{3\beta}, \quad \beta = 1 + \gamma. \tag{1.43}$$

Proof. If follows from (L) that

$$\mathbf{E}(|\xi|^{m/\beta})/m! \le C_\gamma, \quad m = 1, 2, \dots .$$

For $m = (\widehat{k\beta})$, where

$$\hat{x} := \min\{n \ge x \mid n \text{ an integer}\},$$

we have

$$\mathbf{E}|\xi|^k \le (\mathbf{E}|\xi|^{m/\beta})^{k\beta/m} \le (m!C_\gamma)^{k\beta/m} \le 2C_\gamma \beta^{k\beta} e^\beta (k!)^\beta.$$

Consequently,

$$\mathbf{E}|\xi|^k \le (2C_\gamma e^\beta \beta^{3\beta})^{k-2}(k!)^\beta$$

for $m \ge 3$. By Lemma 1.8, this yields

$$|\Gamma_k(\xi)| \le (k!)^\beta (C_\gamma^{(1)})^{k-2}, \quad k = 3, 4, \dots ,$$

where $C_\gamma^{(1)}$ is given by (1.43).

§1.2. Highlights of the Proofs of Lemmas 1.4 and 1.5

Lemma 1.4 occupies a dominant place in the proofs of Lemmas 1.2, 1.3 and 1.6. We now proceed to give a short presentation of its proof.

Let

$$\tilde{\varphi}(z) := \exp\left(\sum_{k=2}^{s} \frac{1}{k!}\Gamma_k(\xi)z^k\right) = 1 + \sum_{k=2}^{\infty} \frac{1}{k!}\tilde{m}_k z^k, \tag{1.44}$$

where

$$\tilde{m}_k = \frac{d^k}{dz^k}\tilde{\varphi}(z)\big|_{z=0},$$

and by condition (S^*), $\tilde{m}_k = m_k = \mathbf{E}\xi^k$, $1 \le k \le s$, $s \le 2\Delta^2$. Noting that $\Gamma_1(\xi) = 0$, one can readily show that

$$\tilde{m}_k = k! \sum_{j=1}^{[k/2]} \frac{1}{j!} \sum_{\substack{k_1+\dots+k_j=h \\ k_i=2,3,\dots,s}} \frac{\Gamma_{k_1}(\xi)\dots\gamma_{k_j}(\xi)}{k_1!\dots k_j!}.$$

By the hypothesis of the lemma, this yields

$$|\tilde{m}_k| \le k! \sum_{j=1}^{[k/2]} \frac{1}{j!} \left(\sum_{i=2}^{s} \frac{1}{i(i-1)} \right)^j \Delta^{2j-k}. \tag{1.45}$$

After some simple computations, we obtain

$$|\tilde{m}_k| \le \frac{k!}{a^k(x)}, \quad a(k) = (\sqrt{k/2e} \wedge \Delta/\sqrt{e}). \tag{1.46}$$

We next determine $h = h(z)$ from the equation

$$x = \tilde{m}(h) := \frac{d}{dh} \ln \tilde{\varphi}(h) = \sum_{k=2}^{s} \frac{1}{(k-1)!} \Gamma_k(\xi) h^{k-1}, \tag{1.47}$$

and we let

$$\tilde{\sigma}^2(h) := \frac{d^2}{dh^2} \ln \tilde{\varphi}(h) = \sum_{k=2}^{s} \frac{1}{(k-2)!} \Gamma_k(\xi) h^{k-2}. \tag{1.48}$$

Assume that $0 \le h \le \delta a$ with $0 < \delta < 1$ and $a = \sqrt{s/(4e)}$. Since $s \le 2\Delta^2$, we have $a \le \Delta/\sqrt{2e}$ and $h < \Delta/\sqrt{2e}$. Applying condition (S^*), we obtain

$$x = h\big(1 + \theta(\sqrt{2e}h/(3\Delta))\big) = h\big(1 + \theta(\delta/3)\big) \tag{1.49}$$

and

$$\tilde{\sigma}^2(h) = 1 + \theta(3\sqrt{2e}h/(4\Delta)) = 1 + \theta(3\delta/4). \tag{1.50}$$

Consequently $\tilde{\sigma}^2(h) > 0$ and corresponding to each value of x in the equation $x = \tilde{m}(h)$ there is a single value of h. To have $0 \le h < a$, we assume x is such that $0 \le x < 2a/3$. Consider the following functions:

$$F_h(y) := \int_{-\infty}^{\tilde{\sigma}(h)y + \tilde{m}(h)} \tilde{\varphi}^{-1}(h) g_h(u) dF_\xi(u), \tag{1.51}$$

$$g_h(y) := \sum_{k=0}^{s} \frac{1}{k!}(hy)^k + y^2 \sum_{k=s+1}^{\infty} \frac{1}{k!} \tilde{m}_k h^k := \tilde{e}(hy) + y^2 \tilde{r}(h). \tag{1.52}$$

Then

$$1 - F_\xi(x) = \tilde{\varphi}(h) \int_0^\infty g_h^{-1}(\tilde{\sigma}(h)y + z) dF_h(y). \tag{1.53}$$

Let us estimate $\sup_y |F_h(y) - \Phi(y)|$. By the definition of $F_h(y)$ (see (1.51)), its *Fourier transform* is

$$\begin{aligned} f_h(t) &= \int_{-\infty}^{\infty} e^{ity} dF_h(y) \\ &= \frac{\exp(-itx/\tilde{\sigma}(h))}{\tilde{\varphi}(h)} \int_{-\infty}^{\infty} \exp\left(\frac{itu}{\tilde{\sigma}(h)}\right) g_h(u) dF_\xi(u). \end{aligned} \tag{1.54}$$

Put

$$\tilde{f}_h(t) = \frac{\exp\left(-\frac{itx}{\tilde{\sigma}(h)}\right)}{\tilde{\varphi}(h)} \int_{-\infty}^{\infty} g_z(u)dF_\xi(u) = \frac{\exp\left(-\frac{itx}{\tilde{\sigma}(h)}\right)\tilde{\varphi}(z)}{\tilde{\varphi}(h)}, \tag{1.55}$$

where $z = h + it/\tilde{\sigma}(h)$. Then

$$\left|f_h(t) - \exp\left(-\frac{1}{2}t^2\right)\right| \le |f_h(t) - \tilde{f}_h(t)| + \left|\tilde{f}_h(t) - \exp\left(-\frac{1}{2}t^2\right)\right|. \tag{1.56}$$

Expanding $\ln \tilde{f}_h(t)$ by Taylor's theorem in a neighborhood of h with $|h + |t|/\tilde{\sigma}(h)| \le \delta_2 a$, $0 < \delta < \delta_2 < 1$, that is, with $|t| \le T = (\delta_2 - \delta)\tilde{\sigma}(h)a$, we obtain

$$\left|\tilde{f}_h(t) - \exp\left(-\frac{1}{2}t^2\right)\right| \le \frac{|t|}{T}\left(\exp\left(-\frac{t^2}{4}\right) - \exp\left(-\frac{t^2}{2}\right)\right). \tag{1.57}$$

We now estimate $|f_h(t) - \tilde{f}_h(t)|$. Recalling the definitions of $f_h(t)$ and $\tilde{f}_h(t)$ in (1.54) and (1.55), we have

$$|f_h(t) - \tilde{f}_h(t)| \le I_1 + I_2, \tag{1.58}$$

where

$$I_1 = \tilde{\varphi}^{-1}(h)\left|\int_{-\infty}^{\infty}\left(\exp\left(\frac{itu}{\tilde{\sigma}(h)}\right)\tilde{e}(hu) - \tilde{e}(zu)\right)dF_\xi(u)\right|$$

and

$$I_2 = \tilde{\varphi}^{-1}(h)\left|\int_{-\infty}^{\infty}\left(\exp\left(\frac{itu}{\tilde{\sigma}(h)}\right)r(h) - r(z)\right)dF_\xi(u)\right|.$$

We next observe that $r(z) = \sum_{s+1}^{\infty}(1/k!)\tilde{m}_k z^k$, $\tilde{\varphi}(h) \ge 1$ and $a(k)$ in the estimate (1.46) for $\tilde{m}_k$ is at least $\sqrt{2}a$ when $k > s$. Thus

$$I_2 \le 2\sum_{k=s+1}^{\infty}\frac{1}{k!}|\tilde{m}_k||z|^k \le 2\sum_{k=s+1}^{\infty}(\delta_2/\sqrt{2})^k. \tag{1.59}$$

We now estimate I_1. We have

$$I_1 \le I_1^{(1)} + I_1^{(2)}, \tag{1.60}$$

where

$$I_1^{(1)} = \left|\int_{-b}^{b}\left(\exp\left(\frac{itu}{\tilde{\sigma}(h)}\right)\tilde{e}(hu) - \tilde{e}(zu)\right)dF_\xi(u)\right|,$$

$$I_2^{(2)} = \left|\int_{|u|>b}\left(\exp\left(\frac{itu}{\tilde{\sigma}(h)}\right)\tilde{e}(hu) - \tilde{e}(zu)\right)dF_\xi(u)\right|,$$

$$b := 4a = 4(s/(4e))^{1/2}, \quad z = h + it/\tilde{\sigma}(h).$$

Since $\tilde{e}(zu) = \exp(zu) - \sum_{k=s+1}^{\infty}(1/k!)(zu)^k$, we have

$$I_1^{(1)} \le 2 \sum_{k=s+1}^{\infty} \frac{|z|^k}{k!} \int_{-b}^{b} |u|^k dF_\xi(u) \le 2 \sum_{k=s+1}^{\infty} \frac{(b|z|)^k}{k!} \le 2 \sum_{k=s+1}^{\infty} \delta_2^k \tag{1.61}$$

because $b|z| \le 4a^2\delta_2 = \delta_2 s/e$ and $k! > (s/e)^k$ when $k > s$. To estimate $I_1^{(2)}$, consider the quantity

$$m_k(b) := \int_{|u|>b} |u|^k dF_\xi(u).$$

Applying the bound (1.46) for $\tilde{m}_k$, we have

$$\frac{m_k(b)}{k!} \le \begin{cases} a^{-k}(k) \le a^{-k}, & s/2 < k \le s, \\ \dfrac{(2k)!}{k! a^{2k}(2k)b^k} \le \dfrac{\sqrt{2}}{a^k}, & b < k \le s/2, \\ \dfrac{m_{[b]}(b)}{[b]!} \le \dfrac{\sqrt{2}}{a^{[b]}}, & 0 \le k \le b. \end{cases} \tag{1.62}$$

Assuming that $\delta_2 \ge 1/\sqrt{a}$, we find that

$$I_1^{(2)} \le 2\sqrt{2} \left(\sum_{k=0}^{[b]} \delta_2^{2[b]-k} + \sum_{k=[b]+1}^{\infty} \delta_2^k \right). \tag{1.63}$$

Relations (1.58)–(1.61) and (1.63) allow one to conclude that

$$|f_h(t) - \tilde{f}_h(t)| \le l(\delta_2), \quad l(\delta) = \frac{4\sqrt{2}\delta^{[4a]}}{1-\delta}, \tag{1.64}$$

for $|t| \le T$. We now study $f_h(t)$ in a neighborhood of zero. We have

$$f_h(t) = 1 + it \int_{-\infty}^{\infty} y dF_h(y) + \theta \frac{t^2}{2} \int_{-\infty}^{\infty} y^2 |dF_h(y)|. \tag{1.65}$$

Furthermore,

$$\begin{aligned} \int_{-\infty}^{\infty} y dF_h(y) &= (\tilde{\varphi}(h)\tilde{\sigma}(h))^{-1} \int_{-\infty}^{\infty} (t-x) g_h(t) dF_\xi(t) \\ &= -\frac{x}{\tilde{\sigma}(h)} + (\tilde{\varphi}(h)\tilde{\sigma}(h))^{-1} \left\{ \sum_{k=1}^{s-1} \frac{1}{(k-1)!} m_k h^{k-1} + m_3 r(h) \right\} \\ &= -\frac{x}{\tilde{\sigma}(h)} + (\tilde{\varphi}(h)\tilde{\sigma}(h))^{-1} \Big\{ -\frac{dr(h)}{dh} + \frac{d\tilde{\varphi}(h)}{dh} \\ &\quad + \frac{m_{s+1}h^s}{s!} + m_3 r(h) \Big\} = \theta(0.02\delta^s/a) \end{aligned} \tag{1.66}$$

when $s \ge 30$. It is not hard to see that

$$|dF_h(y)| \le \left(\tilde{e}(ht) + t^2|r(h)|\right)\tilde{\varphi}^{-1}(h)dF_\xi(t), \quad t = \tilde{\sigma}(h)y + x.$$

Next

$$\int_{-\infty}^{\infty} y^2|dF_h(y)| \le (\tilde{\varphi}(h)\tilde{\sigma}^2(h))^{-1}\int_{-\infty}^{\infty}(t-x)^2(\tilde{e}(ht) + t^2|r(h)|)dF_\xi(t) = J_1 + J_2, \tag{1.67}$$

where

$$J_1 = (\tilde{\varphi}(h)\tilde{\sigma}^2(h))^{-1}\int_{-\infty}^{\infty}(t-x)^2\tilde{e}(ht)dt,$$
$$J_2 = (\tilde{\varphi}(h)\tilde{\sigma}^2(h))^{-1}(m_4 - 2xm_3 + x^2)|r(h)|.$$

Applying the bound (1.46) for $\tilde{m}_k$, we have

$$J_1 = (\tilde{\varphi}(h)\tilde{\sigma}^2(h))^{-1}\Bigg\{\sum_{k=2}^{\infty}\frac{\tilde{m}_k h^{k-2}}{(k-2)!} - 2x\sum_{k=1}^{\infty}\frac{\tilde{m}_k h^{k-1}}{(k-1)!} + x^2\sum_{k=0}^{\infty}\frac{\tilde{m}_k h^k}{k!}$$
$$- \sum_{k=s+1}^{\infty}\frac{\tilde{m}_k h^{k-2}}{(k-2)!} - \sum_{k=s+1}^{s+2}\frac{\tilde{m}_k h^{k-2}}{(k-2)!} + 2x\sum_{k=s+1}^{\infty}\frac{\tilde{m}_k h^{k-1}}{(k-1)!} \tag{1.68}$$
$$- 2x\frac{m_{s+1}h^s}{s!} - x^2\sum_{k=s+1}^{\infty}\frac{\tilde{m}_k h^k}{k!}\Bigg\} = 1 + \theta(2\delta^{s-1}/5)$$

and

$$J_2 \le 4|r(h)|\left(4 + \frac{2x}{\Delta} + x^2\right) \le \frac{\delta^{s+1}}{32}. \tag{1.69}$$

Relations (1.65), (1.66) and (1.69) yield

$$f_h(t) = 1 + \theta\left\{0.02\delta^s|t| + (1 + \delta^{s-1}/2)\frac{t^2}{2}\right\}.$$

From this it follows that

$$\left|f_h(t) - \exp\left(-\frac{1}{2}t^2\right)\right| \le 0.02\delta^s|t| + (1 + \delta^{s-1}/4)t^2. \tag{1.70}$$

Resorting to (1.56), (1.57), (1.67) and (1.70), we find that

$$\left|f_h(t) - \exp\left(-\frac{1}{2}t^2\right)\right| \le \min\Bigg\{0.02\delta^s|t| + (1 + \delta^{s-1}/4)t^2,$$
$$\frac{|t|}{T}\left(\exp\left(-\frac{1}{4}t^2\right) - \exp\left(-\frac{1}{2}t^2\right)\right) + l(\delta_2)\Bigg\}. \tag{1.71}$$

Using Lemma 1.1 (Zolotarev (1982)), we obtain

$$\sup_y|F_h(y) - \Phi(y)| \le \frac{d}{\sqrt{2\pi}}, \tag{1.72}$$

where

$$d = \frac{8.5}{(\delta_2 - \delta)a} + \frac{9.1\delta_2^{[4a]}}{(1-\delta_2)}(\ln 2.1\sqrt{s} - \sqrt{s/e}\ln\delta_2). \tag{1.73}$$

Noting (1.53), we have

$$1 - F_\xi(x) = \tilde{\varphi}(h)\int_0^\infty g_h^{-1}(\tilde{\sigma}(h)y + x)dF_h(y) = \tilde{\varphi}(h)(I_1(h) + I_2(h)), \tag{1.74}$$

with

$$I_1(h) = \int_0^\infty g_h^{-1}(\tilde{\sigma}(h)y + x)d(F_h(y) - \Phi(y))$$

and

$$I_2(h) = \int_0^\infty g_h^{-1}(\tilde{\sigma}(h)y + x)d\Phi(y). \tag{1.75}$$

We now estimate $I_1(h)$. From the definition of $g_h(y)$ (see (1.52)), we obtain

$$g_h(t) = \tilde{e}(ht) + t^2 r(h) = \tilde{e}(ht)\big(1 + \theta(2|r(h)|/h^2)\big), \tag{1.76}$$

with

$$\frac{|r(h)|}{h^2} = \left|\sum_{k=s+1}^{\infty} \frac{\tilde{m}_k h^{k-2}}{k!}\right| = \theta \sum_{k=s+1}^{\infty} \frac{h^{k-2}}{(\sqrt{2}a)^k} = \theta 2(\delta/\sqrt{2})^{s-1}.$$

Therefore as a function of y, $g_h^{-1}(\tilde{\sigma}(h)y + x)$ is non-negative and monotone decreasing in the interval $[0, \infty)$. Consequently,

$$\begin{aligned} I_1(h) &\le 2\sup_y |F_h(y) - \Phi(y)| / \left(\tilde{e}(hx)(1 - (\delta/\sqrt{2})^{s-5})\right) \\ &\le \frac{2d\exp(-hx)}{\sqrt{2\pi}(1 - (\delta/\sqrt{2})^{s-5})(1 - (\delta/6)^{s+1})} \end{aligned}$$

since

$$\tilde{e}(hx) = e^{hx}\big(1 + \theta(hx)^{s+1}/(s+1)!\big) = e^{hx}(1 + \theta(\delta/6)^{s+1}).$$

Taking into account that

$$(1 - \Phi(x))\exp\left(\frac{1}{2}x^2\right) \ge \frac{3.1}{4\sqrt{2\pi}(x+1)}$$

for $x \ge 0$, we find from this that

$$I_1(h) \le \frac{8}{3} d\exp\left(-hx + \frac{x^2}{2}\right)(1 - \Phi(x))(x+1). \tag{1.77}$$

The integral $I_2(h)$ remains to be estimated. According to (1.75),

$$I_2(h) = \int_0^{b_1} g_h^{-1}(\tilde{\sigma}(h)y + x)d\Phi(y) + \int_{b_1}^\infty g_h^{-1}(\tilde{\sigma}(h)y + x)d\Phi(y),$$

where $b_1 = (b-x)/\tilde{\sigma}(h) \geq 5a/3$. Then for $0 \leq y \leq b_1$ and $t = \tilde{\sigma}(h)y + x$,

$$g_h(t) = \frac{\exp(-ht)}{(1+\theta(\delta/\sqrt{2})^{s-1})(1+\theta\delta^{s+1})}$$

and

$$I_2(h) = \frac{\exp\left(-hx + \frac{1}{2}(h\tilde{\sigma}(h))^2\right)\left(1-\Phi(h\tilde{\sigma}(h))\right) + \theta\exp\left(-\frac{1}{2}b_1^2\right)/\sqrt{2\pi}}{(1+\theta\delta^{s+1})(1+\theta(\delta/\sqrt{2})^{s-5})}. \tag{1.78}$$

Furthermore,

$$\begin{aligned} h\tilde{\sigma}^2(h) - x &= \sum_{k=3}^{s}\left(\frac{1}{(k-2)!} - \frac{1}{(k-1)!}\right)\Gamma_k(\xi)h^{k-1} \\ &= \theta\sum_{k=3}^{s}\frac{k-2}{k-1}h(h/\Delta)^{k-2} = \theta(3/5)\delta h. \end{aligned}$$

Since $x = h(1+\theta(\delta/3)$, we find that

$$h\tilde{\sigma}(h) = x(1+\theta(0.09\delta)). \tag{1.79}$$

Put $\psi(y) = \exp(y^2/2)(1-\Phi(y))$ and $q(y) = y\psi(y)$. It is not hard to see that $\psi(y)$ is decreasing and $q(y)$ is increasing in the interval $[0,\infty)$. Thus

$$\psi(y+z) = \psi(y)y(y+\theta z)^{-1}$$

for any positive y and $z \in (-y, y)$. Consequently (1.78), (1.79) and the inequality $4b_1^2 \leq s$ lead to

$$\begin{aligned} I_2(h) &= \exp\left(-hx + \frac{1}{2}x^2\right)(1-\Phi(x)) \\ &\times\left(1+\theta\frac{2\delta + (4/3)(x+1)\exp\left(-\frac{1}{8}s\right)}{1-\delta}\right). \end{aligned} \tag{1.80}$$

Let us now estimate d. To this end, choose $\delta_2 = 1 - ((1-\delta)/2s^{1/4})$. Then $\delta_2 - \delta = 1 - (2s^{1/4})^{-1}$ and $1 - \delta_2 = (1-\delta)(2s^{1/4})^{-1}$ when $s \geq 30$ and $\delta_2 \geq 1/\sqrt{a}$. Consequently,

$$\delta_2^{[4a]} \leq \exp\left(-\frac{1}{2}(1-\delta)s^{1/4}\right)\delta_2^{-1}.$$

Going back to the definition of d (see (1.73)), we have

$$d \leq 40 + 35s\exp\left(-\frac{1}{2}(1-\delta)s^{1/4}\right)/(\sqrt{s}(1-\delta)). \tag{1.81}$$

Since $\delta \leq 3x/2a < 5x/\sqrt{s}$, it follows from (1.77), (1.81) and (1.80) that

$$\frac{1-F_\xi(x)}{1-\Phi(x)} = \exp(\tilde{L}(x))\left(1+\theta_1\tilde{f}(x)\frac{x+1}{\sqrt{s}}\right)$$

for all $0 \le x < (2/3)\sqrt{s/4e}$, where

$$\tilde{L}(x) = \frac{1}{2}x^2 + \ln\tilde{\varphi}(h) - hx \tag{1.82}$$

and $\tilde{f}(x)$ is given by (1.30).

The function $h = h(x)$ may be expanded in a power series $h = x + \sum_{k=2}^{\infty} b_k x^k$ which converges for $|x| < \sqrt{2}\Delta/3\sqrt{e}$. Since $|h(z)|_{|z|=2\Delta/(3\sqrt{2e})} \le \Delta/\sqrt{2e}$, Cauchy's inequality yields

$$b_k = \theta\left(\frac{3}{2}\right)^k\left(\frac{\sqrt{2e}}{\Delta}\right)^{k-1}, \quad k \ge 2. \tag{1.83}$$

The coefficients b_k can be expressed in terms of the first $r_k = \min(k+3, s)$ cumulants of ξ. Furthermore, it is not hard to see that

$$\frac{d\tilde{L}(x)}{dx} = x - h.$$

This gives $\tilde{L}(x) = \sum_{k=0}^{\infty} \tilde{l}_k x^{k+3}$, where

$$\tilde{l}_k = -\frac{b_{k+2}}{k+3} = \theta\frac{1.5^{k+2}(\sqrt{2e})^{k+1}}{(k+3)\Delta^{k+1}}, \quad k = 0, 1, 2, \dots . \tag{1.84}$$

Resorting to relation (1.47), we have $h = h(z) = z/(1+\theta(\delta/3))$, $\delta = \sqrt{2e}|h|/\Delta = 3\sqrt{e}/\sqrt{2}\Delta < 1$ and

$$\begin{aligned}\tilde{L}(z) &= \frac{1}{2}(z-h)^2 + \sum_{k=3}^{s}\frac{1}{k!}\Gamma_k(\xi)z^k = \frac{1}{2}z^2\left(1-(1+\theta(\delta/3))^{-1}\right)\\ &\quad + \frac{9|z|^2}{4}\sum_{k=1}^{\infty}\left(\frac{\delta}{\sqrt{2e}}\right)^k/((k+1)(k+2)),\\ \tilde{L}(z) &= \theta 5|z|^3/(4\Delta).\end{aligned}$$

In addition,

$$\tilde{L}(x) = \inf_{0\le h<\Delta/\sqrt{2e}}\left(\frac{1}{2}(x-h)^2 + \sum_{k=3}^{s}\frac{1}{k!}\Gamma_k(\xi)z^h\right)$$

for $0 \le x < \sqrt{2}\Delta/(3\sqrt{e})$.

Let $h = \Delta x/(x+\Delta)$. Making use of condition (S^*), we obtain the bound

$$|\tilde{L}(x)| \le \frac{1}{2}x^2 \cdot \frac{x}{x+2\Delta}.$$

This allows one to complete the proof of Lemma 1.4 if one takes into account that the lemma is trivial for $s \leq 30$.

A highlight of the proof of Lemma 1.5 is that the condition (S_γ): $|\Gamma_k(\xi)| \leq (k!)^{1+\gamma}/\Delta^{k-2}$, $k = 2, 3, \ldots$, implies the inequality

$$|\Gamma_k(\xi)| \leq (k-2)!/\Delta_s^{k-2}, \quad k = 3, 4, \ldots, s+2, \tag{S^*}$$

with s even and not exceeding $2\Delta_s^2$. The following estimate holds when $s \geq 4$:

$$(k!)^{1+\gamma}/\Delta^{k-2} \leq (k-2)! \left(\frac{6(s+2)^\gamma}{\Delta} \right)^{k-2}, \quad k = 3, \ldots, s+2. \tag{1.85}$$

We therefore take $\Delta_s = \Delta/(6(s+2)^\gamma)$. Then condition (S_γ) implies (S^*) for all even s such that

$$4 \leq s \leq \Delta^2/(18(s+2)^{2\gamma}). \tag{1.86}$$

Let

$$s = 2 \left[\frac{1}{2} \left(\frac{\Delta^2}{18} \right)^{1/(1+2\gamma)} \right] - 2. \tag{1.87}$$

Such a choice of s is even and satisfies (1.86). We assume that $\Delta > 10^{1+2\gamma}$ since otherwise Lemma 1.4 becomes trivial. A straightforward consequence of (1.86) is that

$$0.95 \left(\frac{\sqrt{2}\Delta}{6} \right)^{1/(1+2\gamma)} < \sqrt{s} < (\sqrt{2}\Delta/6)^{1/(1+2\gamma)}. \tag{1.88}$$

Applying Lemma 1.4 in the interval $0 \leq x < \Delta_\gamma = (0.95/3\sqrt{e}) \cdot (\sqrt{2}\Delta/6)^{1/(1+2\gamma)}$, we have

$$\frac{1 - F_\xi(x)}{1 - \Phi(x)} = \exp(\tilde{L}(x)) \left(1 + \theta f(x) \frac{x+1}{\Delta_\gamma} \right), \tag{1.89}$$

where

$$f(x) = \left(24 + 500\Delta_\gamma^2 \exp\left(- \left(1 - \frac{x}{\Delta_\gamma} \right) \sqrt{\Delta_\gamma} \right) \right) \left(1 - \frac{x}{\Delta_\gamma} \right)^{-1}.$$

The proof of Lemma 1.5 is completed on cutting off the series $\tilde{L}(x)$. One may show that (see Saulis and Statulevičius (1989))

$$L_\gamma(x) = \tilde{L}(x) = \sum_{0 \leq k < p} \tilde{l}_k x^{k+3} + \theta 0.95 \left(\frac{x}{\Delta_\gamma} \right)^3, \tag{1.90}$$

where $p = \min([1/\gamma] + 2,\ s = 3)$ and the coefficients $\tilde{l}_k$ coincide with the corresponding coefficients in the Cramér-Petrov series which, in the formulation of the lemma, have been denoted by λ_k. Using Lemma 1.4 and the fact that $\Delta_s \geq 4\Delta_\gamma$, we find that

$$L_\gamma(x) \leq \frac{x^2}{2} \cdot \frac{x}{x + 8\Delta_\gamma}, \quad L_\gamma(-x) \geq -\frac{x^3}{3\Delta_\gamma}. \tag{1.91}$$

The complete proofs of Lemmas 1.2–1.8 are given in the following papers:
Lemma 1.2, 1.3 – in Saulis (1981);
Lemmas 1.4, 1.5 and 1.8 – in Rudzkis, Saulis and Statulevičius (1978);
Lemma 1.7 – in Bentkus and Rudzkis (1980).

Remark 1.1. Saulis (1980) proved a general lemma for the density of an arbitrary r.v. (if it exists) for which condition (S_γ) holds. Saulis and Statulevičius (1989) and Saulis (1986) and (1990) proved theorems on the basis of it for large deviations for the density of sums of independent r.v.'s in Cramér's zone and in Linnik's power zones.

It should be mentioned that Statulevičius' cumulant method for large deviation probabilities can also be applied to the n-dimensional case. Readers interested in these problems are referred to Saulis (1984), (1987a), (1987b), (1988), (1990) and (1991).

Chapter 2
Large Deviation Theorems for Distributions of Sums of Independent Random Variables

Let $\xi_1, \xi_2, \dots, \xi_n$, $n \geq 1$, be independent r.v.'s with $\mathbf{E}\xi_j = 0$ and $\sigma_j^2 = \mathbf{V}\xi_j^2 > 0$, $j = 1, 2, \dots, n$. Put

$$S_n = \sum_{j=1}^n \xi_j, \quad B_n^2 = \sum_{j=1}^n \sigma_j^2, \quad Z_n = S_n / B_n, \tag{2.1}$$

$$F_{Z_n}(x) = \mathbf{P}\{Z_n < x\}, \quad p_{Z_n}(x) = \frac{d}{dx} F_{Z_n}(x),$$

$$L_{k,n} = \sum_{j=1}^n \mathbf{E}|\xi_j|^k / B_n^k. \tag{2.2}$$

$L_{k,n}$ is called the k-th *Lypaunov quotient.*

We shall say that condition (B_γ) holds for r.v.'s ξ_j, with $\mathbf{E}\xi_j = 0$ and $\sigma_j^2 = \mathbf{V}\xi_j^2 > 0$, $j = 1, 2, \dots, n$, if there exist constants $\gamma \geq 0$ and $K > 0$ such that

$$|\mathbf{E}\xi_j^k| \leq (k!)^{1+\gamma} K^{k-2} \cdot \sigma_j^2, \quad k = 3, 4, \dots . \tag{B_γ}$$

Theorem 2.1 (Rudzkis, Saulis and Statulevičius (1978)). *Suppose that (B_γ) holds for the r.v.'s ξ_j, $j = 1, 2, \dots, n$. Then a bound for the k-th cumulant of the r.v. Z_n given by* (2.1) *is*

$$|\Gamma_k(Z_n)| \leq (k!)^{1+\gamma} / \Delta_n^{k-2}, \quad k = 3, 4, \dots , \tag{2.3}$$

where

$$\Delta_n = B_n/K_n, \quad K_n = 2(K \vee \max_{1 \le j \le n} \sigma_j). \tag{2.4}$$

In addition, the relations (1.31), (1.27) *and the bounds* (1.28), (1.39) *hold for* $\xi = Z_n$ *with*

$$\Delta_\gamma = c_\gamma \Delta_n^{1/(1+2\gamma)}, \quad H = 2^{1+\gamma} \quad \text{and} \quad \tilde{\Delta} = \Delta_n,$$

where c_γ *and* Δ_n *are given by* (1.26) *and* (2.4), *respectively.*

Corollary 2.1. *Let the r.v.'s* ξ_j, $j = 1, 2, \ldots, n$, *be subject to condition* (B_γ). *Then*

$$\lim_{n\to\infty} \frac{1 - F_{Z_n}(x)}{1 - \Phi(x)} = 1, \quad \lim_{n\to\infty} \frac{F_{Z_n}(-x)}{\Phi(-x)} = 1 \tag{2.5}$$

as $\Delta_n \to \infty$ *if* $x \ge 0$,

$$x = o(\Delta_n^\nu), \quad \text{and} \quad \nu = \nu(\gamma) = (1 + 2(1 \vee \gamma))^{-1}.$$

If (B_γ) *holds and all the moments of* ξ_j, $j = 1, 2, \ldots$, *up to order* $[1/\gamma] + 2$ *inclusively coincide with the corresponding moments of the normal distributions, then relations* (2.5) *are true for*

$$x \ge 0, \quad x = o(\Delta_n^{1/(1+2\gamma)}).$$

(Of course, the last part of the assertion is meaningful only for $0 < \gamma < 1$.)

Proofs of Theorem 2.1 and Corollary 2.1. Applying condition (B_γ) and Lemma 1.8, we find that

$$|\Gamma_k(\xi_j)| \le (k!)^{1+\gamma} 2(K \vee \sigma_j)^{k-2} \cdot \sigma_j^2, \quad \forall k, \ k \ge 3. \tag{2.6}$$

From the independence of the r.v.'s ξ_j, $j = 1, 2, \ldots, n$, it follows that

$$|\Gamma_k(S_n)| \le (k!)^{1+\gamma} K_n^{k-2} \cdot B_n^2, \quad \forall k, \ k \ge 3, \tag{2.7}$$

where K_n is given by (2.4). Since $\Gamma_k(Z_n) = \Gamma_k(S_n)/B_n^k$, we obtain the bound (2.3).

To complete the proof of Theorem 2.1, it suffices to make use of (2.3) and Lemmas 1.2, 1.5 and 1.3–1.7.

The first part of Corollary 2.1 is deduced at once by using $L_\gamma(x)$ as given by (1.33). Taking into account that the moments of ξ_j up to order $m + 3$ inclusively are the same as those of the normal law as well as (1.36), we find that

$$\begin{aligned} L_\gamma(x) &= \sum_{k=m+1}^{\infty} \lambda_k x^{k+3} + \theta \left(\frac{x}{\Delta_\gamma}\right)^3 \\ &= \theta_1 \frac{6 \cdot 16^{m+2}((m+5)!)^\gamma}{m+4} \cdot \frac{x^{m+4}}{\Delta_n^{m+2}} + \theta_2 c_\gamma^3 \left(\frac{x}{\Delta_n^{1/(1+2\gamma)}}\right)^3, \end{aligned}$$

where Δ_n is given by (2.4). Since $(m+4)/(1+2\gamma) \le m+2$, it follows that $L_\gamma(x) \to 0$ for $x = o(\Delta_n^{1/(1+2\gamma)})$ as $\Delta_n \to \infty$.

We shall say that r.v.'s ξ_j, $j = 1, 2, \dots$, are subject to condition (P) if there are positive constants A, c_1, c_2, ..., such that

$$\left| \frac{\ln \mathbf{E} \exp(z\xi_j)}{z^2} \right| \le c_j^2, \quad |z| < A \quad (j = 1, 2, \dots), \tag{P}$$

and

$$\limsup_{n\to\infty} \frac{1}{B_n^2} \sum_{j=1}^n c_j^2 \le C. \tag{2.8}$$

Theorem 2.2 (Saulis and Statulevičius (1989)). *Suppose that the r.v.'s ξ_j, with $\mathbf{E}\xi_j = 0$ and $\sigma_j^2 = \mathbf{E}\xi_j^2 > 0$, satisfy condition (P). Then*

$$|\Gamma_k(Z_n)| \le Ck!/(AB_n)^{k-2}, \quad \forall k, \ k \ge 3. \tag{2.9}$$

The large deviation relations (1.31), (1.39) *and estimate* (1.28) *hold for Z_n with $\Delta_\gamma = (1/6)(\sqrt{2}/6)AB_n/(1 \vee C)$ and $\bar{\Delta} = AB_n$.*

Proof. By virtue of the relation

$$\Gamma_k(\xi_j) = \frac{d^k}{dz^k} \ln \mathbf{E} \exp(z\xi_j)\big|_{z=0},$$

the condition (P) and Cauchy's inequality for the derivatives of analytic functions, we find that

$$|\Gamma_k(\xi_j)| \le c_j^2 k!/(A^{k-2}), \quad \forall k, \ k \ge 3.$$

By the independence of the ξ_j, $j = 1, 2, \dots, n$, this leads to

$$|\Gamma_k(S_n)| \le k! \sum_{j=1}^n c_j^2/A^{k-2}$$

and

$$|\Gamma_k(Z_n)| \le k! C/(AB_n)^{k-2}, \quad \forall k, \ k \ge 3, \tag{2.10}$$

where C is determined by (2.8). Using (2.10) and resorting to Lemmas 1.3, 1.5 and 1.7, we arrive at the assertion of the theorem.

Let ξ_j, $j = 1, 2, \dots, n$, be independent r.v.'s with $\mathbf{E}\xi_j = 0$ and $\sigma_j^2 = \mathbf{E}\xi_j^2$ and let $\{a_{k,n}, \ 1 \le k \le n, \ 1 \le n < \infty\}$ be a triangular array of non-negative numbers. Put

$$\tilde{S}_n = \sum_{j=1}^n a_{j,n}\xi_j, \quad \tilde{B}_n^2 = \sum_{j=1}^n a_{j,n}^2 \sigma_j^2, \tag{2.11}$$

$$\tilde{Z}_n = \tilde{S}_n/\tilde{B}_n, \quad \gamma_n = \max(a_{j,n}, \ 1 \le j \le n). \tag{2.12}$$

Theorem 2.3 (Saulis (1979)). *Suppose that ξ_j, with $\mathbf{E}\xi_j = 0$ and $\sigma_j^2 = \mathbf{E}\xi_j^2$, $j = 1, 2, \ldots, n$, satisfies condition (B_γ). Then*

$$|\Gamma_k(\tilde{Z}_n)| \leq \frac{(k!)^{1+\gamma}}{\tilde{\Delta}_n^{k-2}}, \quad k = 3, 4, \ldots, \tag{2.13}$$

where

$$\tilde{\Delta}_n = \frac{\tilde{B}_n}{K_n \gamma_n}, \quad K_n = 2(K \vee \max_{1 \leq j \leq n} \sigma_j). \tag{2.14}$$

In addition, the large deviation relations (1.31) *and* (1.27) *and estimates* (1.28) *and* (1.39) *hold for $\xi = \tilde{Z}_n$ with*

$$\Delta_\gamma = c_\gamma \tilde{\Delta}_n^{1/(1+2\gamma)}, \quad H = 2^{1+\gamma} \quad and \quad \tilde{\Delta} = \tilde{\Delta}_n,$$

where c_γ and $\tilde{\Delta}_n$ are given by (1.26) *and* (2.14).

Proof. Making use of the independence of the ξ_j, $j = 1, 2, \ldots, n$, we obtain

$$\Gamma_k(\tilde{S}_n) = \sum_{j=1}^{n} a_{j,n}^k \Gamma_k(\xi_j). \tag{2.15}$$

Since condition (B_γ) holds (see p. 205), the estimate (2.6) yields

$$|\Gamma_k(\tilde{S}_n)| \leq (k!)^{1+\gamma} (K_n \gamma_n)^{k-2} \cdot \tilde{B}_n^2, \tag{2.16}$$

in which (2.12), (2.14) and (2.11) determine γ_n, K_n and $\tilde{B}_n^2$. Noting that $\Gamma_k(\tilde{Z}_n) = \Gamma_k(\tilde{S}_n)/\tilde{B}_n^k$, we can derive the bound (2.13). This together with Lemmas 1.2, 1.5 and the estimate (1.28) gives the results in Theorem 2.3, Only the coefficients λ_k in (1.3) remain to be expressed in terms of the cumulants of $\tilde{Z}_n$, where

$$\Gamma(\tilde{Z}_n) = \sum_{j=1}^{n} a_{j,n}^k \Gamma_k(\xi_j)/\tilde{B}_n^k, \quad \forall k, \ k \geq 2.$$

Let us recall Linnik's condition: There exists a constant C_γ such that

$$\mathbf{E} \exp(|\xi_1|^{1/(1+\gamma)}) \leq C_\gamma < \infty. \tag{L}$$

The relationship between (L) and (S_γ) was indicated on p. 196.

Let

$$\hat{Z}_n = \frac{\tilde{S}_n}{b_n}, \quad b_n^2 = \sum_{j=1}^{n} a_{j,n}^2, \tag{2.17}$$

$$C_\gamma^{(1)} = 4C_\gamma e^\beta \beta^{3\beta}, \quad \beta = 1 + \gamma, \tag{2.18}$$

with $\tilde{S}_n$ given by (2.11).

Theorem 2.4 (Saulis (1979)). *Suppose that the r.v.'s* ξ_j, $j = 1, 2, \ldots, n$, *are identically distributed with* $\mathbf{E}\xi_j = 0$ *and* $\sigma^2 = \mathbf{E}\xi_j^2$. *Let the condition* (L) *hold. Then*

$$|\Gamma_k(\hat{Z}_n)| \leq \frac{(k!)^{1+\gamma}}{\hat{\Delta}_n^{k-2}}, \quad k = 3, 4, \ldots, \tag{2.19}$$

where

$$\hat{\Delta}_n = b_n/(C_\gamma^{(1)}\gamma_n).$$

In addition, the large deviation relations (1.31) *and* (1.27) *and the estimates* (1.28) *and* (1.39) *hold for* $\xi = \hat{Z}_n$ *with*

$$\Delta_\gamma = c_\gamma \hat{\Delta}_n^{1/(1+2\gamma)}, \quad H = 2^{1+\gamma}, \quad \tilde{\Delta} = \tilde{\Delta}_n.$$

The quantities b_n, $C_\gamma^{(1)}$ *and* γ_n *occurring in the expression for* $\tilde{\Delta}_n$ *are given by* (2.17), (2.18) *and* (2.12), *respectively.*

Proof. The fact that the r.v.'s ξ_j have a common distribution function $F_{\xi_1}(x)$ yields

$$\Gamma_k(\tilde{S}_n) = \Gamma_k(\xi_1)\sum_{j=1}^{n} a_{j,n}^k. \tag{2.20}$$

Resorting to (1.42), we obtain

$$|\Gamma_k(\tilde{S}_n)| \leq (k!)^{1+\gamma}(C_\gamma^{(1)})^{k-2}\sum_{j=1}^{\eta} a_{j,n}^k.$$

If follows from this that the bound (2.19) holds for the k-th cumulant of $\hat{Z}_n = \tilde{S}_n/b_n$ (see (2.17) for b_n). To finish the proof of the theorem, it suffices to apply Lemmas 1.2 and 1.5 and the bounds (1.28) and (1.39).

Remark 2.1. In summation theory of weighted r.v.'s, the following condition is sometimes used (Book (1973)): There exist two numbers $0 < \alpha \leq 1$ and $0 < q \leq 1$ for which there are αn numbers among the non-negative $a_{j,n}$, $j = 1, 2, \ldots, n$, such that $a_{j,n} \geq q\gamma_n$, where $\gamma_n = \max\{a_{j,n},\ 1 \leq j \leq n\}$. This condition means that finitely many terms cannot determine the behaviour of the sum $\tilde{S}_n$ as $n \to \infty$. When it is satisfied.

$$b_n^2 = \sum_{j=1}^{n} a_{j,n}^2 \geq \alpha n (q\gamma_n)^2. \tag{2.21}$$

Then according to (2.19), it suffices to take $\hat{\Delta}_n = q(\alpha n)^{1/2}/C_\gamma^{(1)}$ in Theorem 2.4, where $C_\gamma^{(1)}$ is given by (2.18).

It is known that the *Lypunov quotients*

$$L_{k,n} = \frac{1}{B_n^k}\sum_{j=1}^{n} \mathbf{E}|\xi_j|^k, \quad k = 1, 2, \ldots,$$

are natural building blocks for the asymptotic expansions of the distribution function $F_{Z_n}(x)$ and its density $p_{Z_n}(x) = dF_{Z_n}(x)/dx$ (if it exists) (Statulevičius (1965)).

It turns out that large deviation probabilities can be studied in terms of Lyapunov's quotients both in the Cramér-Petrov zone and Linnik's power zones. Thus, the large deviation probabilities in such zones depend primarily not on the individual properties of the terms but on the means (this fact was pointed out by Wolf (1970), (1975)).

Theorem 2.5 (Rudzkis, Saulis and Statulevičius (1978)). *Suppose that there exist $\gamma \geq 0$ and $\tau_n > 0$ such that*

$$L_{k,n} \leq (k!)^{1+\gamma}/\tau_n^{k-2}, \quad k = 3, 4, \ldots . \tag{L^*}$$

Then in the interval $0 \leq x < \tau_n^*$,

$$\tau_n^* = \begin{cases} c\tau_n/|\ln \tau_n|, & \gamma = 0, \\ c_\gamma^* \tau_n^{1/(1+2\gamma)}, & \gamma > 0 \end{cases} \tag{2.22}$$

the relations (1.28) *and* (1.31) *hold for $\xi = Z_n$ with $\Delta_\gamma = \tau_n^*$. Here*

$$\begin{aligned} c_\gamma^* &\geq 96\sqrt{e} \times 3^{1/(1+2\gamma)}(e(1+\gamma))^3 \\ &\quad \times e^{\gamma(1+2\gamma)}\left((3(1+2\gamma)^2/(e\gamma))^{3(1+2\gamma)}\right)^{-1}, \\ c &> \sqrt{6}(36 \cdot 27)^{-1}. \end{aligned} \tag{2.23}$$

Proposition 2.1. *If ξ_j, $j = 1, 2, \ldots, n$, with $\mathbf{E}\xi_j = 0$ and $\sigma_j^2 = \mathbf{E}\xi_j^2$, satisfy condition ($L^*$) with exponent $\gamma = 0$, then*

$$|\Gamma_k(Z_n)| \leq \frac{k!}{(\tau_n/(27|\ln \tau_n|)^{k-2}}, \quad k = 3, 4, \ldots . \tag{2.24}$$

But if (L^*) holds with $\gamma > 0$, then

$$|\Gamma_k(Z_n)| \leq \frac{(k!)^{1+\gamma}}{(\tau_n/C_1^*(\gamma))^{k-2}} \vee \frac{k!}{\tau_n^2(\tau_n^{1/(1+2\gamma)}/C_2^*(\gamma))^{k-2}}, \quad k = 3, 4, \ldots, \tag{2.25}$$

where

$$\begin{aligned} C_1^*(\gamma) &= 48 \exp\left(3(1+\gamma)\right), \\ C_2^*(\gamma) &= 16^{1/(1+2\gamma)}(e(1+\gamma))^3 6^{\gamma/(1+2\gamma)} \left(\frac{3(1+2\gamma)^2}{e\gamma}\right)^{3(1+2\gamma)}. \end{aligned} \tag{2.26}$$

A complete proof of this result is given in Rudzkis, Saulis and Statulevičius (1979) and Saulis and Statulevičius (1989).

We now give a condition due to Sakhanenko (1984) which is equivalent to (L^*) when $\gamma = 0$: Let there exist a positive λ such that

$$L_n := \lambda \sum_{j=1}^n \mathbf{E}|\xi_j|^3 e^{\lambda|\xi_j|} \leq \mathbf{V}S_n. \tag{B^*}$$

Proposition 2.2. *If condition (B^*) is satisfied, then condition (L^*) is true for $\gamma = 0$ and $\tau_n = \lambda B_n$. Conversely, if (L^*) is satisfied for $\gamma = 0$, then (B^*) holds for $\lambda = \tau_n/(12B_n)$.*

Proof. We first prove that (B^*) implies (L^*) with $\gamma = 0$ and $\tau_n = \lambda B_n$. We have

$$\begin{aligned}\lambda^{k-2}\mathbf{E}|\xi_j|^k/(k-3)! &= \lambda\mathbf{E}|\xi_j|^3(\lambda|\xi_j|)^{k-3}/(k-3)! \\ &\leq \lambda\mathbf{E}|\xi_j|^3 e^{\lambda|\xi_j|}, \quad \forall k,\ k \geq 3.\end{aligned}$$

On the basis of (B^*), this yields

$$\lambda^{k-2}L_{k,n}/(k-3)! \leq \lambda B_n^{-k}\sum_{j=1}^n \mathbf{E}|\xi_j|^3 e^{\lambda|\xi_j|} \leq B_n^{2-k}.$$

Consequently,

$$L_{k,n} \leq \frac{(k-3)!}{(\lambda B_n)^{k-2}}, \quad \forall k,\ k \geq 3.$$

Conversely, if (L^*) holds with $\gamma = 0$, then if $\lambda = \tau_n/(12B_n)$,

$$\begin{aligned}L_n(\lambda) &= \lambda\sum_{k=3}^\infty \frac{\lambda^{k-3}}{(k-3)!}L_{k,n}B_n^k \leq B_n^2\sum_{k=3}^\infty k(k-1)(k-2)(\lambda B_n/\tau_n)^{k-2} \\ &\leq B_n^2\sum_{k=3}^\infty\left(\frac{6\lambda B_n}{\tau_n}\right)^{k-2} \leq B_n^2.\end{aligned}$$

We next give an example which shows that when (L^*) holds with $\gamma = 0$, the bound (2.24) for the k-th cumulant of Z_n is unimprovable up to a constant (see Sakhanenko (1984)). More precisely, the example is a sequence satisfying (B^*) and (L^*) with $\tau = \lambda$ when $\gamma = 0$ and we shall show that if

$$|\Gamma_k(Z_n)| \leq k!/R^{k-2}, \quad k = 3, 4, \ldots, \tag{2.27}$$

then necessarily $R < \pi\tau/\ln\tau$.

Let ξ_0 assume two values $\pm\sigma$ with the same probabilities and with $\sigma \leq 1/e$. Let $\xi_1, \ldots, \xi_n$ be r.v.'s with a common normal distribution function. Then $\mathbf{E}\xi_0 = 0$ and $\mathbf{V}\xi_0 = \sigma^2$. Put

$$S = \xi_0 + S', \quad S' = \xi_1 + \cdots + \xi_n.$$

If we take $\mathbf{V}S = 1$, then $\mathbf{V}S' = 1 - \sigma^2$ and $\mathbf{V}\xi_1 = (1-\sigma^2)/n$. In that event,

$$L_{k,n} = \mathbf{E}|\xi_0|^k + \sum_{j=1}^n \mathbf{E}|\xi_j|^k \to \sigma^k \quad (n \to \infty)$$

and

$$L_n(\lambda) = \lambda \mathbf{E}|\xi_0|^3 e^{\lambda|\xi_0|} + \lambda \sum_{j=1}^{0} \mathbf{E}|\xi_j|^3 e^{\lambda|\xi_j|} \to \lambda \mathbf{E}|\xi_0|^3 e^{\lambda|\xi_0|} \quad (n \to \infty).$$

In turn,

$$\begin{aligned} \lambda \mathbf{E}|\xi_0|^3 e^{\lambda|\xi_0|} &= \lambda\sigma^3 e^{\lambda\sigma} = (1/\sigma)\ln(1/\sigma)\sigma^3 \exp\left(\sigma(1/\sigma)\ln(1/\sigma)\right) \\ &= \sigma \ln(1/\sigma) < \sigma(1/\sigma) = 1 \end{aligned}$$

since $\ln(1/\sigma) \leq 1/\sigma - 1 < 1/\sigma$ and $\sigma \leq 1/e$. Consequently, (B^*) holds with $\lambda = (1/\sigma)\ln(1/\sigma)$. Thus, by Proposition 2.2, (L^*) holds with $\tau = \lambda = (1/\sigma)\ln(1/\sigma)$ and $\gamma = 0$.

Next, if z is complex,

$$\begin{aligned} \varphi(z) &:= \ln \mathbf{E}e^{zS} = \ln \mathbf{E}e^{z\xi_0} + \ln \mathbf{E}e^{zS'} = \ln \mathbf{E}e^{z\xi_0} \\ &+ \sum_{k=2}^{\infty} \frac{1}{k!}\Gamma_k(S')z^k = \ln \mathbf{E}e^{z\xi_0} + \Gamma_2(S')\frac{z^2}{2} = \ln \mathbf{E}e^{z\xi_0} + (1-\sigma^2)\frac{z^2}{2}. \end{aligned}$$

In addition,

$$\mathbf{E}e^{z\xi_0} = (e^{\sigma z} + e^{-\sigma z})/2 = \cos(i\sigma z) = 0$$

for $z = \pi i/(2\sigma)$ because $\cos(i\sigma z) = \cos(i^2\pi\sigma/(2\sigma)) = \cos(-\pi/2) = 0$. Let $R_0 = \pi/(2\sigma)$. Then when $z = iR_0$, we have $\mathbf{E}e^{z\xi_0} = 0$ (that is, $\ln \mathbf{E}^{z\xi_0} = -\infty$).

Since (2.27) implies that $\varphi(z)$ is analytic in the disc $|z| < R$, this shows that if (2.27) is true, then $R < R_0$.

Since $\tau = \lambda = (1/\sigma)\ln(1/\sigma)$, we have $R_0 = \pi/(2\sigma) = \pi(1/\sigma)\ln(1/\sigma) \div (2\ln(1/\sigma)) = \pi\tau/(2\ln(1/\sigma))$. Since $\ln(1/\sigma) \leq (1/\sigma) - 1 < 1/\sigma$, we have $\tau = (1/\sigma)\ln(1/\sigma) < (1/\sigma)(1/\sigma) = 1/\sigma^2$ or $\ln\tau \leq \ln(1/\sigma^2)$. In other words, $\ln\tau < 2\ln(1/\sigma)$ and so $R_0 < \pi\tau/\ln\tau$.

Chapter 3
Large Deviation Theorems for Sums of Dependent Random Variables

Let X_t, $t = 1, 2, \ldots$, be a random process which is defined on probability space $(\Omega, \mathcal{F}, \mathbf{P})$ and let $\{\mathcal{F}_s^t,\ 1 \leq s \leq t < \infty\}$ be a family of σ-algebras such that

1. $\mathcal{F}_s^t \subset \mathcal{F}, \quad \forall s \leq t;$
2. $\mathcal{F}_{s_1}^{t_1} \subset \mathcal{F}_{s_2}^{t_2}, \quad \forall [s_1, t_1] \subset [s_2, t_2];$
3. $\mathcal{F}_s^t \supset \sigma\{X_u,\ s \leq u \leq t\}.$

As usual, we consider the *α-mixing*, *φ-mixing* and *ψ-mixing* functions given by the following relations:

$$\alpha(s,t) = \sup_{A\in\mathcal{F}_1^s, B\in\mathcal{F}_t^\infty} |\mathbf{P}\{AB\} - \mathbf{P}\{A\}\mathbf{P}\{B\}|$$

(Rosenblatt (1956));

$$\varphi(s,t) = \sup_{\substack{A\in\mathcal{F}_1^s, B\in\mathcal{F}_t^\infty \\ P(A)>0}} \left| \frac{\mathbf{P}\{AB\} - \mathbf{P}\{A\}\mathbf{P}\{B\}}{\mathbf{P}\{A\}} \right|$$

(Ibragimov (1959)); and

$$\psi(s,t) = \sup_{\substack{A\in\mathcal{F}_1^s, B\in\mathcal{F}_t^\infty \\ P(A)>0 \ P(B)>0}} \left| \frac{\mathbf{P}\{AB\} - \mathbf{P}\{A\}\mathbf{P}\{B\}}{\mathbf{P}\{A\}\mathbf{P}\{B\}} \right|$$

(Blum, Hanson and Koopmans (1963)).

Classes of generalized mixing functions. Let

$$\begin{aligned} \mathcal{K} := \{f \in L_\infty(R^2) \mid 0 \le f(s_1,t_1) \le f(s,t) \le f(s_2,t_2), \\ [s_2,t_2] \subset [s,t] \subset [s_1,t_1]\}. \end{aligned} \tag{3.1}$$

Each $f \in \mathcal{K}$ will be termed a *generalized mixing function.* Let

$$d(f,g) := \sup_{(s,t)\in R^2} |f(s,t) - g(s,t)|$$

be distance in $\mathcal{K}$. Define $\mathcal{K}^{(\le 1)}$ and $\mathcal{K}^{(\ge 1)}$ by

$$\begin{aligned} \mathcal{K}^{(\le 1)} &= \{f \in \mathcal{K} \mid d(0,f) \le 1\}, \\ \mathcal{K}^{(\ge 1)} &= \{f \in \mathcal{K} \mid d(0,f) \ge 1\}. \end{aligned} \tag{3.2}$$

Then

$$\begin{aligned} &\alpha,\ \varphi,\ \psi,\ \overline{m} \in \mathcal{K}, \\ &\alpha,\ \varphi,\ \overline{m} \in \mathcal{K}^{(\le 1)}, \\ &4\alpha,\ \varphi,\ \psi,\ \overline{m} \in \mathcal{K}^{(\ge 1)}, \end{aligned}$$

where $\overline{m}(s,t) = \mathbf{1}_{\{t-s\le m\}}(s,t)$ is the *m-dependence function.*

Let

$$\begin{aligned} \mathfrak{N} &= \{1,2,\ldots,n\}, \\ I &= \{t_1,\ldots,t_k \mid t_j \in \mathfrak{N}\}, \quad t_1 \le \ldots \le t_k. \end{aligned}$$

$\{I_1,\ldots,I_\nu\}$ is a partition of I and $I_p = \{t_1^{(p)},\ldots,t_{k_p}^{(p)}\}$, with $t_1^{(p)} \le \ldots \le t_{k_p}^{(p)}$, $1 \le p \le \nu$ and $k_1 + \ldots + k_\nu = k$.

§3.1. Bounds for the Centered k-th Order Moments of Mixing Random Processes

Playing an essential role in limit theorems for sums $S_n = \sum_{t=1}^n X_t$ of dependent random variables under various mixing conditions are upper bounds for $\hat{\mathbf{E}}X_sX_t = \mathbf{E}X_sX_t - \mathbf{E}X_s\mathbf{E}X_t$ in terms of α, φ and ψ. Let us recall the main ones:

$$|\hat{\mathbf{E}}X_sX_t| \le 4C^2\alpha(s,t) \tag{A}$$

if $|X_s| \le C$ and $|X_t| \le C$ with probability 1 (Volkonskii and Rozanov (1959));

$$|\hat{\mathbf{E}}X_sX_t| \le 6\alpha^{1-(1/u)-(1/v)}(s,t)\mathbf{E}^{1/u}|X_s|^u\mathbf{E}^{1/v}|X_t|^v \tag{B}$$

for any $u \ge 1$, $v \ge 1$ with $1/u + 1/v \le 1$ provided $\mathbf{E}|X_s|^u$ and $\mathbf{E}|X_t|^v$ are finite (Davydov (1968));

$$|\hat{\mathbf{E}}X_sX_t| \le 2\varphi^{1/p}(s,t)\mathbf{E}^{1/p}|X_s|^p\mathbf{E}^{1/q}|X_t|^q \tag{C}$$

for any $p \ge 1$, $q \ge 1$ with $1/p + 1/q = 1$ provided $\mathbf{E}|X_s|^p$ and $\mathbf{E}|X_t|^q$ are finite (Ibragimov (1959));

$$|\hat{\mathbf{E}}X_sX_t| \le \psi(s,t)\mathbf{E}|X_s|\mathbf{E}|X_t| \tag{D}$$

if $\mathbf{E}|X_s|$ and $\mathbf{E}|X_t|$ exist (Philipp (1964)).

These estimates will now be extended to $\hat{\mathbf{E}}X_{t_1}\dots X_{t_k}$.

Theorem 3.1. *If $|X_{t_j}| \le C$ with probability 1, $j = 1,\dots,k$ and $k = 2,3,\dots$, then for all $i = 1,\dots,k-1$*

1. $|\hat{\mathbf{E}}X_{t_1}\dots X_{t_k}| \le 2^kC^k\alpha(t_i,\ t_{i+1})$,
2. $|\hat{\mathbf{E}}X_{t_1}\dots X_{t_k}| \le 2^{k-1}C^k\varphi(t_i,\ t_{i+1})$,
3. $|\hat{\mathbf{E}}X_{t_1}\dots X_{t_k}| \le 2^{k-2}C^k\psi(t_i,\ t_{i+1})$.

Theorem 3.2. *If $\mathbf{E}|X_{t_j}|^{p_j}$, $j = 1,\dots,k$, exist for some collection $p_j \ge 1$, $j = 1,\dots,k$, such that $\sum_{j=1}^k(1/p_j) \le 1$, then for all $i = 1,\dots,k-1$,*

1. $|\hat{\mathbf{E}}X_{t_1}\dots X_{t_k}| \le 3\cdot 2^{k-1}\alpha^{1-\sum_{j=1}^k(1/p_j)}(t_i,\ t_{i+1})\prod_{j=1}^k \mathbf{E}^{1/p_j}|X_{t_j}|^{p_j}$,
2. $|\hat{\mathbf{E}}X_{t_1}\dots X_{t_k}| \le 2^{k-1}\varphi^{\sum_{j=1}^k(1/p_j)}(t_i,\ t_{i+1})\prod_{j=1}^k \mathbf{E}^{1/p_j}|X_{t_j}|^{p_j}$,
3. $|\hat{\mathbf{E}}X_{t_1}\dots X_{t_k}| \le 2^{k-2}\psi(t_i,\ t_{i+1})\prod_{j=1}^k \mathbf{E}^{1/p_j}|X_{t_j}|^{p_j}$.

Let $l_1 < \ldots < l_r$ be the growth points of the sequence $t_1 \le \ldots \le t_k$ and let m_j be the number of elements in $\{t_1, \ldots, t_k\}$ equal to b_j. Henceforth, the numbers m_j will be regarded as having been determined but no clarification of their structure will be given.

Theorem 3.3. *If for some* $k \in \{2, 3, \ldots\}$

$$|\hat{\mathbf{E}}(X_{l_j}^{m_j} \mid \mathcal{F}_1^{l_j - 1})| < \infty \textit{ with probability } 1, \; j = 1, \ldots, r,$$

then

$$|\hat{\mathbf{E}}X_{t_1} \ldots X_{t_k}| \le 2^k \alpha(t_i, \; t_{i+1}) \prod_{j=1}^{r} \text{ess sup}\, |\mathbf{E}(X_{l_j}^{m_j} \mid \mathcal{F}_1^{l_j-1})|$$

for all $i = 1, \ldots, k-1$.

Corollary (to Theorem 3.3). *If*

$$|\mathbf{E}(X_{l_j}^{k} \mid \mathcal{F}_1^{l_j-1})| \le (k!)^{1+\gamma_2} H_2^k \textit{ with probability } 1$$

for some $\gamma_2 \ge 0$ *and* $H_2 > 0$, $j = 1, \ldots, r$, *and* $k = 2, 3, \ldots$, *then*

$$|\hat{\mathbf{E}}X_{t_1} \ldots X_{t_k}| \le \prod_{j=1}^{r} (m_j!)^{1+\gamma_2} 2^k H_2^k \alpha(t_i, \; t_{i+1})$$

for all $i = 1, \ldots, k-1$.

Consider the case where the X_t's are connected in a Markov chain ξ_t (that is, $X_t = g_t(\xi_t)$ with $g_t(x)$ a measurable function for each value of t) having transitions probabilities

$$\mathbf{P}_{st}(x, A) = \mathbf{P}\{\xi_t \in A \mid \xi_s = x\}, \; \mathbf{P}_t(A) = \mathbf{P}\{\xi_t \in A\}.$$

Let $\mathcal{F}_s^t = \sigma\{\xi_u, \; s \le u \le t\}$. Then

$$\varphi(s,t) = \sup_{x, \; A \in \mathcal{F}_s^t} |\mathbf{P}_{st}(x, A) - \mathbf{P}_t(A)| \le 1 - \alpha_{st},$$

where α_{st} is the *ergodicity coefficient* of $\mathbf{P}_{st}(\cdot, \cdot)$ given by

$$\alpha_{st} = 1 - \sup_{x, y, \; A \in \mathcal{F}_s^t} |\mathbf{P}_{st}(x, A) - \mathbf{P}_{st}(y, A)|$$

(Dobrushin (1956a,b)). It is easy to show in this case that

$$\hat{\mathbf{E}}X_{t_1} \ldots X_{t_k} = \underbrace{\int \ldots \int}_{k} g_{t_1}(x_1) P_{t_1}(dx_1) \times \prod_{j=2}^{k} g_{t_j}(x_j)(P_{t_{j-1}t_j}(x_{j-1}, dx_j) - P_{t_j}(dx_j)) \qquad (3.3)$$

if $t_1 < \ldots < t_k$.

Theorem 3.4. *Suppose that the X_t's are connected in a Markov chain ξ_t. If $|X_{t_j}| \le C$ with probability 1, $j = 1, \ldots, r$ and $r = 2, 3, \ldots$, then*

$$1.\quad |\hat{\mathbf{E}} X_{t_1} \ldots X_{t_k}| \le 2^{k-1} C^k \prod_{j=1}^{r-1} \varphi(l_j,\ l_{j+1}),$$

$$2.\quad |\hat{\mathbf{E}} X_{t_1} \ldots X_{t_k}| \le 2^{k-r} \prod_{j=1}^{r-1} \psi(l_j,\ l_{j+1}).$$

Theorem 3.5. *Suppose that the X_t's are connected in a Markov chain ξ_t.*

1. *If $\mathbf{E}|X_{l_j}|^{m_j q_j}$, $j = 1, \ldots, r$, exist for some collection $q_j \ge 1$, $j = 1, \ldots, r$ and $r = 2, 3, \ldots$, such that $\sum_{j=1}^r (1/q_j) = 1$, then*

$$|\hat{\mathbf{E}} X_{t_1} \ldots X_{t_k}| \le 2^{k-1} \prod_{j=1}^{r-1} \varphi^{\sum_{i=1}^j (1/q_i)}(l_j,\ l_{j+1}) \prod_{j=1}^{r} \mathbf{E}^{1/q_j} |X_{t_j}|^{m_j q_j}.$$

2. *If $\mathbf{E}|X_{l_j}|^{m_j}$, $j = 1, \ldots, r$, exist for some $r \in \{2, 3, \ldots\}$, then*

$$|\hat{\mathbf{E}} X_{t_1} \ldots X_{t_k}| \le 2^{k-r} \prod_{j=1}^{r-1} \psi(l_j,\ l_{j+1}) \prod_{j=1}^{r} \mathbf{E}|X_{l_j}|^{m_j}.$$

Theorem 3.6. *Suppose that the X_t's are connected in a Markov chain ξ_t. If*

$$|\mathbf{E}(X_{l_j}^{m_j} \mid \mathcal{F}_1^{l_j - 1})| < \infty \text{ with probability } 1,\ j = 1, \ldots, r,$$

for some $r \in \{2, 3, \ldots\}$, then

$$|\hat{\mathbf{E}} X_{t_1} \ldots X_{t_k}| \le 2^{k-1} \prod_{j=1}^{r-1} \varphi(l_j,\ l_{j+1}) \prod_{j=1}^{r} \operatorname{ess\,sup} |\mathbf{E}(X_{l_j}^{m_j} \mid \mathcal{F}_1^{l_j - 1})|.$$

Corollary (to Theorem 3.6). *Suppose that the X_t's are connected in a Markov chain ξ_t. If*

$$|\mathbf{E}(X_{l_j}^k \mid \mathcal{F}_1^{l_j - 1})| \le (k!)^{1+\gamma_2} H_2^k \text{ with probability } 1$$

for some $\gamma_2 \ge 0$ and $H_2 > 0$, $j = 1, \ldots, r$, $r = 2, 3, \ldots$ and $k = 2, 3, \ldots$, then

$$|\hat{\mathbf{E}} X_{t_1} \ldots X_{t_k}| \le 2^{k-1} H_2^k \prod_{j=1}^{r} (m_j!)^{1+\gamma_2} \prod_{j=1}^{r-1} \varphi(l_j,\ l_{j+1}).$$

It is known that bounds for the cumulants of random variables connected in a Markov chain (Statulevičius (1965), (1969a,b), (1970a)) or satisfying the more general RMT condition (Zhurbenko (1972), (1982), Zuev (1973), Statulevičius (1969a,b), (1970a,b), (1979), Rosenblatt (1956)) can be expressed in terms of a product of mixing functions. But one cannot do this for arbitrary variables. One has to confine oneself to "maximal-interval" bounds. It should be noted that the inequalities

$$\alpha(s,t) \le \varphi(s,t) \le \psi(s,t)$$

justify the immediate passage from bounds in terms of $\alpha(s,t)$ to bounds in terms of $\varphi(s,t)$ and from $\varphi(s,t)$ to $\psi(s,t)$ by making a straightforward replacement of the mixing function (Iosifescu (1978)).

§3.2. Bounds for Mixed Cumulants of Mixing Random Processes

Having the estimates in Theorems 3.1–3.6 and their corollaries, one can use (1.22) and the behavior of $N_\gamma(I_1, \ldots, I_\nu)$ to obtain bounds for the cumulants $\Gamma(X_{t_1}, \ldots, X_{t_n})$.

Theorem 3.7. *If* $|X_{t_j}| \le C$ *with probability* 1, $j = 1, \ldots, k$ *and* $k = 2, 3, \ldots$, *then*

$$\begin{aligned} &1.\ \ |\Gamma(X_{t_1}, \ldots, X_{t_k})| \le (k-1)!2^k C^k \alpha(t_i,\ t_{i+1}),\\ &2.\ \ |\Gamma(X_{t_1}, \ldots, X_{t_k})| \le (k-1)!2^{k-1} C^k \varphi(t_i,\ t_{i+1}),\\ &3.\ \ |\Gamma(X_{t_1}, \ldots, X_{t_k})| \le (k-1)!2^{k-2} C^k \psi(t_i,\ t_{i+1}) \end{aligned}$$

for all $i = 1, 2, \ldots, k-1$. *Here* $t_{i+1} - t_i = \max_{1\le j\le k}(t_{j+1} - t_j)$.

Theorem 3.8. *If* $\mathbf{E}|X_{t_j}|^{p_j}$, $j = 1, \ldots, k$, *exist for some collection* $p_j \ge 1$, *such that* $\sum_{j=1}^n (1/p_j) \le 1$, $k = 2, 3, \ldots$, *then*

$$1.\ \ |\Gamma(X_{t_1}, \ldots, X_{t_k})| \le 3(k-1)!2^{k-1}\alpha^{1-\sum_{j=1}^k (1/p_j)}(t_i,\ t_{i+1}) \prod_{j=1}^k \mathbf{E}^{1/p_j}|X_{t_j}|^{p_j},$$

$$2.\ \ |\Gamma(X_{t_1}, \ldots, X_{t_k})| \le (k-1)!2^{k-1}\varphi^{\sum_{j=1}^k (1/p_j)}(t_i,\ t_{i+1}) \prod_{j=1}^k \mathbf{E}^{1/p_j}|X_{t_j}|^{p_j},$$

$$3.\ \ |\Gamma(X_{t_1}, \ldots, X_{t_k})| \le (k-1)!2^{k-2}\psi(t_i,\ t_{i+1}) \prod_{j=1}^k \mathbf{E}^{1/p_j}|X_{t_j}|^{p_j}$$

for all $i = 1, \ldots, k-1$.

Theorem 3.9. *If*

$$\text{ess sup}\,|\mathbf{E}(X_{l_j}^{m_j} \mid \mathcal{F}_1^{l_j-1}| < \infty \text{ with probability } 1,\ j = 1, \dots, r,$$

for some $k \in \{2, 3, \dots\}$, *then*

$$|\Gamma(X_{t_1}, \dots, X_{t_k})| \le (k-1)!2^k \alpha(t_i,\ t_{i+1}) \prod_{j=1}^{r} \text{ess sup}\,|\mathbf{E}(X_{l_j}^{m_j} \mid \mathcal{F}_1^{l_j-1})|$$

for all $i = 1, \dots, k-1$.

Corollary (to Theorem 3.9). *If*

$$|\mathbf{E}(X_{l_j}^k \mid \mathcal{F}_1^{l_j-1})| \le (k!)^{1+\gamma_2} H_2^k \text{ with probability } 1,$$

$j = 1, \dots, r$, $k = 2, 3, \dots$, *for some* $\gamma_2 \ge 0$ *and* $H_2 > 0$, *then*

$$|\Gamma(X_{t_1}, \dots, X_{t_k})| \le (k!)^{1+\gamma_2} 2^{2k-1} H_2^k \alpha(t_i,\ t_{i+1})$$

for all $i = 1, \dots, k-1$.

Theorem 3.10. *Suppose that the* X_t*'s are connected in a Markov chain* ξ_t. *If with probability* 1, $|X_{t_j}| \le C$, $j = 1, \dots, r$ *and* $k \in \{2, 3, \dots\}$, *then*

$$1.\quad |\Gamma(X_{t_1}, \dots, X_{t_k})| \le (k-1)!2^{k-1} C^k \prod_{j=1}^{r-1} \varphi(l_j,\ l_{j+1}),$$

$$2.\quad |\Gamma(X_{t_1}, \dots, X_{t_k})| \le (k-1)!2^{k-r} C^k \prod_{j=1}^{r-1} \psi(l_j,\ l_{j+1}).$$

Theorem 3.11. *Suppose that the* X_t*'s are connected in a Markov chain.*

1. *If* $\mathbf{E}|X_{l_j}|^{m_j q_j}$, $j = 1, \dots, r$, *exist for some collection* $q_j \ge 1$ *and* $r \in \{2, 3, \dots\}$ *such that* $\sum_{j=1}^{r}(1/q_j) = 1$, *then*

$$|\Gamma(X_{t_1}, \dots, X_{t_k})| \le (k-1)!2^{k-1} \prod_{j=1}^{r-1} \varphi^{\sum_{i=1}^{j}(1/q_i)}(l_j,\ l_{j+1}) \prod_{j=1}^{r} \mathbf{E}^{1/q_j}|X_{l_j}|^{m_j q_j}.$$

2. *If* $\mathbf{E}|X_{l_j}|^{m_j}$, $j = 1, \dots, r$, *exist for some* $r \in \{2, 3, \dots\}$, *then*

$$|\Gamma(X_{t_1}, \dots, X_{t_k})| \le (k-1)!2^{k-r} \prod_{j=1}^{r-1} \psi(l_j,\ l_{j+1}) \prod_{j=1}^{r} \mathbf{E}|X_{l_j}|^{m_j}.$$

Theorem 3.12. *Suppose that the X_t's are connected in a Markov chain ξ_t. If*

$$|\mathbf{E}(X_{l_j}^{m_j} \mid \mathcal{F}_1^{l_j-1})| \le \infty \text{ with probability } 1,\ j = 1, \ldots, r,$$

for some $r \in \{2, 3, \ldots\}$, then

$$|\Gamma(X_{t_1}, \ldots, X_{t_k})| \le (k-1)!2^{k-1} \prod_{j=1}^{r-1} \varphi(l_j,\ l_{j+1}) \prod_{j=1}^{r} \text{ess sup}\, |\mathbf{E}(X_{l_j}^{m_j} \mid \mathcal{F}_1^{l_j-1})|.$$

Corollary (to Theorem 3.12). *Suppose that the X_t's are connected in a Markov chain ξ_t. If*

$$\mathbf{E}(X_{l_j}^{k} \mid \mathcal{F}_1^{l_j-1})| \le (k!)^{1+\gamma_2} H_2^k \text{ with probability } 1,$$

$j = 1, \ldots, r$, $r = 2, 3, \ldots$, and $k = 2, 3, \ldots$, for some $\gamma_2 \ge 0$ and $H_2 > 0$, then

$$|\Gamma(X_{t_1}, \ldots X_{t_k})| \le (k!)^{1+\gamma_2} 4^{k-1} H_2^k \prod_{j=1}^{r-1} \varphi(l_j,\ l_{j+1}).$$

The proofs of Theorems 3.7–3.12 rest on the following inequalities.

Lemma 3.1.

$$|\Gamma(X_{t_1}, \ldots, X_{t_k})| \le (k-1)! \max_{1 \le \nu \le k} \prod_{p=1}^{\nu} |\hat{\mathbf{E}}(X_{I_p})|,$$

$$|\Gamma(X_{t_1}, \ldots, X_{t_k})| \le k!2^{k-1} \max_{1 \le \nu \le k} \prod_{p=1}^{\nu} |\hat{\mathbf{E}}(X_{I_p})|/k_p!.$$

Theorem 3.7 is a straightforward consequence of Theorem 3.1 and Lemma 3.1 since

$$\max_{1 \le \nu \le k} \prod_{p=1}^{\nu} |\hat{\mathbf{E}}(X_{I_p})| \le 2^k C^k \alpha(t_i,\ t_{i+1}),$$

$$\max_{1 \le \nu \le k} \prod_{p=1}^{\nu} |\hat{\mathbf{E}}(X_{I_p})| \le 2^{k-1} C^k \varphi(t_i,\ t_{i+1}),$$

$$\max_{1 \le \nu \le k} \prod_{p=1}^{\nu} |\hat{\mathbf{E}}(X_{I_p})| \le 2^{k-2} C^k \psi(t_i,\ t_{i+1}).$$

To prove part 1 of Theorem 3.8, observe that the bound in part 1 of Theorem 3.2 readily leads to

$$\max_{1 \le \nu \le k} \prod_{p=1}^{\nu} |\hat{\mathbf{E}}(X_{I_p})| \le 3 \cdot 2^k C^k \alpha^{1-\sum_{j=1}^{k} 1/p_j}(t_i,\ t_{i+1}) \prod_{j=1}^{k} \mathbf{E}^{1/p_j} |X_{t_j}|^{p_k}$$

because $\alpha \le 1$. Since the hypotheses of Theorem 3.1 are satisfied, the theorem is proved.

Part 2 of Theorem 3.8 is proved similarly to part 1. We merely mention that because $\varphi \le 1$, one can in this case set

$$\max_{1\le\nu\le k}\prod_{p=1}^{\nu}|\hat{\mathbf{E}}(X_{I_p})| \le 2^{k-1}\varphi^{\sum_{j=1}^{i}1/p_j}(t_i,\ t_{i+1})\prod_{j=1}^{k}\mathbf{E}^{1/p_j}|X_{t_j}|^{p_j}.$$

Theorem 3.9 is a consequence of Theorem 3.3 and Lemma 3.1 by virtue of the estimate

$$\max_{1\le\nu\le k}\prod_{p=1}^{\nu}|\hat{\mathbf{E}}(X_{I_p})| \le 2^{k}\alpha(t_i,\ t_{i+1})\prod_{j=1}^{r}\operatorname{ess\,sup}|\mathbf{E}(X_{l_j}^{m_j}\mid \mathcal{F}_1^{l_{j-1}})|.$$

Theorem 3.10 follows from the estimates

$$1.\quad \max_{1\le\nu\le k}\prod_{p=1}^{\nu}|\hat{\mathbf{E}}(X_{I_p})| \le 2^{k-1}C^k\prod_{j=1}^{r-1}\varphi(l_j,\ l_{j+1}),$$

$$2.\quad \max_{1\le\nu\le k}\prod_{p=1}^{\nu}|\hat{\mathbf{E}}(X_{I_p})| \le 2^{k-r}C^k\prod_{j=1}^{r-1}\psi(l_j,\ l_{j+1}).$$

Part 1 of Theorem 3.11 is a consequence of the estimate

$$\max_{1\le\nu\le k}\prod_{p=1}^{\nu}|\hat{\mathbf{E}}(X_{I_p})| \le 2^{k-1}\prod_{j=1}^{r-1}\varphi^{\sum_{i=1}^{j}1/q_i}(l_j,\ l_{j+1})\prod_{j=1}^{r}\mathbf{E}^{1/q_j}|X_{l_j}|^{m_jq_j}.$$

Part 2 of Theorem 3.11 is a consequence of the estimate

$$\max_{1\le\nu\le k}\prod_{p=1}^{\nu}|\hat{\mathbf{E}}(X_{I_p})| \le 2^{k-1}\prod_{j=1}^{r-1}\psi(l_j,\ l_{j+1})\prod_{j=1}^{r}\mathbf{E}|X_{l_j}|^{m_j}.$$

Theorem 3.12 is a consequence of the estimate

$$\max_{1\le\nu\le k}\prod_{p=1}^{\nu}|\hat{\mathbf{E}}(X_{I_p})| \le 2^{k-1}\prod_{j=1}^{r-1}\varphi(l_j,\ l_{j+1})\prod_{j=1}^{r}\operatorname{ess\,sup}|\mathbf{E}(X_{l_j}^{m_j}\mid \mathcal{F}_1^{m_{j-1}})|.$$

The corollary to Theorem 3.12 is a consequence of the estimate

$$\max_{1\le\nu\le k}\prod_{p=1}^{\nu}|\hat{\mathbf{E}}(X_{I_p})|/k_p! \le (k!)^{\gamma_2}2^{k-1}H_2^k\prod_{j=1}^{r-1}\varphi(l_j,\ l_{j+1}).$$

§3.3. Bounds for the Cumulants of Sums of Dependent Random Variables

Let $S_n = \sum_{t=1}^n X_t$ and let $\Gamma_k(S_n)$ be the k-th cumulant of S_n. Put

$$\Lambda_n(f,u) := 1 \vee \max_{1\le s\le n} \sum_{t=s}^{n} f^{1/u}(s,t),$$

where $f(s,t)$ is one of the mixing functions α, φ or ψ and $u > 0$.

Theorem 3.13. *If $|X_t| \le C$ with probability 1, $t = 1,2,\dots,n$, then for positive β and δ and all $k = 2,3,\dots$*

1. $|\Gamma_k(S_n)| \le 2k!8^{k-1}C^k\Lambda_n^{k-1}(\alpha,\ k-1)n,$
2. $|\Gamma_k(S_n)| \le k!8^{k-1}C^{k-2}\Lambda_n^{k-2}\big(\varphi,\ (1+\beta)(1+1/\delta)(k-2)\big)$
$$\times \sum_{1\le s\le t\le n} \varphi^{\frac{\beta\delta}{(1+\beta)(1+\delta)}}(s,t)\mathbf{E}^{\frac{\delta}{1+\delta}}|X_s|^{1+1/\delta}\mathbf{E}^{\frac{1}{1+\delta}}|X_t|^{1+\delta}.$$

Theorem 3.14. *If $\mathbf{E}|X_t|^{(1+\delta)k}$, $t = 1,\dots,n$, exists for some $k \in \{2,3,\dots\}$ and $\delta > 0$, then for all $\beta > 0$,*

1. $|\Gamma_k(S_n)|$
$$\le 2k!12^{k-1}\Lambda_n^{k-1}\big(\alpha,\ (1+1/\delta)(k-1)\big) \max_{1\le t\le n} \mathbf{E}^{\frac{1}{1+\delta}}|X_t|^{(1+\delta)k}\cdot n,$$
2. $|\Gamma_k(S_n)|$
$$\le k!8^{k-1}\Lambda_n^{k-2}\big(\varphi,\ (1+\beta)(1+1/\delta)(k-2)\big) \max_{1\le t\le n} \mathbf{E}^{\frac{k-2}{(1+\delta)k}}|X_t|^{(1+\delta)k}$$
$$\times \sum_{1\le s\le t\le n} \varphi^{\frac{\beta\delta}{(1+\beta)(1+\delta)}}(s,t)\mathbf{E}^{\frac{1}{(1+\delta)k}}|X_s|^{(1+\delta)k}\mathbf{E}^{\frac{1}{(1+\delta)k}}|X_t|^{(1+\delta)k}.$$

Theorem 3.15. *If*

$$\mathbf{E}|X_t|^k \le (k!)^{1+\gamma_1}H_1^k,\ \ t = 1,\dots,n,\ \ k = 2,3,\dots,$$

for some $\gamma_1 \ge 0$ and $H_1 > 0$, then

$$|\Gamma_k(S_n)| \le 2(k!)^{2+\gamma_1}12^{k-1}H_1^k(\widehat{1+\delta})^{(1+\gamma_1)k}\Lambda_n^{k-1}\big(\alpha,\ (1+1/\delta)(k-1)\big)\cdot n$$

for all positive δ. Here $\hat{u} = \min\{v \ge u \mid v$ an integer$\}$.

Theorem 3.16. *If*

$$|\mathbf{E}(X_t^k \mid \mathcal{F}_1^{t-1})| \le (k!)^{1+\gamma_2}H_2^k \text{ with probability } 1,$$

$t = 1,\dots,n$ and $k = 2,3,\dots$, for some $\gamma_2 \ge 0$ and $H_2 > 0$, then

$$|\Gamma_k(S_n)| \le 2(k!)^{1+\gamma_2}16^{k-1}H_2^k\Lambda_n^{k-1}(\alpha,\ k-1)\cdot n.$$

Theorem 3.17. *Let the X_t's be connected in a Markov chain ξ_t. If $|X_t| \le C$ with probability 1, $t = 1, \dots, n$, then*

$$1.\quad |\Gamma_k(S_n)| \le k! 8^{k-1} C^k \Lambda_n^{k-1}(\varphi,\ 1) n,$$
$$2.\quad |\Gamma_k(S_n)| \le k! 8^{k-1} C^{k-2} \Lambda_n^{k-2}(\varphi,\ 1 + 1/\delta) \times \sum_{1\le s\le t\le n} \varphi^{\frac{\delta}{(1+\delta)}}(s,t) \mathbf{E}^{\frac{\delta}{1+\delta}} |X_s|^{1+1/\delta} \mathbf{E}^{\frac{1}{1+\delta}} |X_t|^{1+\delta}$$

for all $k = 2, 3, \dots$, and positive δ.

Theorem 3.18. *Let the X_t's be connected in a Markov chain ξ_t. If $\mathbf{E}|X_t|^{(1+\delta)k}$, $t = 1, \dots, n$, exist for some $k \in \{2, 3, \dots\}$ and positive δ, then*

$$1.\quad |\Gamma_k(S_n)| \le k! 8^{k-1} \Lambda_n^{k-1}(\varphi,\ 1 + 1/\delta) \max_{1\le t\le n} \mathbf{E}^{\frac{1}{1+\delta}} |X_t|^{(1+\delta)k} \cdot n,$$
$$2.\quad |\Gamma_k(S_n)| \le k! 8^{k-1} \Lambda_n^{k-1}(\varphi,\ 1 + 1/\delta) \max_{1\le t\le n} \mathbf{E}^{\frac{k-2}{(1+\delta)k}} |X_t|^{(1+\delta)k} \times \sum_{1\le s\le t\le n} \varphi^{\frac{\delta}{(1+\delta)}}(s,t) \mathbf{E}^{\frac{1}{(1+\delta)k}} |X_s|^{(1+\delta)k} \mathbf{E}^{\frac{1}{(1+\delta)k}} |X_t|^{(1+\delta)k}.$$

Theorem 3.19. *Let the X_t's be connected in a Markov chain ξ_t. If*

$$\mathbf{E}|X_t|^k \le (k!)^{1+\gamma_1} H_1^k,\quad t = 1, \dots, n,\quad k = 2, 3, \dots,$$

for some $\gamma_1 \ge 0$ and $H_1 > 0$, then

$$1.\quad |\Gamma_k(S_n)| \le (k!)^{2+\gamma_1} 8^{k-1} H_1^k \widehat{(1+\delta)}^{(1+\gamma_1)k} \Lambda_n^{k-1}(\varphi,\ 1 + 1/\delta) n,$$
$$2.\quad |\Gamma_k(S_n)| \le (k!)^{1+\gamma_1} 16^{k-1} H_1^k \Lambda_n^{k-1}(\psi,\ 1) n$$

for all positive δ.

Theorem 3.20. *Let the X_t's be connected in a Markov chain ξ_t. If*

$$|\mathbf{E}(X_t^k \mid \mathcal{F}_1^{t-1})| \le (k!)^{1+\gamma_2} H_2^k \text{ with probability } 1,$$

$t = 1, \dots, n$ and $k = 2, 3, \dots$, for some $\gamma_2 \ge 0$ and $H_2 > 0$, then

$$|\Gamma_k(S_n)| \le (k!)^{1+\gamma_2} 16^{k-1} H_2^k \Lambda_n^{k-1}(\varphi,\ 1) n.$$

Lemma 3.2. *If*

$$|\hat{\mathbf{E}}(X_{I_p})| \le C_0^{k_p - \varepsilon} C_2^{k_p} \min_{1\le i\le k_p} f^{1/u}(t_i^{(p)},\ t_{i+1}^{(p)}), \tag{3.4}$$

with $0 \le \varepsilon \le k_p$, $u \ge 1$, $C_0 \ge 1$, $C_2 > 0$, $\delta \in \mathcal{K}$,[2] $1 \le p \le \nu$ and $1 \le \nu \le k$, then

---

[2] The classes of mixing functions $\mathcal{K}$ and $\mathcal{K}^{(\le 1)}$, $\mathcal{K}^{(\ge 1)}$ are defined by the respective relations (3.1) and (3.2).

$$\left|\sum_{I\in\mathfrak{N}}\Gamma(X_I)\right|\le nk!4^{k-1}C_0^{k-\varepsilon}C_2^k\max\Lambda_n^{k-1}(f,\ su). \tag{3.5}$$

If

$$|\hat{\mathbf{E}}(X_{I_p})|\le C_0^{k_p-\varepsilon}C_2^{k_p}\prod_{j=1}^{r_p-1}f^{1/u}(l_j^{(p)},\ l_{j+1}^{(p)}) \tag{3.6}$$

with $0\le\varepsilon\le k_p$, $u\ge 1$, $C_0\ge 1$, $C_2>0$, $f\in\mathcal{K}^{(\ge 1)}$, $1\le p\le\nu$ *and* $1\le\nu\le k$, *then*

$$\left|\sum_{I\in\mathfrak{N}}\Gamma(X_I)\right|\le nk!4^{k-1}C_0^{k-\varepsilon}C_2^k\Lambda_n^{k-1}(f,u). \tag{3.7}$$

Corollary (to Lemma 3.2). *If*

$$|\hat{\mathbf{E}}(X_{I_p})|\le C_0^{k_p-\varepsilon}C_2^{k_p}\min_{1\le i\le k_p}f^{1/u}(t_i^{(p)},\ t_{i+1}^{(p)}), \tag{3.8}$$

with $0\le\varepsilon\le k_p$, $u\ge 1$, $C_0\ge 1$, $C_2>0$, $f\in\mathcal{K}^{(\le 1)}$, $1\le p\le\nu$ *and* $1\le\nu\le k$, *then*

$$\left|\sum_{I\in\mathfrak{N}}\Gamma(X_I)\right|\le nk!4^{k-1}C_0^{k-\varepsilon}C_2^k\Lambda_n^{k-1}(f,\ (k-1)u). \tag{3.9}$$

Lemma 3.3. *If*

$$|\hat{\mathbf{E}}(X_{I_p})|\le m_1^{(p)}!\dots m_{r_p}^{(p)}!C_0^{k_p-\varepsilon}C_2^{k_p}\min_{1\le i\le k_p}f^{1/u}(t_i^{(p)},\ t_{i+1}^{(p)}) \tag{3.10}$$

with $0\le\varepsilon\le k_p$, $u\ge 1$, $C_0\ge 1$, $C_2>0$, $f\in\mathcal{K}$, $1\le p\le\nu$ *and* $1\le\nu\le k$, *then*

$$\left|\sum_{I\in\mathfrak{N}}\Gamma(X_I)\right|\le nk!8^{k-1}C_0^{k-\varepsilon}C_2^k\max_{1\le s<k}\Lambda_n^{k-1}(f,\ su). \tag{3.11}$$

If

$$|\hat{\mathbf{E}}(X_{I_p})|\le m_1^{(p)}!\dots m_{r_p}^{(p)}!C_0^{k_p-\varepsilon}C_2^{k_p}\prod_{j=1}^{r_p-1}f^{1/u}(l_j^{(p)},\ l_{j+1}^{(p)}) \tag{3.12}$$

with $0\le\varepsilon\le k_p$, $u\ge 1$, $C_0\ge 1$, $C_2>0$, $f\in\mathcal{K}^{(\ge 1)}$, $1\le p\le\nu$ *and* $1\le\nu\le k$, *then*

$$\left|\sum_{I\in\mathfrak{N}}\Gamma(X_I)\right|\le nk!8^{k-1}C_0^{k-\varepsilon}C_2^k\Lambda_n^{k-1}(f,\ u). \tag{3.13}$$

Corollary (to Lemma 3.3). *If*

$$|\hat{\mathbf{E}}X_{I_p}|\le m_1^{(p)}!\dots m_{r_p}^{(p)}!C_0^{k_p-\varepsilon}C_2^{k_p}\min_{1\le i\le k_p}f^{1/u}(t_i^{(p)},\ t_{i+1}^{(p)}) \tag{3.14}$$

with $0\le\varepsilon\le k_p$, $u\ge 1$, $C_0\ge 1$, $C_2>0$, $f\in\mathcal{K}^{(\le 1)}$, $1\le p\le\nu$, $1\le\nu\le k$, *then*

$$\left|\sum_{I\in\mathfrak{N}}\Gamma(X_I)\right|\le nk!8^{k-1}C_0^{k-\varepsilon}C_2^k\Lambda_n^{k-1}(f,\ (k-1)u). \tag{3.15}$$

To prove Theorem 3.13, it suffices to take $f = \alpha$, $C_0 = 2$, $C_2 = C$, $\varepsilon = 0$ and $u = 1$ in (3.8).

To prove Theorem 3.16, we put $f = \alpha$, $C_0 = 2$, $C_2 = H_2$, $\varepsilon = 0$ and $u = 1$ in (3.14).

Theorem 3.17 is proved by making use of inequality (3.6).

To prove Theorem 3.19, it suffices to take $f = \psi$, $C_0 = 2$, $C_2 = H_1$, $\varepsilon = 1$ and $u = 1$ in (3.12).

If we put $f = \varphi$, $C_0 = 2$, $C_2 = H_2$, $\varepsilon = 1$ and $u = 1$ in (3.12), we obtain Theorem 3.20.

§3.4. Theorems and Inequalities on Large Deviations for Sums of Dependent Random Variables

The bounds for $\Gamma_k(S_n)$ in Theorems 3.13–3.17 and the basic lemmas of Chapter 1 enable one to deduce large deviation theorems and inequalities for the distribution $\mathbf{P}\{Z_n < x\}$ of the normalized sum $Z_n = S_n/B_n$, $B_n^2 = \mathbf{E}S_n^2$ (we shall assume throughout that $\mathbf{E}X_t = 0$, $t = 1, \ldots, n$). Large deviation theorems for $\mathbf{P}\{Z_n < x\}$ will be considered just for stationary random sequences X_t, $t = 1, 2, \ldots$. In the general nonstationary case, it is better to estimate $\Gamma_k(Z_n)$ in Theorems 3.13–3.17 with the help of $\Lambda_n^{k-2} L_{k,n}$ instead of $\Lambda_n^{k-2} n \max_{1 \le t \le n} \mathbf{E}|X_t|^k / B_n^k$, where

$$L_{k,n} = \frac{1}{B_n^k} \sum_{t=1}^{n} \mathbf{E}|X_t|^k$$

is the k-th *Lyapunov quotient.* To this end, one needs to express S_n in terms of new amplified summands and to study the behavior of $\Gamma_k(S_n)$ relative to B_n.

We thus consider a stationary sequence X_t with $\mathbf{E}X_t = 0$, $\mathbf{E}X_t^2 = 1$ and $B_n^2 = \mathbf{E}S_n^2 \ge \sigma_0^2 n$, $\sigma_0 > 0$.

Theorem 3.21. *If* $|X_1| \le C$ *with probability* 1 *and*

$$\alpha(s,t) \le K_1 \exp(-b_1(t-s)), \quad K_1 > 0,\ b_1 > 0,$$

then

$$|\Gamma_k(Z_n)| \le (k!)^2 B_1 \left(\frac{8Ce}{b_1 B_n}\right)^{k-2}$$

with $k = 2, 3, \ldots$, $B_1 = 8C^2 K \exp(1+b_1)/(b_1\sigma_0^2)$ *and* $K = (1 \vee K_1)$. *Moreover, the r.v.* $\xi = Z_n$ *obeys the large deviation relations* (1.31), (1.27) *and estimates* (1.39), (1.28) *with*

$$\gamma = 1,\ \Delta_\gamma = c_\gamma (B_n/H_0)^{1/3},\ \tilde{\Delta} = b_1 B_n (8Ce),$$

where

$$H_0 = (8eC/b_1)(1 \vee B_1), \; H = 4B_1$$

and

$$\Delta_\gamma \geq c_\gamma(\sigma_0/H_0)^{1/3}(\sqrt{n})^{1/3}.$$

Theorem 3.22. *If*

$$\mathbf{E}|X_1|^k \leq (k!)^{1+\gamma_1} H_1^k, \quad k = 2, 3, \ldots,$$

for some $\gamma_1 \geq 0$ *and* $H_1 > 0$, *and if*

$$\alpha(s,t) \leq K_1 \exp(-b_1(t-s)), \quad K_1 > 0, \; b_1 > 0,$$

then

$$|\Gamma_k(Z_n)| \leq (k!)^{3+\gamma_1} B_2 \left(\frac{48e2^{\gamma_1} H_1}{b_1 B_n}\right)^{k-2}$$

for $k = 2, 3, \ldots$ *and* $B_2 = 96 H_1^2 4^{\gamma_1} \sqrt{K} \exp(1 + \frac{1}{2} b_1)/(b_1 \sigma_0^2)$. *Moreover the large deviation relations* (1.31), (1.27) *and bounds* (1.28), (1.39) *hold for the r.v.* $\xi = Z_n$ *with*

$$\gamma = 2 + \gamma_1, \; \Delta_\gamma = c_\gamma(B_n/H_0)^{1/(5+2\gamma_1)}, \; \tilde{\Delta} = b_1 B_n/(48e2^{\gamma_1} H_1),$$

where

$$H_0 = (48e2^{\gamma_1} H_1/b_1)(1 \vee B_2), \; H = 4 \cdot 2^{\gamma_1} B_2.$$

Theorem 3.23. *If*

$$|\mathbf{E}(X_t^k \mid \mathcal{F}_1^{t-1})| \leq (k!)^{1+\gamma_2} H_2^k \text{ with probability } 1,$$

$k = 2, 3, \ldots, t = 1, \ldots, n$, *for some* $\gamma_2 \geq 0$ *and* $H_2 > 0$, *and if*

$$\alpha(s,t) \leq K_1 \exp(-b_1(t-s)), \quad K_1 > 0, \; b_1 > 0,$$

then

$$|\Gamma_k(Z_n)| \leq (k!)^{2+\gamma_2} B_3 \cdot \left(\frac{16eH_2}{b_1 B_n}\right)^{k-2}$$

for $k = 2, 3, \ldots$ *and* $B_3 = 16 H_2^2 K \exp(1+b_1)/(b_1 \sigma_0^2)$. *Moreover the r.v.* $\xi = Z_n$ *satisfies the large deviation relations* (1.31), (1.27) *and bounds* (1.28), (1.39) *with*

$$\gamma = 1 + \gamma_2, \; \Delta_\gamma = c_\gamma(B_n/H_0)^{1/(3+2\gamma_2)}, \; \tilde{\Delta} = b_1 B_n/(16eH_2),$$

where

$$H_0 = (16eH_2/b_1)(1 \vee B_3), \quad H = 4 \cdot 2^{\gamma_2} B_3$$

and

$$\Delta_\gamma \geq c_\gamma(\sigma_0/H_0)^{1/(3+2\gamma_2)}(\sqrt{n})^{1/(3+2\gamma_2)}.$$

Theorem 3.24. *Let the r.v.'s X_t be connected in a Markov chain ξ_t. If $|X_t| \le C$ with probability* 1 *and*

$$\varphi(s,t) \le \exp(-b_2(t-s)), \quad b_2 > 0,$$

then

$$|\Gamma_k(Z_n)| \le k! B_4 \left(\frac{8(1+b_2)C}{b_2 B_n}\right)^{k-2}$$

for $k = 2, 3, \ldots$ *and* $B_4 = 8C^2(1+b_2)/(b_2\sigma_0^2)$. *Moreover, the r.v.* $\xi = Z_n$ *satisfies the relations* (1.31), (1.27) *and bounds* (1.28), (1.39) *with*

$$\gamma = 0,\ \Delta_\gamma = c_\gamma B_n/H_0,\ \tilde{\Delta} = b_2 B_n/(8C(1+b_2)),$$

where

$$H_0 = (8C(1+b_2)/b_2)(1 \vee B_4),\ H = 2B_4$$

and

$$\Delta_\gamma \ge c_\gamma(\sigma_0/H_0)\sqrt{n}.$$

Theorem 3.25. *Let the r.v.'s X_t be connected in a Markov chain ξ_t. If*

$$\mathbf{E}|X_1|^k \le (k!)^{1+\gamma_1} H_1^k, \quad k = 2, 3, \ldots,$$

for some $\gamma_1 \ge 0$ and $H_1 > 0$ and if

$$\varphi(s,t) \le \exp(-b_2(t-s)), \quad b_2 > 0,$$

then

$$|\Gamma_k(Z_n)| \le (k!)^{2+\gamma_1} B_5 \left(\frac{16 \cdot 2^{\gamma_1} H_1(2+b_2)}{b_2 B_n}\right)^{k-2}$$

for $k = 2, 3, \ldots$ *and* $B_5 = 32 \cdot 4^{\gamma_1} H_1^2(2+b_2)/(b_2\sigma_0^2)$. *Moreover, the large deviation relations* (1.31), (1.27) *and bounds* (1.28), (1.39) *hold for the r.v.* $\xi = Z_n$ *with*

$$\gamma = 1 + \gamma_1,\ \Delta_\gamma = c_\gamma(B_n/H_0)^{1/(3+2\gamma_1)},\ \tilde{\Delta} = \frac{b_2 B_n}{16 \cdot 2^{\gamma_1} H_1(2+b_2)},$$

where

$$H_0 = (16 \cdot 2^{\gamma_1} H_1(2+b_2)/b_2)(1 \vee B_5), \quad H = 4 \cdot 2^{\gamma_1} B_5,$$

and

$$\Delta_\gamma \ge c_\gamma(\sigma_0/H_0)^{1/(3+2\gamma_1)}(\sqrt{n})^{1/(3+2\gamma_1)}.$$

Theorem 3.26. *Let the r.v.'s X_t be connected in a Markov chain ξ_t. If*

$$\mathbf{E}|X_1|^k \le (k!)^{1+\gamma_1} H_1^k, \quad k = 2, 3, \ldots,$$

for some $\gamma_1 \ge 0$ and $H_1 > 0$ and if

$$\psi(s,t) \le K_3 \exp(-b_3(t-s)), \quad K_3 > 0,\ b_3 > 0,$$

then

$$|\Gamma_k(Z_n)| \le (k!)^{1+\gamma_1} B_6 \left(\frac{16 H_1 K (1+b_3)}{b_3 B_n} \right)^{k-2}$$

with $K = (1 \vee K_3)$ and $B_6 = 16 H_1^2 K (1+b_3)/(b_3 \sigma_0^2)$. Moreover, the relations (1.31), (1.27) and bounds (1.28), (1.39) hold for the r.v. $\xi = Z_n$ with

$$\gamma = \gamma_1,\ \Delta_\gamma = c_\gamma (B_n/H_0)^{1/(1+2\gamma_1)},\ \tilde{\Delta} = \frac{b_3 B_n}{16 H_1 K (1+b_3)},$$

where

$$H_0 = (H_1 B_6/\sigma_0^2)(1 \vee B_6), \quad H = 2^{1+\gamma_1} \cdot B_6,$$

and

$$\Delta_\gamma \ge c_\gamma (\sigma_0/H_0)^{1/(1+2\gamma_1)} (\sqrt{n})^{1/(1+2\gamma_1)}.$$

Theorem 3.27. *Let the r.v.'s X_t be connected in a Markov chain ξ_t. If*

$$|\mathbf{E}(X_t^k \mid \mathcal{F}_1^{t-1})| \le (k!)^{1+\gamma_2} H_2^k \text{ with probability } 1$$

for $k = 2, 3, \ldots$, $t = 1, \ldots, n$, and some $\gamma_2 \ge 0$ and $H_2 > 0$, and if

$$\varphi(s,t) \le \exp(-b_2(t-s)), \quad b_2 > 0,$$

then

$$|\Gamma_k(Z_n)| \le (k!)^{1+\gamma_2} B_7 \left(\frac{16 H_2 (1+b_2)}{b_2 B_n} \right)^{k-2}, \quad k = 2, 3, \ldots,$$

where

$$B_7 = 16 H_2^2 (1+b_2)/(b_2 \sigma_0^2).$$

Moreover, the large deviation relations (1.31), (1.27) and bounds (1.28), (1.39) hold for the r.v. $\xi = Z_n$ with

$$\gamma = \gamma_2,\ \Delta_\gamma = c_\gamma (B_n/H_0)^{1/(1+2\gamma_2)},\ \tilde{\Delta} = \frac{b_2 B_n}{16 H_2 (1+b_2)},$$

where

$$H_0 = (\sigma_0^2 B_7/H_2)(1 \vee B_7), \quad H = 2^{1+\gamma_2} B_7,$$

and

$$\Delta_\gamma \ge c_\gamma (\sigma_0/H_0)^{1/(1+2\gamma_2)} (\sqrt{n})^{1/(1+2\gamma_2)}.$$

Theorems 3.21–3.27 can be proved by straightforward calculation of γ, Δ_γ, $\tilde{\Delta}$ and H and by just applying Theorems 3.13–3.20 and the basic lemmas of Chapter 1. We merely point out that if $f(s,t) \le \exp(-b(t-s))$ and $K \ge 1$, then

$$\begin{aligned} &\Lambda_n(f,1) \le K(1+\exp(-b)+\cdots+\exp(-b(n-s))) \\ &\le K/(1-\exp(-b)) = K(1+1/(\exp(b)-1)) \le K(1+1/b), \\ &\Lambda_n(f,k-1) \le K^{1/(k-1)}(1+(k-1)/b). \end{aligned}$$

Because $k^k \le k!\exp(k)$ and $k \ge 2$, if follows that

$$\begin{aligned} &\Lambda_n^{k-1}(f,k-1) \le K(1+(k-1)/b)^{k-1} \\ &= K((k-1)/b)^{k-1}(1+b/(k-1))^{k-1} \le K(e/b)^{k-1}(k-1)!e^b \\ &\le k!(K/(2b))\exp(1+b)(e/b)^{k-2}. \end{aligned}$$

Chapter 4
Large Deviation Theorems for Polynomial Forms, Pitman Polynomial Estimators, U-Statistics, Multiple Stochastic Integrals and Estimates of the Spectra of Stationary Sequences

§4.1. Bounds for the Cumulants and Large Deviation Theorems for Polynomial Forms, Pitman Polynomial Estimators and U-Statistics

Consider the *polynomial form*

$$\zeta_n^{(p)} = \sum_{1 \le \alpha_1 \le \cdots \le \alpha_p \le n} a_{\alpha_1,\ldots,\alpha_p} X_{\alpha_1} \cdot \ldots \cdot X_{\alpha_p} \tag{4.1}$$

of degree $p \ge 1$. $X_1,\ldots,X_n$ are independent and identically distributed r.v.'s with $\mathbf{E}X_1 = 0$ and $\mathbf{E}X_1^2 = \sigma^2 > 0$ and the coefficients $a_{\alpha_1,\ldots,\alpha_p}$ are invariant under permutations of the indices $\alpha_1,\ldots,\alpha_p$. For brevity, $\max\limits_{\alpha}$ and $\sum\limits_{\alpha}$ will denote these operations over all collection $\alpha = \{\alpha_1,\ldots,\alpha_p\}$, $1 \le \alpha_1 \le \ldots \le \alpha_p \le n$; $\max_{\alpha(1,s)} \sum_{\alpha(s+1,p)} a_{\alpha_1,\ldots,\alpha_p}$ will denote summation over the indices $1 \le \alpha_{s+1} \le \ldots \le \alpha_p \le n$ and then maximization over the remaining indices $1 \le \alpha_1 \le \ldots \le \alpha_s \le n$. Also, by definition, let

$$X_\alpha = X_{\alpha_1} \cdot \ldots \cdot X_{\alpha_p},\ a_\alpha = a_{\alpha_1,\ldots,\alpha_p}.$$

Then

$$\zeta_n^{(p)} = \sum_{\alpha} a_\alpha X_\alpha, \tag{4.2}$$

$$B_n^2 = \mathbf{V}\zeta_n^{(p)} = \sum_{\substack{\alpha,\alpha' \\ \alpha\cap\alpha'\neq\emptyset}} a_\alpha a_{\alpha'} \mathbf{E}(X_\alpha - \mathbf{E}X_\alpha)(X_{\alpha'} - \mathbf{E}X_{\alpha'}). \tag{4.3}$$

Write

$$A_n^2 = \max_{\substack{1\le s_1,s_2\le p \\ s_1+s_2=p}} \left(\max_{\alpha(1,s_1)} \sum_{\alpha(s_1+1,p)} |a_\alpha| \right) \left(\max_{\alpha(1,s_2)} \sum_{\alpha(s_2+1,p)} |a_\alpha| \right), \tag{4.4}$$

$$Z_n^{(p)} := \frac{\zeta_n^{(p)} - \mathbf{E}\zeta_n^{(p)}}{B_n}, \tag{4.5}$$

Theorem 4.1 (Basalykas and Plikusas (1987)). *Let* $\beta_k = \mathbf{E}|X_1|^k < \infty$ *for all* $k \ge 1$. *Then*

$$|\Gamma_k(Z_n^p)| \le k!\beta_{kp}4^k(A_n/B_n)^{k-2} \tag{4.6}$$

for $k \ge 2$.

If positive constants H_0 *and* σ *exist for which Bernstein's condition*

$$|\mathbf{E}X_1^k| \le \frac{k!}{2}H_0^{k-2}\sigma^2, \quad k = 2,3,\ldots,$$

holds, then

$$|\Gamma_k(Z_n^{(p)})| \le (k!)^p c_p^2 H_1^{4k} \left(\frac{c_p A_n}{B_n}\right)^{k-2}, \quad c_p = 2(pH_0)^p, \tag{4.7}$$

and $H_1 = 1 \vee (\sigma/H_0)$. *Moreover, the r.v.* $\xi = Z_n^{(p)}$ *given by* (4.5) *satisfies in the interval*

$$0 \le x < \Delta_\gamma, \quad \gamma = p-1,$$

the large deviation relations (1.31) *with*

$$\Delta_\gamma = \frac{1}{6}\left(\frac{3\sqrt{2}B_n}{c_p^3 H_1^8 A_n}\right)^{1/(2p-1)} \tag{4.8}$$

and the bounds (1.28) *and* (1.39) *with* $\gamma = p-1$, $H = 2^p c_p^2 \sigma^2 H_1^6/H_0^2$ *and* $\tilde{\Delta} = 2B_n/(c_p^2 H_1^4 A_n)$, $H_1 = 1 \vee \sigma/H_0$.

If $|X_1| \le C$ *with probability* 1, *then in the interval*

$$0 \le x < \Delta_\gamma, \quad \gamma = 0,$$

$\xi = \zeta_n^{(p)}$ *satisfies* (1.31) *with*

$$\Delta_\gamma = \frac{\sqrt{2}B_n}{144C^p(1 \vee 16C^{2p-2})A_n} \tag{4.9}$$

and (1.28), (1.39) *with* $\gamma = 0$, $H = 32\sigma^2 C^{2p-2}$ *and* $\tilde{\Delta} = B_n/(4C^p A_n)$.

The theorem is a straightforward consequence of (4.6), (4.7) and Lemmas 1.5, 1.3 and 1.7.

Now let $X_1, X_2, \ldots, X_n$ be sample data of the form $X_i = \theta + \xi_i$, where $\theta \in R$ is a (location) parameter that has to be estimated and $\xi_1, \ldots, \xi_n$ are independent and identically distributed r.v.'s with distribution function $F(x)$ such that

$$\mu_s = \int_{-\infty}^{\infty} x^s dF(x), \quad s = 1, 2, \ldots, 2p.$$

Consider

$$\overline{X} = \frac{1}{n} \sum_{j=1}^{n} X_j, \quad m_j = \frac{1}{n} \sum_{j=1}^{n} (X_i - \overline{X})^j, \tag{4.10}$$

and let V_p be the space of all polynomials in $X_1 - \overline{X}, \ldots, X_n - \overline{X}$ of at most p-th degree. One can estimate θ by the so-called *Pitman polynomial estimator* $t_{n,p}^{(1)}$ or (simpler) *modified Pitman polynomial estimator* $t_{n,p}^{(2)}$ (Kagan (1966), Kagan, Klebanov and Fintushal (1974)). They possess a number of good properties and are given by

$$t_{n,p}^{(1)} = \overline{X} - \tilde{\mathbf{E}}(\overline{X} \mid V_p), \tag{4.11}$$

and

$$t_{n,p}^{(2)} = \overline{X} - A_1 + \sum_{j=2}^{p} A_j m_j, \tag{4.12}$$

where $\tilde{\mathbf{E}}(\cdot \mid V_p)$ is the projection operator on V_p and $A_1, \ldots, A_p$ are expressible in terms of $\mu_1, \mu_2, \ldots, \mu_{2p}$. Observe that

$$\begin{aligned} \mathbf{E}_\theta t_{n,p}^{(1)} &= \theta, \\ \mathbf{E}_\theta t_{n,p}^{(2)} &= \theta + O\left(\frac{1}{n}\right). \end{aligned}$$

Since the estimators $t_{n,p}^{(j)}$, $j = 1, 2$, can be represented as polynomial forms such as (4.1) with coefficients satisfying

$$|a^{(j)}_{\alpha^{(1)}, \ldots, \alpha^{(s)}}| \le C_j(s, \mu_1, \ldots, \mu_{2p})/n^s, \quad s = 1, \ldots, p,$$

an analogue of Theorem 4.1 can be proved for them.

Proposition 4.1 (Basalykas (1988)). *Suppose that* $\mathbf{E}\xi_1 = 0$, $\sigma^2 = \mathbf{E}\xi_1^2$ *and* $|\xi_1| \le L$ *with probability* 1, $i = 1, 2, \ldots, n$, $L > 0$.
Then

$$\left| \Gamma\left((t_{n,p}^{(j)} - \theta)/\sqrt{\mathbf{V} t_{n,p}^{(j)}} \right) \right| \le k! \left(\frac{H_{1,j}}{\sqrt{n}} \right)^{k-2}, \quad j = 1, 2, \tag{4.13}$$

for all $k = 3, 4, \ldots$, *where* $H_{1,j} = H_{1,j}(\sigma, L, p)$ *depends on the arguments within the parentheses. Its explicit form is found in Basalykas (1985).*

If $\mathbf{E}\xi_1 = 0$, $\sigma^2 = \mathbf{E}\xi_1^2 > 0$ *and a positive constant* H_0 *exists such that*

$$|\mathbf{E}\xi_1^k| \le \frac{1}{2} k! H_0^{k-2} \sigma^2, \quad k = 2, 3, \ldots, \tag{B}$$

then

$$\left| \Gamma \left((t_{n,p}^{(j)} - \theta) / \sqrt{\mathbf{V} t_{n,p}^{(j)}} \right) \right| \le (k!)^p \left(\frac{H_{2,j}}{\sqrt{n}} \right)^{k-2} \tag{4.14}$$

for $k = 2, 3, \ldots$, *where* $H_{2,j} = H_{2,j}(H_0, \mu_2, \mu_3, \ldots, \mu_{2p}, p)$, $j = 1, 2$.

Theorem 4.2 (Basalykas (1988)). *Suppose that* $X_1, \ldots, X_n$ *are sample data of the form* $X_j = \theta + \xi_j$, *where* $\xi_1, \ldots, \xi_n$ *are independent and identically distributed r.v.'s with* $\mathbf{E}\xi_j = 0$ *and* $\sigma^2 = \mathbf{E}\xi_1^2 > 0$ *that satisfy condition* (B). *Then in the interval*

$$0 \le x < \Delta_\gamma^{(j)}, \quad \gamma = p - 1,$$

the estimators $t_{n,p}^{(j)}$, $j = 1, 2$, *given by* (4.11) *and* (4.12), *satisfy the large deviation relations* (1.31) *with*

$$\Delta_\gamma = \Delta_\gamma^{(j)} = \frac{1}{6} \left(\frac{\sqrt{2}\sqrt{n}}{6H_{2,j}} \right)^{1/(2p-1)}, \quad j = 1, 2,$$

and the bounds (1.28), (1.39) *with*

$$\gamma = p - 1, \ H = 2^p, \ \tilde{\Delta} = \Delta^{(j)} = \frac{\sqrt{n}}{H_{2,j}}, \ j = 1, 2.$$

If $\mathbf{E}\xi_1 = 0$, $\mathbf{E}\xi_1^2 = \sigma^2 > 0$ *and* $|\xi_i| \le L$ *with probability* 1, $i = 1, \ldots, n$, *then in the interval*

$$0 \le x < \Delta_\gamma^{(j)}, \ \gamma = 0,$$

$t_{n,p}^{(j)}$ *satisfies* (1.31) *with*

$$\Delta_\gamma = \Delta_\gamma^{(j)} = \frac{\sqrt{2}}{36H_1}\sqrt{n}, \quad j = 1, 2,$$

and (1.28), (1.39) *with*

$$\gamma = 0, \ H = 2, \ \tilde{\Delta} = \sqrt{n}/H_{1,j}, \ j = 1, 2.$$

Let $X_1, X_2, \ldots$ be a sequence of independent and identically distributed r.v.'s with common distribution function F and let $\varphi(x_1, x_2)$ be a function symmetric in its arguments. Consider the second-order *U-statistic* with kernel $\varphi(x_1, x_2)$ given by

$$U_n := \binom{n}{2}^{-1} \sum_{1 \le i < j \le n} \varphi(X_i, X_j).$$

We shall assume that

$$\mathbf{E}\varphi(X_1, X_2) = 0 \tag{4.15}$$

and that

$$g(x) = \mathbf{E}(\varphi(X_1, X_2) \mid X_1 = x)$$

satisfies

$$\sigma^2 = \mathbf{V}g(X_1) > 0. \tag{4.16}$$

It is not hard to show that

$$\begin{aligned} \sigma_U^2 &:= \mathbf{V}U_n = \mathbf{E}U_n^2 \\ &= \frac{4\sigma^2}{n} + \frac{2}{n(n-1)}\mathbf{E}\psi^2(X_1, X_2) = \frac{4\sigma^2}{n} + O\left(\frac{1}{n^2}\right), \end{aligned} \tag{4.17}$$

where

$$\psi(X_1, X_2) := \varphi(X_1, X_2) - g(X_1) - g(X_2).$$

The study of large deviation probabilities for U-statistics was begun by Hoeffding (1963). He derived a Bernstein-type inequality for U-statistics with bounded kernel. Malevich and Abdalimov (1979) investigated Cramér and Linnik-type large deviations. They showed that if (4.15) and (4.16) hold and if there exist constants $K > 0$ and $\alpha \geq 0$ such that

$$\mathbf{E}|\varphi(X_1, X_2)|^k \leq K^k k^{\alpha k}, \quad k = 1, 2, \ldots, \tag{4.18}$$

then

$$\mathbf{P}\{\sqrt{n}U_n/(2\sigma) \geq x\} \sim 1 - \Phi(x), \quad n \to \infty, \tag{4.19}$$

uniformly in x in the region

$$0 < x \leq \rho(n)n^{1/(2(5+2\alpha))},$$

where $\rho(n)$ approaches zero as slowly as desired as $n \to \infty$.

We now give bounds for the cumulants of second-order U-statistics. From these it will follow that the large deviation relations (1.31) hold for $\xi = U_n/\sigma_U$ in the interval

$$0 < x \leq cn^{1/(2(1+2\alpha))}, \quad c > 0.$$

Proposition 4.2 (Aleškevičienė (1991a)). *If* (4.14) *and* (4.16) *hold and*

$$\mathbf{E}|\varphi(X_1, X_2)| \leq C^k (k!)^{1+\gamma}, \quad \gamma \geq 0, \tag{4.20}$$

for all $k = 3, 4, \ldots,$ *then*

$$|\Gamma_k(U_n)| \leq 2e^{2(k-2)}\frac{2^k - 1}{k}C^k(k!)^{2+\gamma}n^{-(k-1)}, \quad n \geq 7,$$

for all $k = 3, 4, \ldots, n-1$. *Consequently,*

$$|\Gamma_k(U_n/\sqrt{\mathbf{V}U_n})| \leq (k!)^{2+\gamma}\left(\frac{2\sqrt{2}eC(\sigma)}{\sqrt{n}}\right)^{k-2}, \quad k = 3, \ldots, n-1, \tag{4.21}$$

for $n \geq n_0 \geq 7$; *here* $C(\sigma) = C/\sigma$ *if* $C \leq \sigma$ *and* $C(\sigma) = C^3/\sigma^3$ *if* $C > \sigma$, *and* $n_0 \geq 7$ *is so determined that for all* $n \geq n_0$

$$\mathbf{V}U_n \geq \frac{e^2}{2}\frac{\sigma^2}{n}.$$

Theorem 4.3 (Aleškevičienė (1991b)). *If* (4.15), (4.16) *and* (4.20) *hold with* $\gamma \geq 1$, *then in the interval* $0 \leq x < \varepsilon_\gamma n^{1/(2+4\gamma)}$, *where*

$$\begin{aligned} \varepsilon_\gamma &= \frac{1}{2} \wedge \frac{\sigma}{12B_\gamma} \wedge (3C^{1/(1+\gamma)})^{-(1+\gamma)/(1+2\gamma)}, \\ B_\gamma &= (2.7)^{1+\gamma} 3C, \end{aligned}$$

the following relations are satisfied:

$$\begin{aligned} P\left\{\frac{U_n}{\sqrt{\mathbf{V}U_n}} > x\right\} &= (1-\Phi(x))\left(1 + O\left(\frac{1+x^3+x\ln n}{\sqrt{n}}\right)\right), \\ P\left\{\frac{U_n}{\sqrt{\mathbf{V}U_n}} < -x\right\} &= \Phi(-x)\left(1 + O\left(\frac{1+x^3+x\ln n}{\sqrt{n}}\right)\right). \end{aligned}$$

Large deviations have been studied for L-statistics and more general statistics by Aleškevičienė (1991a), (1991b) and Zitikis (1990).

§4.2. Bounds for the Cumulants and Large Deviation Theorems for Multiple Stochastic Integrals and Spectral Estimates of Stationary Sequences

Consider a complex *Gaussian random measure* $\beta(\Lambda)$, $\Lambda \subset R^1$, possessing the standard properties

(a) $\mathbf{E}\beta(\Lambda) = 0$,
(b) $\beta(\Lambda) = \overline{\beta(-\Lambda)}$,
(c) $\mathbf{E}\beta(\Lambda_1)\overline{\beta(\Lambda_2)} = F(\Lambda_1 \cap \Lambda_2)$.

Λ, Λ_1 and Λ_2 are measurable sets in R^1 and F is the spectral measure of β. We shall assume that F is continuous (it has no atoms).

Consider the space $L_2(F)$ of even complex-valued functions defined in R^m:

$$L_2^{(m)}(F) := \left\{\varphi : ||\varphi||^2 = \int_{R^m} |\varphi(\lambda_1, \ldots, \lambda_m)|^2 \prod_{j=1}^m F(d\lambda_j) < \infty, \right.$$
$$\left. \varphi(-\lambda_1, \ldots, -\lambda_m) = \overline{\varphi(\lambda_1, \ldots, \lambda_m)}, \quad m = 1, 2, \ldots \right\}.$$

One can define a multiple stochastic integral for functions $\varphi \in L_2^{(m)}(F)$ with respect to the measure β as follows (see, for example, Itô (1951)):

$$I^{(m)}(\varphi) = \int \cdots \int \varphi(\lambda_1, \ldots, \lambda_m)\beta(d\lambda_1) \cdots \beta(d\lambda_m).$$

The definition can be built up in the standard way beginning with step-functions vanishing on the diagonals of R^m. Let

$$\tilde{\varphi}(\lambda_1, \ldots, \lambda_m) = \frac{1}{m!} \sum{}' \varphi(\lambda_{i_1}, \ldots, \lambda_{i_m}),$$

where $\sum'$ denotes summation over all permutations of $\{1, \ldots, m\}$. Then $I^{(m)}(\varphi) = I^{(m)}(\tilde{\varphi})$. Some other properties of multiple stochastic integrals are

1. $I^{(m)}(\varphi)$ is a real random variable;
2. $\mathbf{E}I^{(m)}(\varphi) = 0$;
3. $\mathbf{E}I^{(m)}(\varphi)I^{(n)}(\psi) = 0$ if $m \neq n$; if $m = n$,

$$\mathbf{E}I^{(m)}(\varphi)I^{(m)}(\psi) = m! \int \tilde{\varphi}(\lambda_1, \ldots, \lambda_m)\overline{\tilde{\psi}(\lambda_1, \ldots, \lambda_m)} \prod_{j=1}^{m} F(d\lambda_j);$$

4. Let $\varphi_1, \ldots, \varphi_n$ be an orthonormal system of functions in $L_2^{(1)}(F)$ and let $H_k(x) = (-1)^k e^{\frac{1}{2}x^2} \frac{d^k e^{-\frac{1}{2}x^2}}{dx^k}$ be the *Hermite polynomial* of degree k. Then

$$\underbrace{\int \cdots \int}_{n} \varphi_1(\lambda_1) \cdot \ldots \cdot \varphi_1(\lambda_{p_1})\varphi_2(\lambda_{p_1+1}) \ldots \varphi_2(\lambda_{p_1+p_2})$$

$$\ldots \varphi_n(\lambda_{p_1+\cdots+p_{n-1}+1}) \cdot \ldots \cdot \varphi_n(\lambda_{p_1+\cdots+p_n}) \prod_{i=1}^{r} \beta(d\lambda_i)$$

$$= \prod_{j=1}^{n} H_{p_j}\left(\int \varphi_j(x)\beta(d\lambda)\right), \quad r = p_1 + \cdots + p_n;$$

5. Let $X_t = \int e^{it\lambda}\beta(d\lambda)$ and $\mathcal{F} = \sigma\{X_t,\ t \in R^1\}$. Then any $\mathcal{F}$-measurable r.v. ξ with $\mathbf{E}\xi^2 < \infty$ may be expanded in a series

$$\xi = \mathbf{E}\xi + \sum_{m=1}^{\infty} I^{(m)}(\varphi_m),$$

which converges in the mean-square. The set of functions $\varphi_m \in L_2^{(m)}(F)$ is unique if one considers just symmetric functions.

Let D or $D_{k,m}$ denote an array of pairs of indices consisting of k rows and m columns: $D_{k,m} = \{(i,j),\ i = \overline{1,k},\ j = \overline{1,m}\}$. $D = D' \cup D''$ will be called a *row partition* if any row in D belongs either to D' or D''. A partition $D_{k,m} = \bigcup_{j=1}^{r} D_j$ will be called *indecomposable* if no row partition $D = D' \cup D''$ exists for which any D_j belongs to either D' or D''.

Lemma 4.1 (Plikusas (1980)). *Let* $\varphi \in L_2^{(m)}(F)$. *Then* $\Gamma_k(I^{(m)}(\varphi)) = 0$ *if* km *is odd and*

$$\Gamma_k(I^{(m)}(\varphi)) = \sum\nolimits^* \int_{R^r} \prod_{j=1}^{k} \varphi(\lambda_{j,1}, \ldots, \lambda_{j,m}) \prod_{s=1}^{r} F(d\lambda_s) \tag{4.22}$$

if km *is even, where* $r = km/2$. $\sum^*$ *denotes summation over all indecomposable partitions of* $D_{k,m}$ *into subsets of two elements.* $\prod_{j=1}^k \varphi(\lambda_{j,1}, \ldots, \lambda_{j,m})$ *is determined by an indecomposable partition* $D_{k,m} = \bigcup_{j=1}^r D_j$ *as follows: if* $D_j = \{(p,q),(s,t)\}$, *then* $\lambda_{p,q} = \lambda_j$ *and* $\lambda_{s,t} = -\lambda_j$.

Proposition 4.3. *Let* $\varphi_j \in L_2^{(m)}(F)$, $j = 1, 2, \ldots, k$, km *be even and* $r = \frac{1}{2}km$. *Then*

$$\left| \int_{R^r} \prod_{j=1}^{k}{}^* \varphi_j(\lambda_{j_1}, \ldots, \lambda_{j_m}) \prod_{s=1}^{r} F(d\lambda_s) \right| \leq \prod_{j=1}^{k} ||\varphi_j||. \tag{4.23}$$

The product $\prod^*$ *is taken over those collections of different indices* $(j_1, \ldots, j_m)$ *such that each* j_i *is repeated in exactly two collections.*

Corollary 4.1. *Let* $\varphi \in L_2^{(2)}(F)$ *be symmetric. Then*

$$\Gamma_k(I^{(2)}(\varphi)) \tag{4.24}$$
$$= 2^{k-1}(k-1)! \int_{R^k} \varphi(\lambda_1, -\lambda_2)\varphi(\lambda_2, -\lambda_3) \cdots \varphi(\lambda_k, -\lambda_1) \prod_{j=1}^{k} F(d\lambda_j).$$

Proof. All of the terms on the right-hand side of (4.22) are clearly equal in this case. It merely remains to determine the number of indecomposable partitions of $D_{k,2}$, which is simple to do

Corollary 4.2. *Let* $\varphi \in L_2^{(2)}(F)$ *be symmetric. Then*

$$|\Gamma_k(I^{(2)}(\varphi))| \leq 2^{k-1}(k-1)! ||\varphi||^k.$$

This is a consequence of inequality (4.23).

This estimate allows one to say that the distribution of the r.v. $I^{(2)}(\varphi)$ is determined by its moments or cumulants.

It is known that a symmetric $L_2^{(2)}(F)$-function φ, which is nonvanishing F-a.e., determines a self-adjoint operator from $L_2^{(1)}(F)$ into $L_2^{(1)}(F)$ by means of the relation

$$\varphi_1(\lambda_1) = \int \varphi_2(\lambda_2)\varphi(\lambda_1, \lambda_2) F(d\lambda_2), \quad \varphi_2 \in L_2^{(1)}(F).$$

It has nonvanishing eigenvalues $\{\mu_j\}$. If $\{\psi_j\} \subset L_2^{(1)}(F)$ is an orthonormal set of eigenfunctions, then the expansion

$$\varphi(\lambda_1, \lambda_2) = \sum_j \mu_j \psi_j(\lambda_1)\psi_j(\lambda_2) \tag{4.25}$$

holds, where the convergence in (4.25) means convergence in $L_2^{(2)}(F)$.

Proposition 4.4. *Let $\varphi \in L_2^{(2)}(F)$ be symmetric. Then*

$$\Gamma_k(I^{(2)}(\varphi)) = 2^{k-1}(k-1)! \sum_j \mu_j^k, \quad k = 2, 3, \dots . \tag{4.26}$$

This is proved by substituting (4.25) in (4.24).

The expression (4.26) implies that $I^{(2)}(\varphi)$ is distributed the same as $\sum_j \mu_j(X_j^2 - 1)$, where the X_j's are independent standard Gaussian variables.

Lemma 4.2. *The following estimate holds:*

$$|\Gamma_k(I^{(m)}(\varphi))| \le M(k,m)||\varphi||^k, \quad k = 2, 3, \dots, \tag{4.27}$$

where $M(k,m) = \Gamma_k(H_m(X))$ and X is a standard Gaussian r.v. Equality is attained in (4.27) *when $\varphi(\cdot) = \psi(\lambda_1)\dots\psi(\lambda_m)$.*

Proof. Noting (4.22) and (4.23), we obtain (4.27) with $M(k,m)$ the number of indecomposable partitions of the array $D_{k,m}$. When $\varphi = 1$ and $F(R^1) = 1$, we have $I^{(m)}(\varphi) = H_m(X)$. The substitution of this in (4.22) yields $M(k,m) = \Gamma_k(H_m(X))$.

Proposition 4.5. *There exist positive constants H_i and C_i, $i = 1, 2$, that depend only on m such that*

$$H_1 C_1^k (k!)^{m/2} \le M(k,m) \le H_2 C_2^k (k!)^{m/2}, \quad k = 2, 3, \dots .$$

In particular, possible choices are $C_1 = \sqrt{2}$ for $H_1 = 1/8$ and $C_2 = m^{m/2}$ for $H_2 = 1$.

Thus

$$|\Gamma_k(I^{(m)}(\varphi))| \le (m^{m/2}||\varphi||)^k (k!)^{m/2}. \tag{4.28}$$

Applying (4.28) and the exponential bounds (1.40), we arrive at the next result.

Theorem 4.4 (Plikusas (1980)). *The inequality*

$$\mathbf{P}\{I^{(m)}(\varphi) \ge \sigma_m x\} \le \begin{cases} \exp(-c_1 x^2), & x \le A_m, \\ \exp(-c_2 x^{2/m}), & x > A_m, \end{cases}$$

is true, where $\sigma_m^2 = \mathbf{E}\big(I^{(m)}(\varphi)\big)^2$,

$$c_1 = \frac{1}{4}\left(\frac{2}{3}e^2\right)^{-m/2} \sqrt{\pi m}, \quad c_2 = (2\pi m)^{1/(2m)}/(4e)$$

and

$$A_m = 2^{(m^2+1)/(4(4m-1))} 3^{m^2/(4(m-1))} e^{m/2}/(\pi m)^{1/4}.$$

Consider a *stationary Gaussian process*

$$X_t = \int_{R^1} e^{it\lambda}\beta(d\lambda), \quad t \in R^1, \tag{4.29}$$

defined on probability space $(\Omega, \mathcal{F}, \mathbf{P})$. Assume that the measure F is absolutely continuous with respect to Lebesgue measure on the real line. In other words, $F(d\lambda) = f(\lambda)d\lambda$ and $F(R^1) = 1$. By the relation $X_0(S_t(\omega)) = X_t(\omega)$, the process X_t determines a shift operator $S_t: \ \Omega \to \Omega$. It is known that then

$$I^{(m)}(\varphi)(S_t(\omega))$$
$$= \int \cdots \int e^{it(\lambda_1+\cdots+\lambda_m)}\varphi(\lambda_1, \ldots, \lambda_m)\beta(d\lambda_1)\ldots\beta(d\lambda_m) \overset{\text{df}}{=} I_t^{(m)}(\varphi).$$

Put

$$Y_T^{(m)} = Y_T^{(m)}(\varphi) = \int_0^T I_t^{(m)}(\varphi)dt, \quad T > 0.$$

Our aim is to deduce limit theorems for the r.v. $Y_T^{(m)}$ as $T \to \infty$ which establish the rate of convergence to the normal distribution. We also want to study the large deviation probabilities.

We shall now define a polynomial form $\Phi_{k,m}(x_1, \ldots, x_{k-1})$ of $\varphi \in L_2^{(m)}(F)$ in terms of which the k-th cumulant of $Y_T^{(m)}(\varphi)$ can be expressed. Assume that km is even. Let

$$\varphi^*(\lambda_1, \ldots, \lambda_r) = \prod_{s=1}^{k} \varphi(\lambda_{s,1}, \ldots, \lambda_{s,m}), \quad r = km/2.$$

The product on the right-hand side is determined by an indecomposable partition $D_{k,m} = \bigcup_{j=1}^r D_j$ into subsets of two elements. (The variables $\lambda_1, \ldots, \lambda_r$ each appear in the product twice: once with a plus sign and once with a minus sign.) We make a change of variables $\lambda_1, \ldots, \lambda_r$ by means of the linear transformation

$$\begin{aligned} \sum_{j=1}^{m} \lambda_{s,j} &= x_s, \quad s = 1, \ldots, k-1, \\ \lambda_{j_k} &= x_k, \\ &\cdots\cdots\cdots \\ &\cdots\cdots\cdots \\ \lambda_{j_r} &= x_r. \end{aligned}$$

The first $k-1$ equations are linearly independent because of the indecomposability of the partitioning of $D_{k,m}$. The remaining equations may be chosen trivially as long as the transformation is non-singular. It is easy to see that the Jacobian of the transformation has absolute value 1. Consider now the function

$$\varphi^*(\lambda_1, \ldots, \lambda_r) \prod_{j=1}^{r} f(\lambda_j).$$

Let $\varphi_1^*(x_1, \ldots, x_r)$ denote this expression after the change of variables. Put

$$\Phi_{k,m}(x) = \Phi_{k,m}(x_1, \ldots, x_{k-1}) = \sum\nolimits^* \int_{R^{r-k+1}} \varphi_1^*(x_1, \ldots, x_r)dx_1 \ldots dx_r.$$

As earlier, $\sum^*$ denotes summation over all indecomposable partitions of $D_{k,m}$ into doubleton subsets. If km is odd, then $\Phi_{k,m}(x)$ is taken to be 0.

Let

$$\Psi_T^{(n)}(x) = \frac{2}{\pi^n T} \cdot \frac{\sin \frac{Tx_1}{2}}{x_1} \cdot \ldots \cdot \frac{\sin \frac{Tx_n}{2}}{x_n} \cdot \frac{\sin \frac{T(x_1 + \cdots + x_n)}{2}}{x_1 + \cdots + x_n}, \tag{4.30}$$
$$n = 1, 2, \ldots, \quad T > 0, \quad x = (x_1, \ldots, x_n).$$

When $n = 1$, $\Psi_T^{(n)}(x)$ is customarily called *Fejér's kernel.* The functions $\Psi_T^{(n)}(x)$, $n \geq 2$, were introduced by Bentkus (1972a) and are one of the possible generalizations of Fejér's kernel to the n-dimensional case. He derived the basic properties of $\Psi_T^{(n)}(x)$ in that article. We now state the following result.

Lemma 4.3.

$$\Gamma_k \left(\int_0^T I_t^{(m)}(\varphi)dt \right) = (2\pi)^{k-1} T \int_{R^{k-1}} \Psi_T^{(k-1)}(x) \Phi_{k,m}(x)dx, \; \varphi \in L_2^{(m)}(F).$$

The lemma is first proved for step-functions. Then the inequality $|\Psi_T^{k-1}(x)| \leq T^{k-1}$ is applied when taking the limit.

We next state some consequences of this lemma.

Proposition 4.6. *Let $\varphi \in L_2^{(2)}(F)$ be symmetric. Then*

$$\Gamma_k \left(\int_0^T I_t^{(2)}(\varphi)dt \right) = (4\pi)^{k-1}(k-1)!T \int_{R^{k-1}} \Psi_T^{k-1}(x)\varphi^*(x)dx,$$

where

$$\varphi^*(x) = \int_{R^1} \prod_{j=1}^{k} \varphi(y_j, -y_{j+1}) f(y_j)dx_k,$$

$$x = (x_1, \ldots, x_{k-1}), \; y_j = \sum_{s=j}^{k} x_s, \; y_{k+1} = -y_1, k = 2, 3, \ldots .$$

Proposition 4.7.

$$v_T^2 := \Gamma_2 \left(\int_0^T I_t^{(m)}(\varphi)dt \right) = 4m! \int_{R^1} \frac{1}{\lambda} \sin^2 \frac{T\lambda}{2}$$
$$\times \int_{R^{m-1}} \left| \varphi \left(\lambda - \sum_{j=1}^{m-1} \mu_j, \mu_1, \ldots, \mu_{m-1} \right) \right|^2 f \left(\lambda - \sum_{j=1}^{m-1} \mu_j \right) \prod_{j=1}^{m-1} f(\mu_j)d\mu_j d\lambda.$$

In particular, if $\varphi = C$, *then*

$$v_T^2 = 4m!C^2 \int_{R^1} \frac{1}{\lambda} \sin^2 \frac{T\lambda}{2} f^{*m}(\lambda) d\lambda,$$

where f^{*m} *is the* m*-th convolution of the spectral density* $f(\lambda)$.

Proposition 4.8. *Suppose that* $\Phi_{2,m}(\lambda)$ *is continuous at* $\lambda = 0$. *Then*

$$v_T^2 = 2\pi\Phi_{2,m}(0)T + o(T), \quad T \to \infty.$$

This statement follows from Proposition 4.7 and Theorem 18.3.1 in Ibragimov and Linnik (1965).

Proposition 4.9. *Suppose that* $\Phi_{k,m}(x)$, $k, m = 2, 3, \ldots$, *is bounded and continuous at* $x = (x_1, \ldots, x_{k-1}) = (0, \ldots, 0)_{k-1}$. *Then*

$$\Gamma_k \left(\int_0^T I_t^{(m)}(\varphi) dt \right) = (2\pi)^{k-1} \Phi_{k,m}(0)T + o(T), \quad T \to \infty.$$

The proof of this statement may be found in Plikusas (1980).

We now proceed to consider upper bounds for the cumulants. Let $R(t) = \mathbf{E}X_0X_t$ be the *correlation function* of the original Gaussian process.

Lemma 4.4. *Suppose that*

$$\text{ess sup}\, |\varphi(\lambda_1, \ldots, \lambda_m)| = A_\varphi < \infty.$$

Then

$$|\Gamma_k(Y_T^{(m)}(\varphi))| \le A_\varphi^k M(k, m) T \left(\int_{-T}^T |R(t)| dt \right)^{k-2} \int_{-T}^T |R(t)|^2 dt, \; k = 2, 3, \ldots .$$

It is now possible to derive a bound for the cumulants of the requisite form.

Lemma 4.5 (Plikusas (1980)). *Suppose that*

1. ess sup $|\varphi(\lambda_1, \ldots, \lambda_m)| \le A_\varphi$;
2. $\int_{R^1} |R(t)| dt = c_R < \infty$;
3. *a positive constant* c_1 *exists such that*

$$v_T \ge c_1 \sqrt{T}.$$

Then

$$|\Gamma_k(Y_T^{(m)}(\varphi)/v_T)| \le (k!)^{m/2} / (A_1 \sqrt{T})^{k-2}.$$

A_1 *may be taken to be* $c_1(A_\varphi c_R m^{m/2})^{-1}$ *if* $A_\varphi^2 m^m c_R \le c_1^2$. *Otherwise*

$$A_1 = \begin{cases} (c_1(A_\varphi m^{m/2} c_R)^{-1})^3 & \text{if } c_R > 1, \\ (c_1(A_\varphi m^{m/2})^{-1})^3 & \text{if } c_R \le 1. \end{cases}$$

Let $F_T(x)$ be the distribution function of $Y_T^{(m)}(\varphi)/v_T$. The following statement is true.

Theorem 4.5 (Plikusas (1980)). *Under the hypotheses of Lemma* 4.5,

$$\sup_x |F_T(x) - \Phi(x)| \le \frac{c}{T^{1/(2(m-1))}},$$

and the large deviation relations (1.31) *hold in the range*

$$0 \le x < \frac{1}{6}\left(\frac{\sqrt{2}A_1\sqrt{T}}{6}\right)^{1/(m-1)}.$$

Consider now a *Poisson process* $X(t)$, $t \in [0,\infty)$, with mean $\mathbf{E}X(t) = m(t)$ and $X(0) = 0$, or in other words, a nondecreasing process with non-negative values and independent increments. The distribution of the increments is given by

$$\mathbf{P}\{X(t) - X(s) = k\} = \frac{(m(t) - m(s))^k}{k!} e^{-(m(t)-m(s))},$$

$t > s$ and $k = 0, 1, 2, \ldots$. We shall assume that $m(t)$ is continuous. Introduce the space of real functions of q variables,

$$L_T^p := \left\{ a(t_1, \ldots, t_q) : \int \cdots \int_{0 \le t_1 < \cdots < t_q \le T} |a(t_1, \ldots, t_q)|^p dm(t_1) \ldots dm(t_q) < \infty \right\}.$$

It is known that one can define a *multiple stochastic integral* of $a \in L_T^2$ with respect to $X(t)$ as follows (Engel (1982)):

$$Y_T^{(q)} = \int \cdots \int_{0 \le t_1 < \cdots < t_q \le T} a(t_1, \ldots, t_q) dX(t_1) \ldots dX(t_q).$$

Observe that when $a \equiv 1$,

$$Y_T^{(q)} = \frac{1}{q!} X(T)(X(T) - 1) \ldots (X(T) - q + 1).$$

A slightly different approach to defining a multiple stochastic integral with respect to a Poisson process was considered by Surgailis (1981).

Let τ_k be the time of the k-th jump in the process $X(t)$. Then $Y_T^{(q)}$ may be expressed in the form

$$Y_T^{(q)} = \sum_{1 \le i_1 < \cdots < i_q \le X(T)} a(\tau_{i_1}, \ldots, \tau_{i_q}). \tag{4.31}$$

Thus, $Y_T^{(q)}$ is a *U-statistic* of a special form relating to queueing theory. Multiple stochastic integrals of various kinds have also been used to construct new classes of self-similar fields (Surgailis (1981)).

Let $D_{k,q}$ denote the array of pairs of indices

$$D_{k,q} = \{(m,i) : m \in \{1,2,\dots,k\},\ i \in \{1,2,\dots,q\}\}.$$

We shall speak of $(m,i) \in D_{k,q}$ as belonging to the m-th row and i-th column.

A partition $D_{k,q} = \bigcup_{r=1}^{i} D_r$ will be said to be *P-indecomposable* if the following two conditions hold:

1. each subset D_r contains at most one element from a row, $r = 1,\dots,j$;
2. no group of r rows exist, $r = 1,\dots,k-1$, having its own partition formed by $D_{k,q} = \bigcup_{r=1}^{i} D_r$.

Lemma 4.6 (Basalykas, Plikusas and Statulevičius (1987)).

$$\Gamma_k(Y_T^{(q)}) = \sum_{j=q}^{k(q-1)+1} \sum{}^{(j)} \int\cdots\int_{0\le t_{m1}<\cdots<t_{mq}\le T} \prod_{m=1}^{q} a(t_{m1},\dots,t_{mq}) \prod_{i=1}^{j} dm(t_i),$$

$$T>0,\ q\ge 2,\ k\ge 1, \tag{4.32}$$

for $a \in L_T^{(k)}$, *where* $\sum^{(j)}$ *denotes summation over all P-indecomposable partitions* $D_{k,q} = \bigcup_{r=1}^{j} D_r$ *into* j *subsets. When* $(m,i) \in D_r$, $r = 1,2,\dots,j$, t_{mi} *is taken to be* t_r *in the product in the integrand.*

Let us clarify (4.32) by way of examples. We first find the variance of $Y_T^{(2)}$. Put $A_{ij} = \{(t_i,t_j) \in R^2 :\ 0 \le t_i < t_j \le T\}$. Then

$$\begin{aligned}\Gamma_2(Y_T^{(2)}) &= \int_{A_{12}} a^2(t_1,t_2)dm(t) + \int_{A_{12}\cap A_{13}} a(t_1,t_2)a(t_1,t_3)dm(t)\\ &+ \int_{A_{12}\cap A_{31}} a(t_1,t_2)a(t_3,t_1)dm(t) + \int_{A_{12}\cap A_{23}} a(t_1,t_2)a(t_2,t_3)dm(t)\\ &+ \int_{A_{12}\cap A_{32}} a(t_1,t_2)a(t_3,t_2)dm(t).\end{aligned}$$

Here $dm(t) = \prod_i dm(t_i)$. Let $\tilde a(t_1,\dots,t_q) = a(t_{i_1},\dots,t_{i_q})$, where $(i_1,\dots,i_q)$ is a permutation of $(1,\dots,q)$ such that $t_{i_1} < \dots < t_{i_q}$; if $t_i = t_j$ for some i,j, then we take $\tilde a = 0$. Thus

$$\Gamma_2(Y_T^{(2)}) = \int_{A_{12}} a^2(t_1,t_2)dm(t) + \int_{[0,T]^3} \tilde a(t_1,t_2)\tilde a(t_2,t_3)dm(t).$$

Applying Hölder's inequality to the second integral, we obtain

$$\Gamma_2(Y_T^{(2)}) \le ||\tilde a_T||^2(1/2 + m(T)).$$

Here $||\tilde a_T||^2 = \int_{[0,T]^2} \tilde a^2(t_1,t_2)dm(t)$.

It is not hard to show in the same fashion that

$$\Gamma_3(Y_T^{(2)}) = \int_{A_{12}} a^3(t_1,t_2)dm(t) + 3\int_{[0,T]^3} \tilde{a}^2(t_1,t_2)\tilde{a}^2(t_1,t_3)dm(t)$$
$$+ \int_{[0,T]^3} \tilde{a}(t_1,t_2)\tilde{a}(t_2,t_3)\tilde{a}(t_3,t_1)dm(t) + \int_{[0,T]^4} \tilde{a}(t_1,t_2)\tilde{a}(t_1,t_3)$$
$$\times \tilde{a}(t_1,t_4)dm(t) + 3\int_{[0,T]^4} \tilde{a}(t_1,t_2)\tilde{a}(t_2,t_3)\tilde{a}(t_3,t_4)dm(t).$$

By once more applying Hölder's inequality, we can establish that the next-to-last integral does not exceed

$$\left(\int_{[0,T]}\left(\int_{[0,T]} |\tilde{a}(t_1,t_2)|^3 dm(t_1)\right)^{1/3} dm(t_2)\right)^3 \leq m^2(T)\int_{[0,T]^2} |\tilde{a}(t_1,t_2)|^3 dm(t).$$

Proceeding in similar fashion with the remaining integrals, we arrive at the estimate

$$|\Gamma_3(Y_T^{(2)})| \leq \frac{1}{2}\int_{[0,T]^2} |\tilde{a}(t_1,t_2)|^3 dm(t) + 3m(T)\int_{[0,T]^2} |\tilde{a}(t_1,t_2)|^3 dm(t)$$
$$+ \left(\int_{[0,T]^2} \tilde{a}^2(t_1,t_2)dm(t)\right)^{3/2} + m^2(T)\int_{[0,T]^2} |\tilde{a}(t_1,t_2)|^3 dm(t)$$
$$+3m(T)\left(\int_{[0,T]^2} \tilde{a}^2(t_1,t_2)dm(t)\right)^{3/2}.$$

By virtue of (4.32), it is not hard to see that for any k the k-th cumulant $\Gamma_k(Y_T^{(q)})$ is bounded by

$$\int_{[0,T]^q} |\tilde{a}(t_1,\ldots,t_q)|^k dm(t).$$

A complete proof of the present lemma is given in Saulis and Statulevičius (1970).

Theorem 4.6 (Basalykas, Plikusas and Statulevičius (1987)). *Suppose that* $|a| \leq c_a$ *and that a positive constant* c_0 *exists such that*

$$\mathbf{V}Y_T^{(q)} \geq c_0(m(T))^{2q-1}, \quad T > 0.$$

Then the large deviation relation (1.31) *holds for the r.v.* $\xi = Z_T^{(q)} := (Y_T^{(q)} - \mathbf{E}Y_T^{(q)})\big/\sqrt{\mathbf{V}Y_T^{(q)}}$ *in the range*

$$0 \leq x < C_q(m(T))^{1/2}.$$

Here

$$C_q = \frac{c_0^2}{c_a^2 e^q} \inf_{0\leq\alpha\leq q-1}\left\{\frac{1}{6}\left(\frac{\sqrt{2}}{6}\right)^{1/(2q-2\alpha-1)} \cdot \alpha^{q-\alpha}\right\}.$$

For the standard Poisson process, $m(T) = aT$ *and the result is true in the range* $0 \le x < c\sqrt{T}$.

We now consider a multiple stochastic integral with respect to a centered Poisson process $\tilde{X}(t) = X(t) - m(t)$. Similarly to the integral $Y_T^{(q)}$, one can define the multiple stochastic integral

$$\tilde{Y}_T^{(q)} = \int_{0 \le t_1 < \ldots < t_q \le T} \cdots \int a(t_1, \ldots, t_q) d\tilde{X}(t_1) \ldots d\tilde{X}(t_q),$$
$$T > 0, \ q \ge 2, \ a \in L_T^2.$$

When $a \equiv 1$, we have $\tilde{Y}_T^{(q)} = K_q(X(T), m(T))$, where

$$K_q(u,t) = \frac{1}{q!} \sum_{r=0}^{q} \binom{q}{r} (-t)^r u_{(q-r)},$$
$$u_{(n)} = u(u-1) \cdots (u-n+1).$$

It is known that the *Poisson-Charlier polynomials* $K_q(u,t)$, $q = 1, 2, \ldots$, are orthogonal with respect to the Poisson measure:

$$\sum_{r=0}^{\infty} K_n(r,t) K_m(r,t) \frac{t^r e^{-r}}{r!} = \frac{t^n}{n!} \delta_{nm}, \ \delta_{nm} = \begin{cases} 1, & n = m, \\ 0, & n \ne m. \end{cases}$$

In contrast to $Z_T^{(q)}$, the normalized r.v. $\tilde{Z}_T^{(q)} := \tilde{Y}_T^{(q)} (\mathbf{V} \tilde{Y}_T^{(q)})^{-1/2}$ does not converge to a Gaussian r.v. as $m(T) \to \infty$.

Lemma 4.7 (Basalykas, Plikusas and Statulevičius (1987)). *Let* $a \in L_T^{(q)}$. *Then*

$$\Gamma_k(\tilde{Y}_T^{(q)})$$
$$= \sum_{j=q}^{[kq/2]} \sum\nolimits^{(j)} \int_{0 \le t_{m1} < \cdots < t_{mq} \le T} \cdots \int \prod_{m=1}^{k} a(t_{m1}, \ldots, t_{mq}) \prod_{i=1}^{j} dm(t_i), \quad (4.33)$$

with $k \ge 2$, $T > 0$, $q \ge 2$ *and* $[kq/2]$ *the integral part of* $kq/2$. $\sum^{(j)}$ *denotes summation over all P-indecomposable partitions* $D_{k,q} = \bigcup_{r=1}^{j} D_r$ *into* j *parts. It suffices to sum over the P-indecomposable partitions into subsets containing at least two elements of* $D_{k,q}$.

We merely point out that $\Gamma_1(\tilde{X}(\Delta)) = 0$, $\Delta \subset R^1$, and consequently one can drop the terms corresponding to partitions containing singletons. Clearly, $\Gamma_k(\tilde{X}(\Delta)) = m(\Delta)$, $k = 2, 3, \ldots$, $\mathbf{E}\tilde{Y}_T^{(q)} = 0$ and

$$\mathbf{E}(\tilde{Y}_T^{(q)})^2 = \Gamma_2(\tilde{Y}_T^{(q)}) = \int_{0 \le t_1 < \cdots < t_q \le T} \cdots \int a^2(t_1, \ldots, t_q) dm(t).$$

This means that the order of growth of the variance is $(m(T))^q$ when $a =$ const. In that event, the even cumulants of $\tilde{Z}_T^{(q)}$ do not converge to zero as $m(T) \to \infty$ by virtue of Lemma 4.7.

If we estimate the cumulants as in Theorem 4.6, then the main contribution is from the terms corresponding to the partition of the array into $[kq/2]$ parts and the cumulants of $\tilde{Y}_t^{(q)}$ will be bounded by $C^k(k!)^{q/2}$. The exponential inequalities for the distribution function (1.39) can be used to deduce the next assertion.

Theorem 4.7 (Basalykas, Plikusas and Statulevičius (1987)). *Let there exist a positive constant* $\tilde{c}_0$ *such that*

$$\mathbf{E}(\tilde{Y}_T^{(q)})^2 \geq \tilde{c}_0(m(T))^q, \quad |a| \leq c_a, \quad m(T) > 1.$$

Then

$$\mathbf{P}\{\tilde{Z}_T^{(q)}) \geq x\} \leq \exp(-hx^{2/q})$$

for $x > 2e^{-q}\tilde{c}_0/c_a$ *and* $h \leq \tilde{c}_0/(2e^q c_a)$.

Consider the estimator

$$\hat{A}(\varphi) = \int_{-\pi}^{\pi} \varphi(x) I_N(x)dx, \tag{4.34}$$

where

$$I_N(x) := \frac{1}{2\pi N} \sum_{s,t=1}^{N} X_s X_t \exp(-i(s-t)x) \tag{4.35}$$

is the *second-order periodogram* formed from a sample of size N of a widely stationary sequence $\{X(t),\ t = \ldots, -1, 0, 1, \ldots\}$ with spectral density $f(\lambda)$ and φ is some function. If $\varphi = W_N(\cdot - \lambda)$, where W_N is a kernel (an asymptotically delta-form function), then $\tilde{A}(\varphi)$ is a (second-order) estimator of the spectral density f at point λ.

It is known that

$$\mathbf{E}\hat{A}(\varphi) = \int_{-\pi}^{\pi} \Psi_N(u)du \int_{-\pi}^{\pi} \varphi(x) f(x+u)dx, \tag{4.36}$$

where

$$\Psi_N(x) := \frac{1}{2\pi N} \cdot \frac{\sin^2 \frac{Nx}{2}}{\sin^2 \frac{x}{2}}$$

is Fejér's kernel. The Leonov-Shiryaev formula (1.17) can be used to derive the following representation for a *mixed cumulant* (see Bentkus (1976)):

$$\Gamma(\hat{A}(\varphi_1), \ldots, \hat{A}(\varphi_k)) = \frac{1}{N^{k-1}} \int_{\Pi^{2k-1}} G(u) \Psi_N^{(2k)}(u)du. \tag{4.37}$$

Here $\Psi_N^{(n)}(x)$ is the generalized Fejér kernel (Bentkus (1976)) given by (4.30), $\Pi := [-\pi, \pi]$ and G is determined by the functions $\varphi_1, \ldots, \varphi_k$ and spectral densities up to order $2k$ inclusively (of course, if they exist).

Lemma 4.8 (Bentkus and Rudźkis (1980)). *Let* $\varphi \in L_{p_1}$ *and* $f \in L_{p_2}$ *with* p_1 *and* $p_2 \in [1, \infty]$. *Then*

$$|\Gamma_k(\hat{A}(\varphi))| \le (k-1)!\mathbf{V}\hat{A}(\varphi)\left(4\pi||\varphi||_{p_1}||f||_{p_2}N^{-1+\frac{1}{p_1}+\frac{1}{p_2}}\right)^{k-2} \tag{4.38}$$

for all $k = 3, 4, \ldots$ *and* $N = 1, 2, \ldots$.

It is easily seen that

$$\hat{A}(\varphi) = \frac{1}{2\pi N}\sum_{s,t=1}^{N} a(s-t)X_sX_t = \frac{1}{2\pi N}(T_\varphi X, X), \tag{4.39}$$

where

$$a(u) = \int_{-\pi}^{\pi} \varphi(x)\cos ux\,dx,$$

$$T_\varphi = ||a(s-t)||_{s,t=\overline{1,N}}, \quad X = (X_1, \ldots, X_n).$$

Diagonalizing T_φ and the covariance matrix T_f of X simultaneously, we find that the r.v. $\hat{A}(\varphi)$ is distributed the same as $1/(2\pi N)\sum_{j=1}^{N}\mu_j\eta_j^2$, where the η_j's are standard normal r.v.'s and the μ_j's are the eigenvalues of the matrix $T_\varphi T_f$. Consequently,

$$\Gamma_k(\hat{A}(\varphi)) = \frac{(k-1)!}{2(\pi N)^k}\sum_{j=1}^{N}\mu_j^k, \quad k = 2, 3, \ldots. \tag{4.40}$$

Noting that $\mathbf{V}\hat{A} = \Gamma_2(\hat{A}(\varphi))$, we find from this that

$$|\Gamma_k(\hat{A}(\varphi))| \le (k-1)!\mathbf{V}\hat{A}\left(\frac{\max_j|\mu_j|}{\pi N}\right)^{k-2}. \tag{4.41}$$

In Bentkus-Rudzkis (1980), it is shown that

$$\max_j|\mu_j| \le (2\pi)^{2-1/p_1-1/p_2}||\varphi||_{p_1}||f||_{p_2}N^{1/p_1+1/p_2}. \tag{4.42}$$

Lemma 4.8 follows from (4.40)–(4.42).

Lemma 4.9 (Bentkus and Rudzkis (1976)). *If* φ *and* f *are bounded functions, then*

$$\mathbf{V}\hat{A}(\varphi) \le 4\pi||\varphi||_1||\varphi||_\infty||f||_\infty^2\frac{1}{N} \tag{4.43}$$

for all $N = 1, 2, \ldots$.

Put

$$\sigma_N^2 = N\mathbf{V}\hat{A}(\varphi)$$

and

$$Z_N = \sqrt{N}\big(\hat{A}(\varphi) - \mathbf{E}\hat{A}(\varphi)\big)/\sigma_N.$$

Then $\mathbf{E}Z_N = 0$, $\mathbf{E}Z_N^2 = 1$ and by Lemma 4.8.

$$|\Gamma_k(Z_N)| \le \frac{(k-1)!}{\Delta_N^{k-2}}, \quad \forall k \ge 1, \tag{4.44}$$

where

$$\Delta_N = \frac{\sqrt{N}\sigma_N}{4\pi||\varphi||_\infty||f||_\infty}. \tag{4.45}$$

Theorem 4.8 (Bentkus and Rudzkis (1980)). *If $\sigma_N > 0$ and φ and f are bounded functions, then in the interval*

$$0 \le x \le \frac{\sqrt{2}}{36}\Delta_N$$

the following large deviation relations hold:

$$\begin{aligned} &\frac{\mathbf{P}\{\sqrt{N}(\hat{A}(\varphi) - \mathbf{E}\hat{A}(\varphi)) \ge \sigma_N x\}}{1-\Phi(x)} \\ &\qquad = \exp(L_0(x))\left(1 + \tilde{\theta}_1 f(x)\frac{x+1}{\Delta_N}\right), \\ &\frac{\mathbf{P}\{\sqrt{N}(\hat{A}(\varphi) - \mathbf{E}\hat{A}(\varphi)) \le -\sigma_N x\}}{\Phi(-x)} \\ &\qquad = \exp(L_0(-x))\left(1 + \tilde{\theta}_2 f(x)\frac{x+1}{\Delta_N}\right). \end{aligned} \tag{4.46}$$

Moreover,

$$\mathbf{P}\{\pm\sqrt{N}(\hat{A}(\varphi) - \mathbf{E}\hat{A}(\varphi)) \ge \sigma_N x\} \le \exp\left(-\frac{x^2}{2(1+x/\Delta_N)}\right) \tag{4.47}$$

for all $x \ge 0$. Here $|\tilde{\theta}_i| \le 36/\sqrt{2}$, $i = 1, 2$, and the functions f and L_0 are given by (1.32) *and* (1.33).

An estimator for the spectral density $f(\lambda)$ is customarily chosen to be $\hat{f}(\lambda) = \hat{A}(\varphi_N(\cdot - \lambda))$, where φ_N is some kernel. As before, φ_N and f are assumed to be bounded and not to vanish almost everywhere. To investigate large deviations for the distribution of the r.v.

$$Z_N = \big(\hat{f}(\lambda) - \mathbf{E}\hat{f}(\lambda)\big)\Big/\sqrt{\mathbf{V}\hat{f}(\lambda)}$$

one can apply the theorem just considered with

$$\hat{A}(\varphi) := \hat{A}(\varphi_N(\cdot - \lambda)) := \hat{f}(\lambda). \tag{4.48}$$

In this instance

$$\Delta_N = \frac{N\sqrt{\mathbf{V}\hat{f}(\lambda)}}{4\pi||\varphi_N||_\infty||f||_\infty}. \tag{4.49}$$

For most of the estimators of the spectral density utilized in practice, $||\varphi_N||_\infty/N \to 0$ as $N \to \infty$ and $\sup_N ||\varphi_N||_1 < \infty$.

If the spectral density f is continuous at the point λ and the kernel φ_N satisfies (see Bentkus (1976))

1. $\int \varphi_N(x)dx = 1, \forall N$,
2. $\sup_N ||\varphi_N||_1 < \infty$,
3. $\int\limits_{[-\pi,\pi]-[-\delta,\delta]} |\varphi_N(x)|dx \to 0 \ (N \to \infty), \forall \delta > 0$,
4. $||\varphi_N||_\infty \to \infty$, $||\varphi_N||_\infty/N \to 0 \ (N \to \infty)$,
5. $$\Lambda_1 = \lim_{N\to\infty} \frac{2\pi}{||\varphi_N||_\infty} \int \varphi_N^2(x)dx$$

and

$$\Lambda_2 = \lim_{N\to\infty} \frac{2\pi}{||\varphi_N||_\infty} \int \varphi_N(x)\varphi_N(-x)dx$$

exist, then

$$\varrho^2(\lambda) = \lim_{N\to\infty} N\mathbf{V}\hat{f}(\lambda)/||\varphi_N||_\infty = \begin{cases} \Lambda_1 f^2(\lambda), & \lambda \notin \{-\pi, 0, \pi\}, \\ (\Lambda_1 + \Lambda_2) f^2(\lambda), & \lambda \in \{-\pi, 0, \pi\}. \end{cases}$$

Consequently, according to (4.49), Δ_N behaves like $(\sqrt{N}\varrho(\lambda))/(4\pi||f||_\infty)$. More details about the asymptotic behavior of spectral estimators may be found in the literature (see Bentkus, Rudzkis and Statulevičius (1975), Bentkus (1976), and Bentkus and Rudzkis (1980), (1982)).

Consider now the deviation probabilities for the estimator (4.48). The deviations are measurable in the uniform metric and X_t will not be assumed to be Gaussian. Define the kernel φ_N by $\varphi_N(x) = \psi(x/h)/h$, where ψ is a given continuously differentiable function having compact support, $\int \psi(x)dx = 1$, and $h = h(N) \to 0$ as $N \to \infty$. Let $0 \le x_1 < x_2 \le \pi$ be fixed numbers.

Theorem 4.9 (Rudzkis (1992)). *Let X_t be a strictly stationary sequence, with mean 0 and spectral density $f(\lambda)$, satisfying*

$$||f_k||_\infty \le C^k (k!)^\beta, \quad k = 2, 3, \ldots,$$

where f_k is the k-th order spectral density and β and C are constants. Let g_1 and g_2 be functions independent of N, which are positive and satisfy along with f a Lipschitz condition with exponent $\alpha > 0$ in a neighborhood of the interval (x_1, x_2). Then

$$\sup_{u>0}\left|\mathbf{P}\left\{-ug_1(x) < \frac{\hat{f}(x)-\mathbf{E}\hat{f}(x)}{\sigma_N f(x)} < ug_2(x),\ x_1 < x < x_2\right\}\right.$$
$$\left. - \prod_{j=1}^{2} \exp\left(-c_N \int_{x_1}^{x_2} \exp(-u^2 g_j^2(x)/2)dx\right)\right| \to 0,\ N \to \infty. \qquad (4.50)$$

Here $\sigma_N = ||\psi||_2\sqrt{2\pi/Nh}$, $c_N = ||\psi'||_2/(2\pi||\psi||_2 h)$ *and* $||\cdot||_p$ *is the* L_p *norm.*

The following corollary results when $g_1(x) \equiv g_2(x) \equiv 1$.

Corollary 4.3. *If the hypotheses of Theorem* 4.9 *hold, then*

$$\mathbf{P}\left\{\sup_{x_1<x<x_2} \frac{|\hat{f}(x)-\mathbf{E}\hat{f}(x)|}{\sigma_N f(x)} \le \sqrt{2\ln(2c_N(x_2-x_1))+2y}\right\}$$
$$= \exp(-e^{-y}) + o(1)$$

as $N \to \infty$ *uniformly in* y.

If we denote $\hat{f}(\lambda) - \mathbf{E}\hat{f}(\lambda)$ by $b_N(\lambda)$, then we have

$$\{|\hat{f}(x) - f(x)| < u\} = \left\{-g_1(x,u) < \frac{\hat{f}(x)-\mathbf{E}\hat{f}(x)}{\sigma_N f(x)} < g_2(x,u)\right\}$$

with $g_j(x,u) = (u + (-1)^j b_N(x))/\sigma_N f(x)$. However Theorem 4.9 cannot be used to analyze the distribution of the maximum error since the functions $g_j(x,u)$ depend on N. Nevertheless, (4.50) is also true in this case under additional assumptions.

Theorem 4.10 (Rudzkis (1992)). *Suppose that the sequence* X_t *satisfies the hypotheses of Theorem* 4.9 *and that the spectral density* f *has a derivative of order* $r \ge 1$ *in a neighborhood of the interval* (x_1, x_2) *satisfying a Lipschitz condition with exponent* $\alpha > 0$. *Let* $\int \psi(x)x^k dx = 0$, $k = 1,\dots,r$, *and let* $h = O(N^{-1/(2r+2\alpha+1)})$. *Then* (4.50) *holds with* $ug_j(x)$ *replaced by* $g_j(x,u)$, *or in other words, as* $N \to \infty$

$$\sup_{u>0}\left|\mathbf{P}\left\{\sup_{x_1<x<x_2} |\hat{f}(x) - f(x)| < u\right\}\right.$$
$$\left. - \prod_{j=1}^{2} \exp\left(-c_N \int_{x_1}^{x_2} \exp\{-g_j^2(x,u)/2\}dx\right)\right| \to 0.$$

Chapter 5
The Cumulant Method in the Central Limit Theorem for Sums of Dependent Random Variables

We now clarify how the cumulant method can be used to prove the central limit theorem, to estimate the rate of convergence, and to derive asymptotic expansions for distributions of sums $S_n = X_1 + \ldots + X_n$ of r.v.'s X_t, $t = 1, \ldots, n$, connected in a general Markov chain (for the definition, see §3.1) with $\mathbf{E}X_t = 0$, $\mathbf{E}X_t^2 = \sigma_t^2 < \infty$ and chain ergodic coefficient $\alpha^{(n)} = \min_{1<t\leq n} \alpha_{t-1,t}$. Let $Z_n = S_n/B_n$ and $B_n^2 = \mathbf{E}S_n^2$.

Theorem 5.1. *If $\alpha^{(n)}$ and B_n are positive and*

$$\frac{1}{\alpha^{(n)} B_n^2} \sum_{j=1}^{n} \int_{|x|>\varepsilon\alpha^{(n)} B_n} x^2 dF_{X_j}(x) \to 0 \quad (n \to \infty) \tag{5.1}$$

for each positive ε, then

$$\sup_x |F_{Z_n}(x) - \Phi(x)| \to 0 \quad (n \to \infty).$$

Proof. It is clear that ε may be replaced in (5.1) by ε_n if $\varepsilon_n \to 0$ slowly enough $(n \to \infty)$. Take $C_n = \varepsilon_n \alpha^{(n)} B_n$ and let $X_j = X_j' + X_j''$, where

$$X_j' = \begin{cases} X_j, & \text{if } |X_j| \leq C_n, \\ 0, & \text{if } |X_j| > C_n, \end{cases} \quad j = 1, \ldots, n.$$

Then $S_n = S_n' + S_n''$, where $S_n' = \sum_{j=1}^n X_j'$. Furthermore, let $Z_n' = S_n'/B_n$ and $Z_n'' = S_n''/B_n$. Consequently,

$$\begin{aligned} f_{Z_n}(t) &= \mathbf{E}e^{itZ_n} = \mathbf{E}e^{itZ_n'}(1 + e^{itZ_n''} - 1) = f_{Z_n'}(t) + \theta \frac{|t|}{B_n} \mathbf{E}|S_n''| \\ &= f_{Z_n'}(t) + \theta \frac{|t|}{B_n C_n} \sum_{j=1}^{n} \int_{|x|>C_n} x^2 dF_{X_j}(x) = f_{Z_n'}(t) + o(1) \end{aligned} \tag{5.2}$$

for each fixed value of t. To show that $f_{Z_n'}(t) \to \exp(-\frac{1}{2}t^2)$ $(n \to \infty)$ for each t, it is sufficient that $\Gamma_k(Z_n') \to 0$ $(n \to \infty)$ for each $k \geq 3$ and that

$$\Gamma_2(Z_n') \to 1 \quad (n \to \infty). \tag{5.3}$$

Moreover, one can show that absolute positive constants H_1 and H_2 exist such that

$$|\Gamma_k(Z_n')| \leq \frac{k! H_1 H_2^{k-2} C_n^{k-2}}{\alpha^{(n)k-2} B_n^{k-2}} \tag{5.4}$$

for all $k \geq 3$.

Let

$$r_i = i\left[\frac{1}{\alpha^{(n)}}+1\right], \quad i = 0,1,\ldots,N-1,$$

where N is determined by the inequality

$$r_{N-2} < n \le r_{N-2} + \left[\frac{1}{\alpha^{(n)}}+1\right]$$

($[x]$ denotes the integral part of x).

Put

$$\gamma_i = \mathbf{E}(S'_{r_i} \mid F^{r_i}_{r_i}),$$

$$\tilde{\gamma}_i = \begin{cases} \gamma_i \wedge 2C_n\left[\frac{1}{\alpha^{(n)}}+1\right] & \text{if } \gamma_i \ge 0, \\ \gamma_i \vee -2C_n\left[\frac{1}{\alpha^{(n)}}+1\right] & \text{if } \gamma_i < 0, \end{cases}$$

and

$$\tilde{\varphi}_i = \tilde{\gamma}_{i-1} + S'_{r_{i-1},r_i}, \quad i = 1,\ldots,N-1, \quad \gamma_0 = 0,$$

where

$$S'_{kl} = \sum_{j=k+1}^{l} X_j, \quad 0 \le k < l \le n.$$

Furthermore, let

$$\varphi_i = \tilde{\varphi}_i - \mathbf{E}\tilde{\varphi}_i, \quad i = 1,\ldots,N-1,$$
$$\varphi_N = \tilde{\gamma}_{N-1}.$$

Then

$$S_n = \sum_{j=1}^{N} \varphi_j,$$

and it is not hard to show that (Dobrushin (1956a,b))

$$\mathbf{V}S'_n \ge \beta \sum_{j=1}^{N} \mathbf{V}\varphi_j,$$

where β is an absolute constant $\ge 10^{-4}$. The r.v.'s $\varphi_1,\ldots,\varphi_N$ are connected in a Markov chain with N moments of time and with ergodic coefficients β_{kl} satisfying

$$1-\beta_{kl} \le \begin{cases} (1-\alpha^{(n)})^{(l-k)/\alpha^{(n)}} & \text{for } l-k \ge 2, \\ 1-\alpha^{(n)} \le e^{-\alpha^{(n)}} & \text{for } l-k = 1, \end{cases}$$

$1 \le k < l \le N$. This is so because $1-\alpha(s,t) \le (1-\alpha(s,u))(1-\alpha(u,t))$ for $s \le u \le t$. Therefore by (3.3).

$$|\hat{\mathbf{E}}\varphi_{l_1}\ldots\varphi_{l_{j_k}}| \le \left(\frac{10C_n}{\alpha^{(n)}}\right)^{k-2} e^{k-2}e^{-(l_{j_k}-l_{j_2})}\mathbf{E}|\varphi_{l_{j_1}}\varphi_{l_{j_2}}|$$

when $l_{j_2} - l_{j_1} < 2$ and

$$|\hat{\mathbf{E}}\varphi_{l_1}\dots\varphi_{l_{j_k}}| \leq \left(\frac{10C_n}{\alpha^{(n)}}\right)^{k-2} e^{(k-2)/2} e^{-(l_{j_k}-l_{j_1})/2}\sqrt{\mathbf{E}\varphi_{l_1}^2\mathbf{E}\varphi_{l_2}^2}$$

when $l_{j_2} - l_{j_1} \geq 2$ with $l_1 \leq l_2 \leq \ldots \leq l_{j_k}$. Next

$$\sum_{1\leq l_1<l_2\leq N} \mathbf{E}|\varphi_{l_1}\varphi_{l_2}| \leq \frac{1}{2}\sum_{1\leq l_1<l_2\leq N}(\mathbf{V}\varphi_{l_1}+\mathbf{V}\varphi_{l_2}) \leq \sum_{l=1}^{N}\mathbf{V}\varphi_l,$$

$$\sum_{\substack{1\leq l_1<l_2\leq N\\ l_2-l_1>2}} e^{-(l_2-l_1)/2}\sqrt{\mathbf{E}\varphi_{l_1}^2\mathbf{E}\varphi_{l_2}^2}$$

$$\leq \frac{1}{2}\sum_{1\leq l_1<\leq l_2\leq N} e^{-(l_2-l_1)/2}(\mathbf{V}\varphi_{l_1}+\mathbf{V}\varphi_{l_2}) \leq \frac{1}{2(1-e^{-(1/2)})}\sum_{l=1}^{N}\mathbf{V}\varphi_l$$

and $\sum_{l=1}^{N}\mathbf{V}\varphi_l \leq \beta^{-1}\mathbf{V}S_n'$. Then by procedures such as in the proof of Theorem 3.15, we can conclude that there are absolute positive constants H_1' and H_2' such that

$$|\Gamma_k(S_n')| \leq \frac{k!H_1'H_2'^{k-2}C_n^{k-2}\mathbf{V}S_n'}{\alpha^{(n)k-2}}. \tag{5.5}$$

For the r.v.'s in the Markov chain,

$$\frac{1}{8}\sum_{j=1}^{n}\min\{a_j,\ a_{j+1}\}\mathbf{V}X_j \leq \mathbf{V}S_n \leq \frac{8}{\alpha^{(n)}}\sum_{j=1}^{n}\sum_{j=1}^{n}\mathbf{V}X_j,$$

where $\alpha_j = \alpha(j-1,\ j)$, $j = 1,\ldots,n$. Therefore

$$\mathbf{V}S_n'' \leq \frac{16}{\alpha^{(n)}}\sum_{j=1}^{n}\mathbf{V}X_j'' \leq \frac{16}{\alpha^{(n)}}\sum_{j=1}^{n}\int_{|x|>C_n} x^2 dF_{X_j}(x) = o(B_n^2).$$

Thus, $\mathbf{V}S_n'/B_n^2 \to 1$ $(n\to\infty)$ and so (5.3) holds. Referring to (5.5), we see that so does (5.4). The theorem has been proved.

Theorem 5.2. *If the r.v. X_j has finite moments $\mathbf{E}|X_j|^s$, $j=1,\ldots,n$, for some integer $s\geq 3$ and if $\alpha^{(n)}>0$, then there exists an absolute positive constant C such that*

$$\sup_x |F_{Z_n}(x)-\Phi(x)| \leq C\left(L_{sn}^{1/(s-2)} + L_{sn}\ln^{s/2}(1+L_{sn}^{-1/(s-2)})\right) \tag{5.6}$$

with

$$L_{sn} = \frac{\sum_{j=1}^{n}\mathbf{E}|X_j|^s}{\alpha^{(n)s-1}B_n^s}.$$

Theorem 5.3. *If $|X_j| \le C^{(n)}$ with probability 1, $j = 1, \dots, n$, and if $\alpha^{(n)} > 0$, then there is an absolute positive constant C' such that*

$$\sup_x |F_{Z_n}(x) - \Phi(x)| \le C' \frac{C^{(n)}}{\alpha^{(n)} B_n}. \tag{5.7}$$

Moreover, there exist absolute positive constants H_3 and H_4 such that

$$|\Gamma_k(Z_n)| \le \frac{k! H_3 H_4^{k-2} {C^{(n)}}^{k-2}}{{\alpha^{(n)}}^{k-2} B_n^{k-2}}, \tag{5.8}$$

and thus F_{Z_n} satisfies (1.31) and the large deviation inequality (1.39) with $\Delta = \alpha^{(n)} B_n / (H_5 C^{(n)})$ and $\gamma = 0$.

The proof of Theorem 5.2 is similar to that of Theorem 5.1 if X_j is truncated at the level

$$C_n = 64^{1/(s-2)} \alpha^{(n)} B_n L_{sn}^{1/(s-2)}.$$

Use is made of the representation $S_n' = \sum_{j=1}^N \varphi_j$ such that $\mathbf{V} S_n' \ge \beta \sum_{j=1}^N \mathbf{V} \varphi_j$ with β an absolute positive constant and $|\varphi_k| \le 10 C_n / \alpha^{(n)}$. Then

$$\mathbf{V} S_n'' \le \frac{16}{\alpha^{(n)}} \sum_{j=1}^n \mathbf{V} X_j'' \le \frac{16}{\alpha^{(n)} C_n^{s-2}} \sum_{j=1}^n \mathbf{E}|X_j|^s = \frac{1}{4} B_n^2,$$

$$\frac{1}{2} B_n^2 \le \mathbf{V} S_n' \le \frac{3}{2} B_n^2,$$

$$\frac{|\Gamma_k(S_n')|}{B_n^k} \le k! H_0^{k-2} L_{sn}^{(k-2)/(s-2)}.$$

Therefore

$$\ln f_{Z_n'}(t) = \sum_{k=2}^s \frac{\Gamma_k(S_n')}{k!} \left(\frac{it}{B_n} \right)^k + 4 s! H_0^{s-2} H_6^s |t|^{s+1} \rho_n^{-1} L_{sn},$$

and

$$f_{Z_n}(t) = f_{Z_n'}(t) + s! H_0^{s-2} |t| \rho_n^{s-1} L_{sn}$$

with

$$|t| \le \frac{\rho_n}{2\sqrt{2} H_0'},$$

where

$$\rho_n = 2\sqrt{2} H_6 a \sqrt{\ln(1 + L_{sn}^{-1/(s-2)})}$$

for any $a > 0$ and H_6 an absolute constant.

We finally obtain

$$|f_{Z_n}(t) - \exp(-t^2/2)| \le H_s H_0 |t| L_{sn}^{1/(s-2)} + H_s' |t| \rho_n^{s-1} L_{sn} \tag{5.9}$$

providing

$$|t| \le \frac{\rho_n}{2\sqrt{2}H_6} \wedge \frac{L_{sn}^{-1/(s-2)}}{2H_0}.$$

Here H_s is a quantity depending on s.

It is also possible to show that

$$|f_{Z_n}(t)| \le \exp\left(-\frac{1}{12\pi^2}t^2\right) \text{ for } |t| \le cL_{sn}^{1/(s-2)}, \tag{5.10}$$

where c is an absolute positive constant.

From (5.9) and (5.10) it follows by Esseen's theorem that

$$|F_{Z_n}(x) - \Phi(x)| \le \frac{\varepsilon}{\pi} + \frac{24A}{\pi T}$$

for any positive T with

$$\varepsilon = \int_{-T}^{T} (|f_{Z_n}(t) - e^{-t^2/2}|)/|t| dt$$

and $A = \sup_x |\Phi'(x)|$. This leads to the proof of (5.6) with $T = cL_{sn}^{-1/(s-2)}$. Then $\varepsilon \le C_1(L_{sn}^{1/(s-2)} + \rho_n^s L_{sn})$, where C_1 is an absolute constant. Inequality (5.8) is derived similarly to (5.4). Resorting to the general large deviation lemmas, we obtain Theorem 5.3 from this.

A more detailed presentation of the material as well as asymptotic expansions for F_{Z_n} may be found in the papers by Statulevičius (1965), (1969a,b), (1970).

We conclude with several remarks on the case where the approximating law is not Gaussian.

Suppose that the approximating distribution is for some r.v. η and that $z_1(z) = \ln \mathbf{E} \exp(z\eta)$ is analytic at $z = 0$. Introduce the k-th order cumulant $\Gamma_k(\xi \,||\, \eta)$ of r.v. ξ relative to η as follows:

$$\ln \mathbf{E} e^{it\xi} = \sum_{l=1}^{k} \frac{1}{l!} \Gamma_l(\xi \,||\, \eta) z_1^l(it) + o(|t|^k)$$

if $\mathbf{E}|\xi^k| < \infty$. Then one can often succeed in estimating the large deviation probabilities $\mathbf{P}\{\xi \ge x\}$ by means of $\mathbf{P}\{\eta \ge x\}$ if bounds are known for $\Gamma_k(\xi \,||\, \eta)$ such as

$$|\Gamma_k(\xi \,||\, \eta)| \le (k!)^{1+\gamma}/\Delta_n^{k-1} \tag{5.11}$$

for all $k \ge 2$. Obviously,

$$\Gamma_k(\eta \,||\, \eta) = \begin{cases} 1, & k = 1, \\ 0, & k > 1. \end{cases}$$

Sometimes the analyticity of $z_1(z)$ at $z = 0$ is optional and it is enough to know an upper bound for moments, as for example, $|\mathbf{E}\eta^k| \le (k!)^{1+\gamma'} H_0^k$ for all $k \geq 1$. Let us state some results when η has a Poisson distribution with parameter λ. There is no loss of generality in assuming that $\lambda = 1$. Then

$$z_1 = z_1(t) = \exp(it) - 1$$

and

$$\mathbf{E}\exp(it\xi) = \mathbf{E}(1+z_1)^{\xi} = 1 + \sum_{l=1}^{k} \frac{1}{l!}\mathbf{E}(\xi)_l z_1^l + o(|t|^k),$$

where $\mathbf{E}(\xi)_l$ is the l-th order factorial moment. Here

$$(x)_l := x(x-1)\dots(x-(l-1)).$$

It is not hard to see that

$$\begin{aligned}\tilde{\Gamma}_k(\xi) :&= \Gamma_k(\xi\,||\,\eta)\\ &= \sum_{\nu=1}^{k} \frac{(-1)^{\nu-1}}{\nu} \sum_{k_1+\dots+k_\nu=k} \frac{k!}{k_1!\dots k_\nu!}\mathbf{E}(\xi)_{k_1}\dots\mathbf{E}(\xi)_{k_\nu}. \quad (5.12)\end{aligned}$$

Let $s(k,l)$ be the Stirling numbers of the first kind, that is,

$$(x)_k = \sum_{l=1}^{k} s(k,l)x^l.$$

Then resorting to the relations

$$(1+y)^x = \sum_{k=0}^{\infty}\sum_{l=0}^{k} s(k,l)\frac{1}{k!}y^k x^l$$

and

$$\frac{1}{l!}(\ln(1+y))^l = \sum_{k=l}^{\infty} s(k,l)\frac{1}{k!}y^k,$$

we obtain

$$\tilde{\Gamma}_k(\xi) = \sum_{l=1}^{k} s(k,l)\Gamma_l(\xi).$$

Recall that $\Gamma_k(\eta) = \lambda$ and $\tilde{\Gamma}_k(\eta) = 0$ for all $k \geq 2$. The numbers $s(k,l)$ have the following combinatorial interpretation.

If

$$\sum_{l=1}^{k} C(k,l)x^l = x(x+1)\dots(x+k-1),$$

then

$$s(k,l) = (-1)^{k+l}C(k,l)$$

where $C(k,l)$ is the number of permutations of order k having l cycles. $\tilde{\Gamma}_k(\xi)$ is known as the k-th factorial cumulant of ξ. Many of the bounds obtained for $\Gamma_k(\xi)$ also hold for $\tilde{\Gamma}_k(\xi)$. For example, it follows from (5.12) that if

$$|\mathbf{E}\xi| \le p, \quad p \le 1/\Delta,$$

and

$$\mathbf{E}|(\xi)_k| \le (k!)^{1+\gamma} p/\Delta^{k-1}$$

for all $k \ge 2$ and some $\gamma \ge 0$, then

$$|\tilde{\Gamma}_k(\xi)| \le (k!)^{1+\gamma} 2^{k-1} p/\Delta^{k-1}.$$

Let $X_t^{(n)}$, $t = 1, 2, \ldots$, $n = 1, 2, \ldots$, be a double array of r.v.'s defined on probability space $(\Omega, \mathcal{F}, \mathbf{P})$ with a family of σ-algebras $\mathcal{F}_s^t$.

Theorem 5.4. *If* $\mathbf{E}X_t^{(n)} = p_t^{(n)}$,

$$|\mathbf{E}(X_t^{(n)})_k| \le k! p_t^{(n)}/\Delta_n^{k-1}, \quad p_t^{(n)} \le 1/\Delta_n,$$

for all $t = 1, 2, \ldots, n$, $k \ge 2$ *and a sequence* $\Delta_n \uparrow \infty$ $(n \to \infty)$, *and if* $\mathcal{F}_s^t$ *is* ψ*-mixing with*

$$\psi(s,t) \le C \exp(-b(t-s)),$$

$C > 0$, $b > 0$, *then*

$$\mathbf{P}\{S_n \ge x\} = \mathbf{P}\{\eta_n \ge x\}(1 + o(1))$$

in the interval

$$0 \le x < \varepsilon\sqrt{\Delta_n},$$

where $S_n = \sum_{t=1}^n X_t^{(n)}$, $\sum_{t=1}^n p_t^{(n)} = \lambda_n$ *and* η_n *is Poisson-distributed with parameter* λ_n; ε *is a positive constant depending on* b *and* C. *Moreover,*

$$|\tilde{\Gamma}_k(S_n)| \le \frac{(k!)^2 \lambda_n H^{k-1}}{\Delta_n^{k-1}},$$

where H *depends on* b *and* C.

Let the approximating law η be infinitely divisible:

$$\log \mathbf{E}e^{it\eta} = \lambda \left(\int_{-\infty}^0 (e^{itx} - 1) dM(x) + \int_0^\infty (e^{itx} - 1) dN(x) \right)$$

(see Gnedenko and Kolmogorov (1949)). If $\mathbf{E}|\eta^k| < \infty$ for all $k \ge 1$, then it is advantageous to use

$$\tilde{\tilde{\Gamma}}_k(\xi) = \int_{-\infty}^0 \tilde{\Gamma}_k(\xi/x) dM(x) + \int_0^\infty \tilde{\Gamma}_k(\xi/x) dN(x)$$

when investigating large deviation probabilities.

A general lemma and how it is used to derive an asymptotic expansion for the distribution function of an integer-valued r.v. ξ in the case of Poisson approximation are found in Bikelis and Žemaitis (1974), (1976) and Aigner (1979). Aleškivičienė (1988) investigated large deviations for double arrays of independent and identically distributed integer-valued r.v.'s.

References*

Aigner, M. (1979): Combinatorial Theory. Springer-Verlag, Heidelberg New York. Zbl. 415.05001

Akhmedov, S.A. (1990): Nonuniform estimates in the central limit theorem for dependent variables. Liet. Mat. Rink. **30**, No. 4, 623–629. English transl.: Lith. Math. J. **30**, No. 4, 285–295. Zbl. 723.60021

Aleškevičienė, A. (1988): Probabilities of large deviations with Poisson approximation. Liet. Mat. Rink. **28**, No. 1, 3–13. English transl.: Lith. Math. J. **28**, No. 1, 1–8. Zbl. 714.60022

Aleškevičienė, A. (1991a): Large and moderate deviations for L-statistics. Liet. Mat. Rink. **31**, No. 2, 145–156. Zbl. 787.60035

Aleškevičienė, A. (1991b): Large deviations for U- and L-Statistics. Teor. Veroyatn. Primen. **36**, No. 4, 774–775. English transl.: Theory Probab. Appl. **36**, No. 4, 806–808. Zbl. 776.60001

Aleškevičienė, A., and Statulevičius, V. (1995): Large deviations in power zones in the approximation by the Poisson law. Usp. Mat. Nauk **50**, 63–82. English transl.: Russ. Math. Sov. **50**, No. 5, 905–924. Zbl. 867.60008

Aleškevičienė, A., and Statulevičius, V. (1997): Probabilities of large deviations in the approximation by $\mathcal{X}^2$-law. Lith. Math. J. **37**, No. 4.

Aleškevičienė, A., and Statulevičius, V. (1998): Theorems on large deviations for non-Gaussian approximation. Probab. Theory and Math. Statist., Proceedings of the Seventh Vilnius Conf., 5–14.

Aleškevičienė, A., and Statulevičius, V. (1999): Theorems on large deviations in the approximation by an infinitely divisible law. Acta Applicandae Mathematicae **58**, 1–13.

Amosova, N.N. (1989): On the necessity of Cramér's condition in local limit theorems. Probab. Theory and Math. Statistics. Proc. of the Fifth Vilnius Conference, 1. VSP, Utrecht, Mokslas, Vilnius, pp. 52–56. Zbl. 731.60021

Andrews, G.E. (1976): The Theory of Partitions. Addison-Wesley (Reprint 1998), Reading, Mass. Zbl. 371.10001

Arak, T.V., and Zaitsev, A.Yu. (1986): Uniform limit theorems for sums of independent random variables. Tr. Mat. Inst. Steklova Akad. Nauk SSSR, Nauka, Leningrad. English transl.: Proc. Steklov Math. Inst. vol. 174. Zbl. 606.60028

* For the convenience of the reader, references to reviews in Zentralblatt für Mathematik (Zbl.), compiled by means of the MATH database, and Jahrbuch über die Fortschritte der Mathematik (Jbuch) have, as far as possible, been included in this bibliography.

Bahadur, R.R. (1960): On the asymptotic efficiency of tests and estimates. Sankhya **22**, No. 3–4, 229–252. Zbl. 109.12503

Bahr, B. von (1967): Multidimensional integral limit theorems for large deviations. Ark. Mat. **7**, No. 1, 89–99. Zbl. 221.60015

Basalykas, A. (1984): Some asymptotic properties of Pitman-Linnik polynomial estimators. Liet. Mat. Rink. **24**, No. 2, 16–29 (Russian). Zbl. 581.62029

Basalykas, A. (1985): Some asymptotic properties of Pitman-Linnik polynomial and modified polynomial estimators. Dokl. Akad. Nauk SSSR **280**, No. 5, 1037–1039. English transl.: Sov. Math., Dokl. **31**, No. 5, 178–180. Zbl. 596.62037

Basalykas, A. (1988): Some asymptotic properties of distributions of multinomial forms. Liet. Mat. Rink. **28**, No. 4, 644–654 (Russian). Zbl. 673.60032

Basalykas, A., Plikusas, A., and Statulevičius, V. (1987): Theorems of large deviations for multinomial forms and multiple stochastic integrals. Proc. of the 1st World Congress of the Bernoulli Soc., 1987. VNU Science Press, Utrecht, pp. 629–639. Zbl. 679.60038

Bentkus, R. (1972a): On the error in estimating the spectral function of a stationary process. Liet. Met. Rink. **12**, No. 1, 55–71 (Russian). Zbl. 253.62050

Bentkus, R. (1972b): On asymptotic normality in estimating a spectral function. Liet Mat. Rink. **12**, No. 3, 5–18 (Russian). Zbl. 253.62051

Bentkus, R. (1976): Cumulants of spectral estimates of a stationary time series. Liet. Mat. Rink. **16**, No. 4, 37–61. English transl.: Lith. Math. J. **16**, No. 4, 501–518. Zbl. 373.62056

Bentkus, R., and Rudzkis, R. (1976): Large deviations for spectral estimates of a stationary Gaussian sequence. Liet. Mat. Rink. **16**, No. 4, 63–77. English transl.: Lith. Math. J. **16**, No. 4, 519–529. Zbl. 401.62075

Bentkus, R., and Rudzkis, R. (1980): On exponential bounds for distributions of random variables. Liet. Mat. Rink. **20**, No. 1, 15–30 (Russian). Zbl. 428.60027

Bentkus, R., and Rudzkis, R. (1982): On the distribution of some statistical estimates of spectral density. Teor. Veroyatn. Primen. **27**, No. 4, 739–756. English transl.: Theory Probab. Appl. **27**, No. 4, 795–814. Zbl. 507.62077

Bentkus, R., Rudzkis, R., and Statulevičius, V. (1975): Exponential inequalities for spectral estimates of a stationary Gaussian sequence. Liet. Mat. Rink. **15**, No. 3, 25–39. English transl.: Lith. Math. J. **15**, No. 3, 392–402. Zbl. 333.60045

Bentkus, V.Yu. (1986): Large deviations in Banach spaces. Teor. Veroyatn. Primen. **31**, No. 4, 710–716. English transl.: Theory Probab. Appl. **31**, No. 4, 627–632. Zbl. 623.60012

Bhattacharya, R.N., and Puri, M.L. (1983): On the order of magnitude of cumulants of von Mises functionals and related statistics. Ann. Probab. **11**, No. 2, 346–354. Zbl. 527.62025

Bhattacharya, R.N., and Ranga Rao, R. (1976): Normal Approximation and Asymptotic Expansions. John Wiley, New York London Toronto. Zbl. 331.41023

Bikelis, A., and Žemaitis, A. (1974): Asymptotic expansions for large deviation probabilities, LL. Liet. Mat. Rink. **14**, No. 1, 45–52. English transl.: Lith. Mat. J. **14**, No. 1, 567–572. Zbl. 326.60027

Bikelis, A., and Žemaitis, A. (1976): Asymptotic expansions for large deviation probabilities, III. Liet. Mat. Rink. **16**, No. 3, 31–50. English transl.: Lith. Math. J. **16**, No. 3, 332–348. Zbl. 353.60033

Blum, J.R., Hanson, D.L., and Koopmans, L.H. (1963): On the strong law of large numbers for a class of stochastic processes. Z. Wahrscheinlichkeitstheorie Verw. Geb. **2**, No. 1, 1–11. Zbl. 117.35603

Book, S.A. (1973): A large deviation theorem for weighted sums. Z. Wahrscheinlichkeitstheorie Verw. Geb. **26**, No. 1, 43–49. Zbl. 252.60013

Borovkov, A.A. (1964a): Analysis of large deviations in boundary problems with arbitrary boundaries, I. Sib. Mat. Zh. **2**, 253–289 (Russian). Zbl. 135.19301

Borovkov, A.A. (1964b): Analysis of large deviations in boundary problems with arbitrary boundaries, II. Sib. Mat. Zh. **5**, 750–767 (Russian). Zbl. 178.53801

Borovkov, A.A. (1967): Boundary problems for random walks and large deviations in function spaces. Teor. Veroyatn. Primen. **12**, No. 4, 635–654. English transl.: Theory Probab. Appl. **12**, No. 4, 575–595. Zbl. 178.20004

Borovkov, A.A. (1983): Boundary problems, invariance principle, and large deviations. Usp. Mat. Nauk **38**, No. 4, 227–254. English transl.: Russ. Math. Surv. **38**, No. 4, 259–290. Zbl. 533.60026

Borovkov, A.A., and Mogul'skii, A.A. (1978): Probabilities of large deviations in topological spaces, I. Sib. Mat. Zl. **19**, No. 5, 988–1004. English transl.: Sib. Math. J. **19**, No. 5, 697–709. Zbl. 397.60029

Borovkov, A.A., and Mogul'skii, A.A. (1980): Probabilities of large deviations in topological spaces, II. Sib. Mat. Zh. **21**, No. 5, 12–26. English transl.: Sib. Math. J. **21**, No. 1, 653–664. Zbl. 446.60016

Borovkov, A.A., and Mogul'skii, A.A. (1985): Uniform theorems on large deviations of sums of random vectors. Preprint, Akad. Nauk SSSR, Sibirsk. Otdel., Inst. Math. **21**. Novosibirsk (Russian).

Borovskikh, Yu.V. (1980): Problem of approximating the distributions of U-statistics and von Mises functionals. II. Preprint, Inst. Math., Ukrainsk. SSR, Kiev, pp. 31–36 (Russian).

Cramér, H. (1938): Sur un nouveau théorème limite de la théorie de probabilités. Act. Sci. Ind. **736**.

Dasgupta, R. (1984): On large deviation probabilities of U-statistics in the non i.i.d. case. Sankhya **A46**, No. 1, 110–116. Zbl. 565.60023.

Davydov, Yu.A. (1968): Convergence of distributions generated by stationary stochastic processes. Teor. Veroyatn. Primen. **13**, No. 4, 730–737. English transl.: Theory Probab. Appl. **13**, No. 4, 691–696. Zbl. 174.49201

Dobrushin, R.L. (1953): Limit laws for Markov chains. Izv. Akad. Nauk SSSR, Ser. Mat. **17**, 291–330 (Russian). Zbl. 052.14301

Dobrushin, R.L. (1956a): Central limit theorems for nonstationary Markov chains. I. Teor. Veroyatn. Primen. **1**, No. 1, 72–89. English transl.: Theory Probab. Appl. **1**, No. 1, 65–80. Zbl. 093.15001

Dobrushin, R.L. (1956b): Central limit theorems for nonstationary Markov chains, II. Teor. Veroyatn. Primen. **1**, No. 4, 365–425. English transl.: Theory Probab. Appl. **1**, No. 4, 329–383. Zbl. 093.15001

Donsker, M.D., and Varadhan, S.R.S. (1975a): Asymptotic evaluation of certain Markov process expectations for large time. Commun. Pure Appl. Math. **28**, No. 1, 1–47. Zbl. 323.60069

Donsker, M.D., and Varadhan, S.R.S. (1975b): Asymptotic evaluation of certain Markov process expectations for large time. Commun. Pure Appl. Math. **28**, No. 2, 279–301. Zbl. 348.60031

Donsker, M.D., and Varadhan, S.R.S. (1976): Asymptotic evaluation of certain Markov process expectations for large time. Commun. Pure Appl. Math. **29**, No. 4, 389–461. Zbl. 348.60032

Engel, D. (1982): The multiple stochastic integral. Mem. Am. Math. Soc. **38**, No. 265, 1–82. Zbl. 489.60064

Feller, W. (1969): Limit theorems for probabilities of large deviations. Z. Wahrscheinlichkeitstheorie Verw. Geb. **14**, No. 1, 1–20. Zbl. 183.47402

Formanov, Sh.K. (1973): Some limit theorems on large deviations for homogeneous Markov chains. Stochastic Process and Statist. Inference, vol. 3, Tashkent, pp. 173–185 (Russian). Zbl. 276.60031

Fortus, M.I. (1957): A uniform limit theorem for distributions attracted to a stable law with index less than one. Teor. Veroyatn. Primen. **2**, No. 4, 486–487. English transl.: Theory Probab. Appl. **2**, No. 4, 478–479.

Fuk, D.Kh., and Nagaev, S.V. (1971): Probability inequalities for sums of independent random variables. Teoriya Veroyat. Primen. **16**, No. 4, 660–675. English transl.: Theory Probab. Appl. **16**, No. 4, 643–660. Zbl. 259.60024

Gnedenko, B.V., and Kolmogorov, A.N. (1949): Limit Distributions for Sums of Independent Random Variables. Nauka, Moscow-Leningrad. English transl.: Addison-Wesley, Reading, Mass. 1954. Zbl. 056.36001

Heinrich, L. (1985a): Some estimates of the cumulant-generating function of a sum of m dependent random vectors and their application to large deviations. Math. Nachr. **120**, 91–101. Zbl. 559.60006

Heinrich, L. (1985b): Non-uniform estimates, moderate and large deviations in the central limit theorem for m-dependent random variables. Math. Nachr. **121**, 107–121. Zbl. 572.60031

Heyde, C.C. (1968): On large deviation probabilities in the case of attraction to a nonnormal stable law. Sankhya **30**, No. 3, 253–258. Zbl. 182.22903

Hoeffding, W. (1963): Probability inequalities for sums of bounded random variables. J. Am. Stat. Assoc. **58**, 13–30. Zbl. 127.10602

Hoeffding, W. (1967): On probabilities of large deviations. Proc. 5th Berkeley Symp. on Math. Statist. and Probab. I, Univ. Calif. Press, Berkeley-Los Angeles, 203–219. Zbl. 211.50401

Ibragimov, I.A. (1959): Some limit theorems for strictly stationary probability processes. Dokl. Akad. Nauk. SSSR **125**, No. 4, 711–714 (Russian). Zbl. 087.13302

Ibragimov, I.A. (1966): On the accuracy of Gaussian approximation to the distribution functions of sums of independent variables. Teor. Veroyatn. Primen. **11**, No. 4, 632–655. English transl.: Theory Probab. Appl. **11**, No. 4, 559–579. Zbl. 161.15207

Ibragimov, I.A. (1967): On the Chebyshev-Cramér asymptotic expansions. Teor. Veroyatn. Primen. **12**, No. 3, 506–519. English transl.: Theory Probab. Appl. **12**, No. 3, 455–469. Zbl. 201.51001

Ibragimov, I.A., and Linnik, Yu.V. (1965): Independent and Stationary Sequences of Random Variables. Nauka, Moscow. English transl.: Wolters-Noordhoff, Groningen, 1971. Zbl. 154.42201

Iosifescu, M. (1980): Recent advances in mixing sequences of random variables. Proc. 3rd Intern. Summer School on Probab. Theory and Math. Statist., 1978, Sofiar-Varna, 111–138. Zbl. 435.60024

Itô, K. (1951): Multiple Wiener integral. J. Math. Soc. Japan **3**, No. 1, 157–169. Zbl. 044.12202

Jakimavičius, D. (1988): Estimates of cumulants and centered moments of mixing stochastic processes. Liet. Mat. Rink. **28**, No. 3, 614–626. English transl.: Lith. Math. J. **28**, No. 3, 308–317. Zbl. 659.60050

Jakimavičius, D., and Statulevičius, V. (1987): Estimates of Cumulants and Centered Moments of Mixing Random Processes. Liet. Mat. Rink. **30**, No. 1, 112–119; No. 2, 360–375. English transl.: Lith. Math. J. **30**, No. 1, 67–80; No. 2, 179–190. Zbl. 666.60027, Zbl. 666.60028

Jakševičius, Š. (1983a): Asymptotic expansions for probability distributions, I. Liet. Mat. Rink. **23**, No. 3, 196-213. English transl.: Lith. Math. J. **23**, No. 3, 341–353. Zbl. 356.60033

Jakševičius, Š. (1983b): Asymptotic expansions for probability distributions, II. Liet. Mat. Rink. **23**, No. 4, 73–83. English transl.: Lith. Math. J. **23**, No. 4, 410–418. Zbl. 571.60032

Jakševičius, Š. (1984): Asymptotic expansions for probability distributions, III. Liet. Mat. Rink. **24**, No. 4, 216–213 (Russian). Zbl. 571.60033

Jakševičius, Š. (1985): Asymptotic expansions for probabilities distributions, IV. Liet. Mat. Rink. **25**, No. 1, 194–208 (Russian). Zbl. 571.60034

Kagan, A.M. (1966): On the estimation theory of location parameters. Sankhya **28**, 335–352. Zbl. 156.39207

Kagan, A.M., Klebanov, L.B., and Fintushal, S.M. (1974): Asymptotic behavior of Pitman polynomial estimators. Zap. Nauch. Semin. LOMI **43**, 30–39 (Russian). Zbl. 358.62028

Kolmogorov, A.N. (1953): Some recent papers in the area of probability theory. Vestn., Mosk. Univ. **8**, No. 10, 29–38 (Russian). Zbl. 052.36202

Kolmogorov, A.N. (1974): Foundations of the Theory of Probability. Nauka, Moscow. English transl. of 1933 German ed.: Chelsea Publishing Co., New York, 2nd edn., 1956. Zbl. 007.21601

Kolmogorov, A.N., and Fomin, S.V. (1976): Elements of Function Theory and Functional Analysis. Nauka, Moscow (Russian). Zbl. 672.46001

Kubilius, J. (1962): Probabilistic Method in Number Theory. Gospolitnauchizdat, Vilnius, English transl.: American Math. Society, Providence, 1964. Zbl. 127.27402

Leonov, V.P. (1964): Applications of Higher Cumulants to the Theory of Stochastic Processes. Nauka, Moscow (Russian). Zbl. 142.14102

Leonov, V.P., and Shiryaev, A.N. (1959): On a method of calculating cumulants. Teor. Veroyatn. Primen. **4**, No. 3, 342–355. English transl.: Theory Probab. Appl. **4**, No. 3, 319–329. Zbl. 087.33701

Linnik, Yu.V. (1960): New limit theorems for sums of independent random variables. Dokl. Akad. Nauka SSSR **133**, No. 6, 1291–1293. English transl.: Sov. Math., Dokl. **1**, No. 6, 972–974. Zbl. 094.32402

Linnik, Yu.V. (1961a): Limit theorems for sums of independent variables taking large deviations into account, I. Theor. Veroyatn. Primen. **6**, No. 2, 145–162. English transl.: Theory Probab. Appl. **6**, No. 2, 131–148. Zbl. 107.13202

Linnik, Yu.V. (1961b): Limit theorems for sums of independent variables taking large deviations into account, II. Teor. Veroyatn. Primen. **6**, No. 4, 377–391. English transl.: Theory Probab. Appl. **6**, No. 4, 345–360. Zbl. 107.13202

Linnik, Yu.V. (1962): Limit theorems for sums of independent variables taking large deviations into account, III. Teor. Veroyatn. Primen. **7**, No. 2, 121–134. English transl.: Theory Probab. Appl. **7**, No. 2, 115–129. Zbl. 107.13202

Liptser, R.Sh., and Shiryaev, A.N. (1989): The Theory of Martingales. Kluwer, Dordrecht-Bosston. Zbl. 728.60048

Malevich, T.L., and Abdalimov, B. (1979): Large deviation probabilities for *U*-statistics. Teor. Veroyatn. Primen. **24**, No. 1, 215–220. English transl.: Theory Probab. Appl. **24**, No. 1, 215–219. Zbl. 396.60031

McDonald, D. (1979): A local limit theorem for large deviations of sums of independent and nonidentically distributed random variables. Ann. Probab. **7**, No. 3, 526–531. Zbl. 421.60026

McLeish, D.L. (1975): Invariance principles for dependent variables. Z. Wahrscheinlichkeitstheorie Verw. Geb. **32**, No. 3, 165–178. Zbl. 305.60010

Mogulskii, A.A. (1975): Large deviations in trajectory space for a sequence of processes with stationary increments. Sib. Mat. Zh. **16**, No. 2, 314–327. English transl.: Sib. Math. J. **16**, No. 2, 242–252. Zbl. 342.60024

Nagaev, A.V. (1967): Local limit theorems taking large deviations into account. Limit Theorems and Stochastic Processes, Tashkent, pp. 71–88 (Russian).

Nagaev, A.V. (1969a): Integral limit theorems taking large deviations into account with Cramér's condition unfulfilled, I. Teor. Veroyatn. Primen. **14**, No. 1, 51–63. English transl.: Theory Probab. Appl. **14**, No. 1, 51–64. Zbl. 172.21901

Nagaev, A.V. (1969b): Integral limit theorems taking large deviations into account with Cramér's condition unfulfilled, II. Teor. Veroyatn. Primen. **14**, No. 2, 203–216. English transl.: Theory Probab. Appl. **14**, No. 2, 193–208. Zbl. 196.21003

Nagaev, S.V. (1961): Refinements of limit theorems for homogeneous Markov chains. Teor. Veroyatn. Primen. **6**, No. 1, 67–86. English transl.: Theory Probab. Appl. **6**, No. 1, 62–81. Zbl. 116.10602

Nagaev, S.V. (1963): An integral limit theorem for large deviations. Dokl. Akad. Nauk. SSSR **148**, No. 2, 280. English transl.: Sov. Math., Dokl. **4**, 79–80. Zbl. 128.38102

Nagaev, S.V. (1965): Some limit theorems for large deviations. Teoriya Veroyat. Primen. **10**, No. 2, 231–251. English transl.: Theory Probab. Appl. **10**, No. 2, 214–235. Zbl. 144.18704

Nagaev, S.V. (1979): Large deviations of sums of independent random variables. Ann. Probab. **7**, No. 5, 745–789. Zbl. 418.60033

Nagaev, S.V., and Pinelis, I.F. (1977): Some inequalities for distributions of sums of independent random variables. Teor. Veroyatn. Primen. **22**, No. 2, 254–263. English transl.: Theory Probab. Appl. **22**, No. 2, 248–256. Zbl. 378.60036

Nagaev, S.V., and Sakoyan, I.F. (1976): On a bound for large deviation probabilities. Limit Theorems and Math. Statist., Tashkent, pp. 132–140 (Russian).

Osipov, L.V. (1978): Large deviation probabilities for sums of independent random vectors. Teor. Veroyatn. Primen. **29**, No. 3, 510–526. English transl.: Theory Probab. Appl. **29**, No. 3, 490–506. Zbl. 437.60005

Osipov, L.V. (1982): Large deviation probabilities of sums of independent random vectors for certain classes of sets. Mat. Zametki **32**, No. 1, 147–153. English transl.: Math. Notes **31**, 75–79. Zbl. 488.60036

Padvelskis, K., and Statulevičius, V. (1998): Theorems on large deviations for sums of random variables connected in a Markov chain, I. Liet. Mat. Rink. **38**, No. 4, 456–471.

Padvelskis, K., and Statulevičius, V. (1999): Theorems on large deviations for sums of random variables connected in a Markov chain, II. Liet. Mat. Rink. **39**, No. 1, 64–85.

Petrov, V.V. (1953): Extension of Cramér's limit theorem to nonidentically distributed independent variables. Vestn. Leningr. Univ. **8**, 13–25 (Russian).

Petrov, V.V. (1954): A generalization of Cramér's limit theorem. Usp. Mat. Nauk. **11**, No. 4, 195–202 (Russian). Zbl. 056.36002

Petrov, V.V. (1965): On large deviation probabilities of sums of independent random variables. Teor. Veroyatn. Primen. **10**, No. 2, 310–322. English transl.: Theory Probab. Appl. **10**, No. 2, 287–298. Zbl. 235.60028

Petrov, V.V. (1968): Asymptotic behavior of probabilities of large deviations. Teor. Veroyatn. Primen. **13**, No. 3, 432–444. English transl.: Theory Probab. Appl. **13**, No. 3, 408–420. Zbl. 181.45003

Petrov, V.V. (1972): Sums of Independent Random Variables. Nauka, Moscow. English transl.: Springer-Verlag, New York (1975). Zbl. 267.60055

Petrov, V.V. (1978): Limit Theorems for Sums of Independent Random Variables, Nauka, Moscow (Russian). (see also Oxford: Clarendon Press 1995. Zbl. 826.60001). Zbl. 621.60022

Philipp, W. (1969): The central limit theorem for mixing sequences of random variables. Z. Wahrscheinlichkeitstheorie Verw. Geb. **12**, No. 2, 155–171. Zbl. 174.49904

Plikusas, A. (1980): Estimation of cumulants and large deviations for certain nonlinear transformations of a stationary Gaussian process. Liet. Mat. Rink. **20**, No. 2, 119–128. English transl.: Lith. Math. J. **20**, No. 2, 150-156. Zbl. 446.60017

Plikusas, A. (1981): Properties of multiple Itô integrals. Liet. Mat. Rink. **21**, No. 2, 163–179. English transl.: Lith. Math. J. **21**, No. 2, 184–191. Zbl. 479.60060

Prokhorov, Yu.V. (1956): Convergence of random processes and limit theorems in probability theory. Teor. Veroyatn. Primen. **1**, No. 2, 177–233. English transl.: Theory Probab. Appl. **1**, No. 2, 157–214. Zbl. 075.29001

Prokhorov, Yu.V. (1968): S.N. Bernstein's inequalities in the multi-dimensional case. Theor. Veroyatn. Primen. **13**, No. 3, 462–470. English transl.: Theory Probab. Appl. **13**, No. 3, 438–447. Zbl. 169.21001

Prokhorov, Yu.V., and Rozanov, Yu.A. (1973): Probability Theory: Basic Concepts, Limit Theorems, Random Process. Nauka, Moscow. English transl. earlier edn.: Springer-Verlag, New York, 1969. Zbl. 186.49501

Ramachandran, B. (1967): Advanced Theory of Characteristic Functions. Statistical Publishing Soc., Calcutta. Zbl. 189.18102

Richter, W. (1957a): A local limit theorem for large deviations. Dokl. Akad. Nauk. SSSR **115**, No. 1, 53–56 (Russian). Zbl. 084.13803

Richter, W. (1957b): Local limit theorems for large deviations. Teoriya Veroyatn. Primen. **2**, No. 2, 214–229. English transl.: Theory Probab. Appl. **2**, No. 2, 206–220. Zbl. 080.34302

Richter, W.-D. (1978): Über Wahrscheinlichkeiten grosser Abweichungen standartisierter Summen unabhängiger Zufallsvektoren. Math. Nachr. **84**, 345–358. Zbl. 391.60033

Richter, W.-D. (1982): Large deviations in finite-dimensional parallelepipeds, Liet. Mat. Rink. **22**, No. 3, 162–169 (Russian). Zbl. 512.60016

Rosenblatt, M. (1956): A central limit theorem and a strong mixing condition. Proc. Na. Acad. Sci. USA **42**, 43–47. Zbl. 070.13804

Rosenblatt, M. (1979): Some remarks on a mixing condition. Ann. Probab. **7**, No. 1, 170–172.

Rozovskii, L.V. (1982): On probabilities of large deviations in convex Borel sets in R^k. IV USSR-Japan Symposium on Probab. Theory and Math. Statisti., Tbilisi, Abstracts of Commun. **191**.

Rubin, H., and Sethuraman, J. (1965): Probabilities of moderate deviations. Sankhya **A27**, No. 2–4. 325–346. Zbl. 178.53802

Rudzkis, R. (1977): A lemma of V.A. Statulevičius. Liet. Mat. Rink. **17**, No. 2, 179–185. English transl.: Lith. Math. J. **17**, No. 2, 263–268. Zbl. 379.60027

Rudzkis, R. (1992): On the distribution of supremum-type functionals of nonparametric estimates of probability and spectral densities. Teor. Veroyatn. Primen. **37**, No. 2, 254–267. English transl.: Theory Probab. Appl. **37**, 236–249. Zbl. 787.62046

Rudzkis, R., Saulis, L., and Statulevičius, V. (1978): A general lemma on large deviation probabilities. Liet. Mat. Rink. **18**, No. 2 99–116. English transl.: Lith. Math. J. **18**, No. 2, 226–238. Zbl. 405.60025

Rudzkis, R., Saulis, L., and Statulevičius, V. (1979): On large deviations of sums of independent random variables. Liet. Mat. Rink. **19**, No. 1, 169–179. English transl.: Lith. Math. J. **19**, No. 1, 118–125. Zbl. 401.60024

Sakhanenko, A.I. (1984): Rate of convergence in an invariance principle for differently distributed variables with exponential moments. Tr. Inst. Mat., Sibirsk. Otd. Akad. Nauk, SSSR, Novosibirsk, 4–49 (Russian). English transl.: Transl. Ser. Math. Eng., 2–73 (1986). Zbl. 541.60024

Sanov, I.N. (1957): On large deviations probabilities of random variables. Mat. Sbo., Nor Ser. **42**(84), No. 1, 11–44 (Russian). Zbl. 078.31202

Saulis, L. (1969): An asymptotic expansion for large deviation probabilities. Liet. Mat. Rink. **9**, No. 3, 605–625 (Russian). Zbl. 188.23701

Saulis, L. (1973): Limit theorems taking account large deviations when Linnik's condition holds. Liet. Mat. Rink. **13**, No. 4, 173–196. English transl.: Lith. Math. J. **13**, No. 4, 646–664. Zbl. 292.60045

Saulis, L. (1979): Large deviations for sums of weighted random variables. Liet. Mat. Rink. **19**, No. 2, 179–187. English transl.: Lith. Math. J. **19**, No. 2, 277–287. Zbl. 433.60029

Saulis, L. (1980): A general lemma for a density taking large deviations into account. Liet. Mat. Rink. **20**, No. 4, 165–185. English transl.: Lith. Math. J. **20**, No. 4, 346–359. Zbl. 456.60024

Saulis, L. (1981): General lemmas on approximating by the normal distribution. Liet. Mat. Rink. **21**, No. 2, 175-189 (Russian). Zbl. 476.60025

Saulis, L. (1984): On large deviations in R^k. Dokl. Akad. Nauk. SSSR **276**, No. 1, 42–45. English transl.: Sov. Math., Dokl. **29**, No. 1, 443–446. Zbl. 584.60039

Saulis, L. (1986): On large deviations for the probability density of sums of independent random variables. Probab. Theory and Math. Statist. – Proceedings of the Fourth Vilnius Conf. **2**, 541–559. Zbl. 646.60033

Saulis, L. (1987a): General lemmas on large deviations for a random vector with regularly behaving cumulants, I. Liet. Mat. Rink. **27**, No. 3, 535–549. English transl.: Lith. Math. J. **27**, No. 3, 263–273. Zbl. 678.60027

Saulis, L. (1987b): General lemmas on large deviations for a random vector with regularly behaving cumulants, II. Liet. Mat. Rink. **27**, No. 4, 747–758. English transl.: Lith. Math. J. **27**, No. 4, 350–356. Zbl. 698.60030

Saulis, L. (1988): General lemmas on large deviations for a random vector with regularly behaving cumulants, III. Liet. Math. Rink. **28**, No. 1, 99–111. English transl.: Lith. Math. J. **28**, No. 1, 58–66. Zbl. 698.60031

Saulis, L. (1990a): On large deviations for the probability density of sums of independent random variables. Probab. Theory and Math. Statist. – Proceedings of the Fifth Vilnius Conf. **2**, 383–393. Zbl. 733.60045

Saulis, L. (1990b): Asymptotic expansions in large deviation zones for the distribution density of sums of independent random variables. New Trends in Probab. and Statist. 1, Proceedings on the Bakuriani Colloquium in Honor of Yu.V. Prokhorov, Bakuriani, Georgia, USSR, 43–56. Zbl. 783.60030

Saulis, L. (1991): Probabilities of large deviations for random variables. Teor. Veroyatn. Primen. **36**, No. 3, 482–493. English transl.: Theory Probab. Appl. **36**, No. 3, 494–507. Zbl. 776.60039

Saulis, L. (1998): Asymptotic expansions, general lemmas and their application to the distribution of sums of random variables. 22nd European Meeting of Statisticians, 7th Vilnius Conference on Prob. Th. and Math. Statistics, Abstracts of Communications, pp. 403–404, 1998, Vilnius

Saulis, L. (1999): Asymptotic expansions of large deviations for sums of nonidentically distributed random variables. Acta Applicandae Mathematicae **58**, 212–231.

Saulis, L.I., and Statulevičius, V. (1970): An asymptotic expansion for large deviation probabilities of sums of random variables in Markov chain. Liet. Mat. Rink. **10**, No. 2, 359–366 (Russian). Zbl. 205.45202

Saulis, L.I., and Statulevičius, V. (1989): Limit Theorems on Large Deviations. Mokslas, Vilnius. English transl.: Kluwer, Dordrecht, 1991. Zbl. 744.60028

Sazonov, V.V. (1974): On the estimation of moments of sums of random variables. Teor. Veroyatn. Primen. **19**, No. 2, 383–386. English transl.: Theory Probab. Appl. **19**, No. 2, 371–374. Zbl. 321.60046

Serfling, R.J. (1968): Contributions to central limit theorem for dependent variables. Ann. Math. Statist. **39**, No. 4, 1158–1175. Zbl. 176.48004

Sethuraman, J. (1964): On the probability of large deviations of families of sample means. Ann. Math. Statist. **35**, No. 4, 1304–1316. Zbl. 147.18803

Sethuraman, J. (1970): Probabilities of deviations. S.N. Roy Memorial Volume. Univ. of North Carolina Press, 655–672. Zbl. 298.60020

Shiryaev, A.N. (1960): Some problems in the spectral theory of higher-order moments. Teor. Veroyatn. Primen. **5**, No. 3, 293–313. English transl.: Theory Probab. Appl. **5**, No. 3, 265–284. Zbl. 109.36001

Shorgin, S.Ya. (1982): Nonclassical estimates of the rate of convergence in the central limit theorem taking large deviations into account. Teor. Veroyatn. Primen. **27**, No. 2, 308–318. English transl.: Theory Probab. Appl. **27**, No. 2, 324–337. Zbl. 565.60018

Sievers, G.L. (1669): On the probabilities of large deviations and exact slopes. Ann. Math. Statist. **40**, 1908–1921. Zbl. 193.46701

Sievers, G.L. (1975): Multivariate probabilities of large deviations. Ann. Math. Statist. **46**, 897–905. Zbl. 313.60023

Statulevičius, V. (1965): Limit theorems for densities and asymptotic expansions for distributions of sums of independent random variables. Teoriya Veroyat. Primen. **10**, No. 4, 645–659. English transl.: Theory Probab. Appl. **10**, No. 4, 582–595. Zbl. 178.53803

Statulevičius, V. (1966): On large deviations. Z. Wahrscheinlichkeitstheorie Verw. Geb. **6**, No. 2, 133–144.

Statulevičius, V. (1969a): Limit theorems for sums of random variables in a Markov chain, I. Liet. Mat. Rink. **9**, No. 2, 345–362 (Russian). Zbl. 203.50206

Statulevičius, V. (1969b): Limit theorems for sums of random variables in a Markov chain, II. Liet. Mat. Rink. **9**, No. 3, 635–672 (Russian). Zbl. 203.50206

Statulevičius, V. (1970a): Limit theorems for sums of random variables in a Markov chain, III. Liet. Mat. Rink. **10**, No. 1, 161–169 (Russian). Zbl. 203.50206

Statulevičius, V. (1970b): Limit theorems for random functions, I. Liet. Mat. Rink. **10**, No. 3, 583–592 (Russian). Zbl. 266.60014

Statulevičius, V. (1979): Large deviation theorems for sums of dependent random variables. Liet. Mat. Rink. **19**, No. 2, 199–208. English transl.: Lith. Math. J. **19**, No. 2, 289–296. Zbl. 442.60028

Statulevičius, V. (1983): On a condition of almost Markov regularity. Teoriya Veroyat. Primen. **28**, No. 2, 358–362. English transl.: Theory Probab. Appl. **28**, No. 2, 379–383. Zbl. 513.60064

Statulevičius, A., and Aleškevičienė, A. (1998): Theorems of large deviations for non-Gaussian approximation. 22nd European Meeting of Statisticians, 7th Vilnius Conference on Prob. Th. and Math. Statistics, Abstracts of Communications, pp. 91–92, 1998, Vilnius.

Statulevičius, V., and Jakimavičius, A. (1988a): Bounds for cumulants and centered moments of mixing random processes, I. Liet. Mat. Rink. **28**, No. 1, 112–129. English transl.: Lith. Math. J. **28**, No. 1, 67–80. Zbl. 666.60027

Statulevičius, V., and Jakimavičius, A. (1988b): Bounds for cumulants and centered moments of mixing random processes, II. Liet. Mat. Rink. **28**, No. 2, 360–375. English transl.: Lith. Math. J. **28**, No. 2, 179–190. Zbl. 666.60028

Stroock, D.W. (1984): An Introduction to the Theory of Large Deviations. Springer-Verlag, New York. Zbl. 552.60022

Surgailis, D. (1981): On infinitely divisible self-similar fields. Z. Wahrscheinlichkeitstheorie Verw. Geb. **58**, 453–477. Zbl. 469.60050

Survila, P. (1966): Large deviations for densities. Liet. Mat. Rink. **6**, No. 4, 591–600 (Russian). Zbl. 158.36301

Svetulevičienė, V. (1981): Large deviation probabilities for sums of random vectors. Liet. Mat. Rink. **21**, No. 2, 191–199. English transl.: Lith. Math. J. **21**, No. 2, 192–197. Zbl. 481.60033

Tkachuk, S.G. (1975): A theorem on large deviations for distributions with tamely varying tails. Random Processes and Statist. Inference, Tashkent, vol. 5, 164–174 (Russian). Zbl. 322.60026

Vandemaele, M. (1982): Large deviation probabilities for U-statistics. Teor. Veroyatn. Primen. **27**, No. 3, 573–574. Also in English: Theory Probab. Appl. **27**, No. 3, 614. Zbl. 488.60037

Ventsel, A.D. (1976): Rough limit theorems on large deviations for Markov stochastic processes, II. Teoriya Veroyat. Primen. **21**, No. 3, 512–526. English transl.: Theory Probab. Appl. **21**, No. 3, 499–512. Zbl. 361.60006

Ventsel, A.D. (1986): Limit Theorems on Large Deviations for Markov Stochastic Processes. Nauka, Moscow. English transl.: Kluwer, Dordrecht-Boston (1990). Zbl. 589.60024

Volkonskii, V.A., and Rozanov, Yu.A. (1959): Some limit theorems for random functions, I. Theor. Veroyatn. Primen. **4**, No. 2, 186–207. English transl.: Theory Probab. Appl. **4**, No. 2, 178–197. Zbl. 092.33502

Wolf, W. (1970): Some limit theorems for large deviations of sums of independent random variables. Dokl. Akad. Nauk. SSSR **191**, No. 6, 1209–1211. English transl.: Sov. Math., Dokl. **11**, 509–512 (1970). Zbl. 235.60027

Wolf, W. (1974): Über Wahrscheinlichkeiten grosser Abweichungen. Math. Nachr. **62**, 216–278. Zbl. 302.60017

Wolf, W. (1975): Über Wahrscheinlichkeiten grosser Abweichungen bei Nichterfüllung der Cramérschen Bedingung. Math. Nachr. **70**, 197–215. Zbl. 324.60028

Wolf, W. (1977): Asymptotische Entwicklungen für Wahrscheinlichkeiten grosser Abweichungen. Z. Wahrscheinlichkeitstheorie Verw. Geb. **40**, 239–256. Zbl. 365.60033

Wolf, W., and Mikosch, T. (1983): Large deviation probabilities for a double array. Liet. Mat. Rink. **23**, No. 2, 43–48. English transl.: Lith. Math. J. **23**, No. 2, 155–159. Zbl. 527.60025

Yurinskii, V.V. (1976): Exponential inequalities for sums of random vectors. J. Multivariate Anal. **6**, No. 4, 473–499. Zbl. 346.60001

Zaitsev, A.Yu. (1984a): On approximation by Gaussian distributions under multivariate analogues of Bernstein's conditions. Dokl. Akad. Nauk. SSSR **276**, No. 5, 1046–1048. English transl.: Sov. Math., Dokl. **29**, No. 5, 624–626. Zbl. 583.60042

Zaitsev, A.Yu. (1984b): On Gaussian approximation of convolutions under multivariate analogues of Bernstein's conditions. Preprint. Akad. Nauk. SSSR LOMI **9**, 3–54 (Russian).

Zhurbenko, I.G. (1972): On strong estimates of mixed cumulants. Sib. Mat. Zh. **13**, No. 2, 293–308. English transl.: Sib. Math. J. **13**, No. 2, 203–213. Zbl. 246.60040

Zhurbenko, I.G. (1982): The Spectral Analysis of Time Series. Moscow Univ. Press, Moscow. English transl.: North Holland, Amsterdam, 1986. Zbl. 498.62080

Zitikis, R. (1990): On large deviations for L-estimates. New Trends in Probab. and Statist. **1**, Proceedings of the Bakuariani Colloquium in Honor of Yu.V. Prokhorov, Baluriani, Georgia, pp. 137–164. Zbl. 778.60021

Zolotarev, V.M. (1962): On a new point of view about limit theorems taking large deviations into account. Trudy VI All-Union Conf. on Probab. and Statist., Vilnius, pp. 43–48 (Russian). Zbl. 133.41201

Zolotarev, V.M. (1965): On proximity of the distributions of two sums of independent random variables. Teoriya Veroyat. Primen. **10**, No. 3, 519–526. English transl.: Theory Probab. Appl. **10**, No. 3, 472–479. Zbl. 214.17402

Zolotarev, V.M. (1986): The Modern Theory of Summation of Independent Random Variables. Nauka, Moscow. Zbl. 64960016

Zuev, N.M. (1973): On bounds for mixed cumulants of random processes. Mat. Zmetki **13**, No. 4, 581–586. English transl.: Utrecht VSP (1997). Zbl. 283.60031

Zuev, N.M. (1981): Estimation of mixed cumulants of strong mixing almost Markov random processes. Liet. Mat. Rink. **21**, No. 2, 81–85 (Russian). Zbl. 468.60084

Zykov, A.A. (1987): Foundations of Graph Theory. Nauka, Moscow. Zbl. 645.05001. English transl.: Moscow: BCS Associated (1990). Zbl. 707.05001

Name Index

Subject Index

MIX
Papier aus verantwortungsvollen Quellen
Paper from responsible sources
FSC® C105338

If you have any concerns about our products,
you can contact us on
ProductSafety@springernature.com

In case Publisher is established outside the EU,
the EU authorized representative is:
Springer Nature Customer Service Center GmbH
Europaplatz 3, 69115 Heidelberg, Germany

Printed by Libri Plureos GmbH
in Hamburg, Germany